Universitext

Springer

*New York
Berlin
Heidelberg
Barcelona
Hong Kong
London
Milan
Paris
Singapore
Tokyo*

Universitext

Editors (North America): S. Axler, F.W. Gehring, and K.A. Ribet

(continued after index)

Heydar Radjavi   Peter Rosenthal

Simultaneous Triangularization

 Springer

Heydar Radjavi
Department of Mathematics
Dalhousie University
Halifax, Nova Scotia B3H 3J5
Canada

Peter Rosenthal
Department of Mathematics
University of Toronto
Toronto, Ontario M5S 1A1
Canada

Mathematics Subject Classification (1991): 15-02, 15A21, 15A30, 15A48, 47-02, 47A15, 47A65, 47C05, 47D03

Library of Congress Cataloging-in-Publication Data
Radjavi, Heydar.
    Simultaneous triangularization / Heydar Radjavi, Peter Rosenthal.
        p.   cm. — (Universitext)
    Includes bibliographical references and index.
    ISBN 0-387-98467-4 (hardcover : alk. paper). — ISBN 0-387-98466-6
(softcover : alk. paper)
        1. Triangularization (Mathematics)   I. Rosenthal, Peter, 1941–   .
    II. Title. III. Series.
    QA197.R33   2000
    512.9′434—dc21   99-23772

Printed on acid-free paper.

Production managed by Allan Abrams; manufacturing supervised by Jeffrey Taub.
Photocomposed copy prepared from the authors' $\mathcal{A}_{\mathcal{M}}S$-T$_{\!E}$X files.
Printed and bound by Maple-Vail Book Manufacturing Group, York, PA.
Printed in the United States of America.

9 8 7 6 5 4 3 2 1

ISBN 0-387-98467-4 Springer-Verlag New York Berlin Heidelberg   SPIN 10659788   (hardcover)
ISBN 0-387-98466-6 Springer-Verlag New York Berlin Heidelberg   SPIN 10659754   (softcover)

To
　　Ursula, Marjan,
　　Shirin, Carrie,

　　Carol, Alan,
　　Jeffrey, Michael,
　　Daniel, Esther,
　　Laurel, Jeremy,
　　Aaron, Margaret

PREFACE

A matrix over the field of complex numbers (or over any other algebraically closed field) is similar to a matrix in upper triangular form. For individual matrices, this is not as important as the Jordan canonical form (partially because it is not canonical), but it is nonetheless useful in many situations. In particular, the eigenvalues of a matrix in upper triangular form are visible: They are just the elements on the main diagonal of the matrix. Moreover, as is illustrated throughout this book, many collections of matrices can be simultaneously put in upper triangular form, whereas it is very rare that two matrices have "simultaneous Jordan forms."

When are two matrices simultaneously similar to matrices in upper triangular form? Equivalently, given two linear transformations on a finite-dimensional complex vector space, when is there a basis for the space with respect to which the matrices of both transformations are upper triangular? More generally, when is a collection of linear transformations "simultaneously triangularizable"?

It is easily shown that commutative sets of matrices are simultaneously triangularizable. Simultaneous triangularizability can be regarded as a kind of generalized commutativity, and implies certain consequences of commutativity. In particular, a set $\{A_1, A_2, \dots, A_m\}$ of linear transformations is simultaneously triangularizable if and only if, for every polynomial p, each eigenvalue of $p(A_1, A_2, \dots, A_m)$ has the form $p(\lambda_1, \lambda_2, \dots, \lambda_m)$ where λ_j is an eigenvalue of A_j for each j. Also, simultaneous triangularizability is equivalent to the existence of simultaneous similarities that are close to commuting transformations.

There are many beautiful classical theorems, associated with names such as Engel, McCoy, Levitzki, Kolchin, and Kaplansky, giving sufficient conditions that collections of linear transformations be simultaneously triangularizable. There are also a number of more recent results by many researchers. Some of the work on triangularizability intersects other areas of linear algebra. For example, triangularization theorems for collections of nonnegative matrices relate to the Perron-Frobenius theory. Triangularizability is also linked to various partial spectral mapping theorems, and to properties of spectral radii and traces.

Beginning around 1980, the theory has been extended to operators on infinite-dimensional Banach spaces. A collection of bounded linear operators is said to be simultaneously triangularizable (or, simply, triangularizable) if there is a maximal chain of subspaces, each of which is left invariant by all the operators in the collection. Many of the finite-dimensional results have satisfactory infinite-dimensional generalizations to collections of compact operators; the basis for such theorems is the lemma established by Lomonosov in his famous paper of 1973, together with Ringrose's Theorem on computing the spectrum of a compact operator from a "triangular" representation. These fundamental results have been supplemented by Turovskii's just-discovered (1998) extension of Lomonosov's work from algebras to semigroups.

There is now a great deal known about triangularizability of collections of compact operators. In particular, simultaneous triangularizability is equivalent to a spectral mapping property for collections of compact operators. As in the finite-dimensional situation, triangularizability generalizes commutativity, and is implied by several other kinds of generalized commutativity. Also, there are many sufficient conditions that collections of compact operators be triangularizable.

Very few of the above results generalize further to arbitrary bounded operators: There are even collections containing just a single operator that have only the trivial invariant subspaces (as shown by Per Enflo), and which are therefore very far from triangularizable. It is not known whether every operator on Hilbert space has a nontrivial invariant subspace, and therefore it is not known whether every operator on Hilbert space is triangularizable. However, even in the Hilbert space case there are counterexamples to most of the natural generalizations of the finite-dimensional results. On the other hand, there are several affirmative results.

In this book we have attempted to give a fairly complete treatment of the classical and recent results in both the finite- and infinite-dimensional settings. We have reworked much of the material to provide a more cohesive and readable treatment than can be obtained by simply taking the union of the published research papers. Moreover, we aspired to make the exposition as elementary and self-contained as possible. In addition, we have included a number of new results.

We hope that this book will be found useful by graduate students and mathematicians working in or contemplating work in either the finite- or infinite-dimensional aspects of this topic. We also hope that making these results more easily available will increase the applications of simultaneous triangularizability to other areas, such as representations of groups and semigroups. The finite-dimensional results are treated independently in the first five chapters; readers who are not interested in operators on Banach spaces may restrict their attention to these chapters. However, we have written the infinite-dimensional sections with a view to making them accessible to those with minimal backgrounds in functional analysis. In particular, Chapter 6 includes a discussion of the basic material required; we hope this encourages all readers to at least peruse the latter part of the book as well. On the other hand, those primarily interested in operator theory could begin with Chapter 7 and read only those earlier sections that are directly referred to thereafter. However, we would suggest that such readers would benefit by at least skimming the earlier chapters first.

In addition, we have written this book so as to make it suitable for students to read as a text. In fact, a very preliminary version formed the basis for graduate courses at Dalhousie University and the University of Toronto. Parts of the book might be used in various courses. The only prerequisite for the first five chapters is a solid course in linear algebra, and all that is required for the balance is an introductory course in functional analysis. Instructors who do not want to spend an entire semester on simultaneous triangularization might use parts of the book as a source of topics in courses covering other material.

Each chapter ends with a section entitled "Notes and Remarks" in which we attempt to outline the development of the material and to discuss other interesting results that could not be included in the main text. Some of these discussions and references could provide direction for reading projects for students.

We use the following scheme for numbering definitions, lemmas, theorems and corollaries throughout the book: chapter.section.number. For example, 7.2.3 is the third numbered item in Section 2 of Chapter 7.

We are grateful to several mathematicians who made suggestions and caught errors (we hope they caught most of them) after reading preliminary versions of the manuscript, especially Marjeta Kramar, Bill Longstaff, Mitja Mastnak, M. H. Shirdarreh, Reza Yahaghi, and Yong Zhong. We are particularly grateful to Ruben Martinez, who found a large number of mistakes, made several excellent suggestions, and was of great assistance in compiling the references and indices. We are also grateful to the three people who did a wonderful job of transforming our sloppy scrawls into beautiful type: Maria Fe Elder, Lucile Lo, and Karin Smith. After this process should have ended, we made an almost infinite number of additions and corrections, which Lucile Lo handled with infinite patience and skill.

The older we grow, the more we appreciate those who taught and encouraged us when we were young, especially Ali Afzalipour, Arthur B. Brown, Robert Cameron, Chandler Davis, Taghi Fatemi, Israel Halperin, Gerhard Kalish, Mohammad Ali Nourghalitchi, Harold Rosenthal, Allen Shields, Manoutchehr Vessal, and Leo Zippin. We are particularly grateful to Paul Halmos for providing inspiration and guidance throughout our careers.

Heydar Radjavi
Dalhousie University

Peter Rosenthal
University of Toronto

CONTENTS

CHAPTER 1
Algebras of Matrices

We begin with a discussion of the fundamental concepts of the subject in the finite-dimensional situation; infinite-dimensional analogues are presented in Chapter 7 below.

There are many known sufficient conditions that a collection of linear transformations be triangularizable. The most satisfactory results are obtained when the collection is an algebra, in which case there are necessary and sufficient conditions for triangularizability (see Theorems 1.3.2, 1.4.6, 1.5.5, and 1.6.6 below). An important preliminary result is Burnside's Theorem (1.2.2) on existence of invariant subspaces for algebras of linear transformations.

Throughout the first five chapters we restrict our attention to collections of linear transformations on a finite-dimensional vector space over an algebraically closed field.

1.1 The Triangularization Lemma

Triangularization is equivalent to the existence of certain chains of invariant subspaces.

Definition 1.1.1. A subspace $\mathcal{M}$ is *invariant* for a collection $\mathcal{C}$ of linear transformations if $Ax \in \mathcal{M}$ whenever $x \in \mathcal{M}$ and $A \in \mathcal{C}$. A subspace is *nontrivial* if it is different from $\{0\}$ and from the entire space. A collection of linear transformations is *reducible* if it has a nontrivial invariant subspace and is *irreducible* otherwise.

The central definition is the following.

Definition 1.1.2. A collection of linear transformations is *triangularizable* if there is a basis for the vector space such that all transformations in the collection have upper triangular matrix representations with respect to that basis.

It is clear that triangularizability is equivalent to the existence of a chain of invariant subspaces

$$\{0\} = \mathcal{M}_0 \subset \mathcal{M}_1 \subset \mathcal{M}_2 \subset \cdots \subset \mathcal{M}_n = \mathcal{V}$$

with the dimension of $\mathcal{M}_j$ equal to j for each j and with $\mathcal{V}$ the entire vector space. (If the collection is triangularizable with respect to the basis $\{e_1, e_2, \dots, e_n\}$, let $\mathcal{M}_j$ be the linear span of $\{e_1, \dots, e_j\}$ for each j.) Any such chain is a *triangularizing chain* for the collection.

Quotient spaces will play an important role in this study.

Definition 1.1.3. If $\mathcal{V}$ is a vector space and $\mathcal{N}$ is a subspace of $\mathcal{V}$, then the *quotient space* $\mathcal{V}/\mathcal{N}$ is the collection of cosets $[x] = x + \mathcal{N} = \{x + z : z \in \mathcal{N}\}$ for $x \in \mathcal{V}$, with $[x] + [y]$ defined as $[x + y]$ and $\lambda[x]$ defined as $[\lambda x]$ for scalars λ. If A is a linear transformation on $\mathcal{V}$ and $\mathcal{N}$ is invariant under A, then the quotient transformation $\tilde{A}$ on $\mathcal{V}/\mathcal{N}$ is defined by $\tilde{A}[x] = [Ax]$ for each $x \in \mathcal{V}$ (the invariance of $\mathcal{N}$ under A ensures that $\tilde{A}$ is well-defined on the cosets). If $\mathcal{C}$ is a collection of linear transformations on $\mathcal{V}$, and if $\mathcal{M}$ and $\mathcal{N}$ are invariant subspaces for $\mathcal{C}$ with $\mathcal{N}$ properly contained in $\mathcal{M}$, then the *collection of quotients of $\mathcal{C}$ with respect to* $\{\mathcal{M}, \mathcal{N}\}$ is the set of all quotient transformations $\tilde{A}$ on $\mathcal{M}/\mathcal{N}$. A property is *inherited by quotients* if every collection of quotients of a collection satisfying the property also satisfies the property. (Note that this implies, in particular, that the property is inherited by restrictions since a restriction to $\mathcal{M}$ is a quotient with respect to $\{\mathcal{M}, 0\}$.)

The following lemma will be used very frequently.

Lemma 1.1.4. (The Triangularization Lemma) *Let $\mathcal{P}$ be a set of properties, each of which is inherited by quotients. If every collection of transformations on a space of dimension greater than 1 that satisfies $\mathcal{P}$ is reducible, then every collection of transformations satisfying $\mathcal{P}$ is triangularizable.*

Proof. Let $\mathcal{C}$ be any collection satisfying $\mathcal{P}$. Choose a maximal chain of invariant subspaces of $\mathcal{C}$:

$$\{0\} = \mathcal{M}_0 \subset \mathcal{M}_1 \subset \cdots \subset \mathcal{M}_m = \mathcal{V}.$$

It suffices to show that each quotient $\mathcal{M}_k/\mathcal{M}_{k-1}$ is one-dimensional, for then $\{\mathcal{M}_j\}$ will be a triangularizing chain for $\mathcal{C}$. Fix any k; if the dimension of $\mathcal{M}_k/\mathcal{M}_{k-1}$ were greater than 1, then the collection of quotients of $\mathcal{C}$ with respect to $\{\mathcal{M}_k, \mathcal{M}_{k-1}\}$ would have a nontrivial invariant subspace L, by hypothesis. But then $\{x \in \mathcal{M}_k : [x] \in L\}$ would be an invariant subspace of $\mathcal{C}$ and would be properly between $\mathcal{M}_{k-1}$ and $\mathcal{M}_k$, contradicting the maximality of the chain $\{\mathcal{M}_j\}$. $\square$

Our first application of the Triangularization Lemma is very well known.

Theorem 1.1.5. *Every commutative collection of linear transformations is triangularizable.*

Proof. Commutativity is a property that is inherited by quotients, so it suffices (by the Triangularization Lemma) to show that commutative collections of transformations on spaces of dimension greater than one have nontrivial invariant subspaces. Let $\mathcal{C}$ be such a collection. If all the transformations in $\mathcal{C}$ are multiples of the identity, then every subspace is invariant under $\mathcal{C}$. On the other hand, if there is an A in $\mathcal{C}$ that is not a multiple of the identity, let λ be any eigenvalue of A and let $\mathcal{M}$ be the corresponding eigenspace. If $B \in \mathcal{C}$ and $x \in \mathcal{M}$, then

$$ABx \; = \; BAx \; = \; \lambda Bx,$$

so $\mathcal{M}$ is invariant under $\mathcal{C}$. □

As we shall see, there are several respects in which triangularizability is a generalization of commutativity.

Corollary 1.1.6. (Schur's Theorem) *Each linear transformation is triangularizable.*

Proof. This is a special case of Theorem 1.1.5. □

If a matrix is in upper triangular form, then its eigenvalues are the entries on the main diagonal. This observation leads immediately to a spectral m apping theorem for triangularizable collections of transformations.

Definition 1.1.7. A *noncommutative polynomial* in the linear transformations $\{A_1, \dots , A_k\}$ is any linear combination of words in the transformations.

We use $\sigma(A)$ to denote the spectrum (which in the present, finite-dimensional, case is simply the set of eigenvalues) of A.

Theorem 1.1.8. (Spectral Mapping Theorem) *If $\{A_1, \dots , A_k\}$ is a triangularizable collection of linear transformations, and if p is any noncommutative polynomial in $\{A_1, \dots , A_k\}$, then*

$$\sigma(p(A_1, \dots , A_k)) \subset p(\sigma(A_1), \dots , \sigma(A_k)),$$

where $p(\sigma(A_1), \dots , \sigma(A_k))$ denotes the set of all $p(\lambda_1, \dots , \lambda_k)$ such that $\lambda_j \in \sigma(A_j)$ for all j.

Proof. This follows immediately from the facts that

   (i) the eigenvalues of a triangular matrix are the entries on the main diagonal,
   (ii) each of the diagonal entries of a product of given triangular matrices is a product of diagonal entries of the given matrices, and
   (iii) each of the diagonal entries of a sum of given matrices is a sum of diagonal entries of the given matrices.

□

1.2 Burnside's Theorem

Definition 1.2.1. An *algebra* of linear transformations is a collection of linear transformations that is closed under addition, multiplication, and multiplication by scalars. An algebra is *unital* if it contains the identity transformation. In a unital algebra with identity I, we use the notation λ as an abbreviation for λI. The notation $\mathcal{B}(\mathcal{V})$ is used to denote the algebra of all linear transformations mapping $\mathcal{V}$ into $\mathcal{V}$. (The notation $\mathcal{L}(\mathcal{V})$ is also common.)

If $\mathcal{A}$ is an algebra of linear transformations and x is any given vector, then $\{Ax : A \in \mathcal{A}\}$ is easily seen to be an invariant subspace for $\mathcal{A}$. However, for some $\mathcal{A}$ and x, $\{Ax : A \in \mathcal{A}\}$ is the entire space (in which case x is said to be a *cyclic vector* for $\mathcal{A}$). The question of which algebras of linear transformations have nontrivial invariant subspaces has the following beautiful classical answer; it forms the foundation for many theorems on triangularization.

Theorem 1.2.2. (Burnside's Theorem) *The only irreducible algebra of linear transformations on the finite-dimensional vector space $\mathcal{V}$ of dimension greater than 1 is the algebra of all linear transformations mapping $\mathcal{V}$ into $\mathcal{V}$.*

Proof. Let $\mathcal{A}$ be an irreducible algebra. We first show that $\mathcal{A}$ contains a transformation of rank 1. For this, let T_0 be a transformation in $\mathcal{A}$ with minimal nonzero rank. We must show that this rank is 1.

If T_0 had rank greater than 1, there would be vectors x_1 and x_2 such that $\{T_0x_1, T_0x_2\}$ is a linearly independent set. Since $\{AT_0x_1 : A \in \mathcal{A}\} = \mathcal{V}$, there is an $A_0 \in \mathcal{A}$ such that $A_0T_0x_1 = x_2$. Then $\{T_0A_0T_0x_1, T_0x_1\}$ is linearly independent. There is a scalar λ such that the restriction of $(T_0A_0 - \lambda)$ to $T_0\mathcal{V}$ is not invertible. Then $(T_0A_0 - \lambda)T_0$ is not 0, since $T_0A_0T_0x_1 - \lambda T_0x_1 \neq 0$. But $(T_0A_0 - \lambda)T_0$ has rank less than that of T_0, since its range is properly contained in that of T_0. This contradicts the minimality of the rank of T_0. We conclude that T_0 has rank 1.

Pick a nonzero vector y_0 in the range of T_0. Then there is a linear functional ϕ_0 on $\mathcal{V}$ such that $T_0x = \phi_0(x)y_0$ for all $x \in \mathcal{V}$. Since every linear transformation of rank 1 has the form $x \mapsto \phi(x)y$ for a vector y in $\mathcal{V}$ and a linear functional ϕ, and since every linear transformation on a finite-dimensional space is the sum of transformations of rank 1, it suffices to show that $\mathcal{A}$ contains every T of the form $Tx = \phi(x)y$.

Note that T_0 in $\mathcal{A}$ implies that the transformation T_0A is in $\mathcal{A}$ for each A in $\mathcal{A}$. But $T_0Ax = \phi_0(Ax)y_0$, so the set of ϕ such that the linear transformation T defined by $Tx = \phi(x)y_0$ is in $\mathcal{A}$ includes all ϕ defined by $\phi(x) = \phi_0(Ax)$ for some A in $\mathcal{A}$. This set of ϕ is a subspace of $\mathcal{V}^*$ (the space of linear functionals on $\mathcal{V}$); if it were not all of $\mathcal{V}^*$, there would be an $x_0 \neq 0$ such that $\phi(x_0) = 0$ for all such ϕ (finite-dimensional spaces are reflexive). But $\phi_0(Ax_0) = 0$ for all A in $\mathcal{A}$ implies $x_0 = 0$, since $\{Ax_0 : A \in \mathcal{A}\} = \mathcal{V}$ when $x_0 \neq 0$. Thus there is no such x_0, and therefore the transformation T defined by $Tx = \phi(x)y_0$ is in $\mathcal{A}$ for every ϕ in $\mathcal{V}^*$.

Since $y_0 \neq 0$, $\{Ay_0 : A \in \mathcal{A}\} = \mathcal{V}$. Given any y in $\mathcal{V}$, choose A in $\mathcal{A}$ such that $Ay_0 = y$. Then $ATx = \phi(x)y$, so $\mathcal{A}$ contains all rank-one transformations. $\qquad \square$

Burnside's Theorem can be rephrased: Every proper subalgebra of $\mathcal{B}(\mathcal{V})$ is reducible. As shown below, this can be combined with the Triangularization Lemma to give many sufficient conditions that algebras of matrices be triangularizable.

Burnside's Theorem provides the simplest proof of the following (which can also be proven directly).

Theorem 1.2.3. *The only two-sided ideals of $\mathcal{B}(\mathcal{V})$ are $\{0\}$ and $\mathcal{B}(\mathcal{V})$.*

Proof. Let $\mathcal{I}$ be a two-sided ideal of $\mathcal{B}(\mathcal{V})$ other than $\{0\}$. Then it is easily seen that $\mathcal{I}$ is an irreducible subalgebra of $\mathcal{B}(\mathcal{V})$. For if $A \neq 0$ is in $\mathcal{I}$ and $x \neq 0$ is in $\mathcal{V}$, there is a $B \in \mathcal{B}(\mathcal{V})$ such that $ABx \neq 0$. Let $y \in \mathcal{V}$. There is a $C \in \mathcal{B}(\mathcal{V})$ such that $CABx = y$. Then $CAB \in \mathcal{I}$, so every nonzero vector x is cyclic for $\mathcal{I}$, and $\mathcal{I}$ is irreducible. By Burnside's Theorem, $\mathcal{I} = \mathcal{B}(\mathcal{V})$.$\square$

The next theorem is useful in several contexts, including the establishing of a "block triangularization theorem" (Theorem 1.5.1) below.

Theorem 1.2.4. *If $\mathcal{V}$ is a finite-dimensional vector space over an algebraically closed field, then every algebra automorphism of $\mathcal{B}(\mathcal{V})$ is spatial (i.e., if $\phi : \mathcal{B}(\mathcal{V}) \to \mathcal{B}(\mathcal{V})$ is an algebra isomorphism, then there is an $S \in \mathcal{B}(\mathcal{V})$ such that $\phi(A) = SAS^{-1}$ for every $A \in \mathcal{B}(\mathcal{V})$).*

Proof. Given such a ϕ, note first that ϕ takes idempotents into idempotents, since $\phi(A^2) = (\phi(A))^2$. Suppose that A_0 is an idempotent of rank 1. Then $\{A_0BA_0 : B \in \mathcal{B}(\mathcal{V})\}$ is a one-dimensional subspace of $\mathcal{B}(\mathcal{V})$, so its image under ϕ, $\{\phi(A_0)C\phi(A_0) : C \in \mathcal{B}(\mathcal{V})\}$, also has dimension 1. Thus $\phi(A_0)$ is a rank-one idempotent whenever A_0 is.

Fix any idempotent A_0 of rank 1. Any two rank-one idempotents are similar (they obviously have the same Jordan canonical form), so $\phi(A_0)$ is similar to A_0. Compose that similarity with ϕ, so that we can assume that $\phi(A_0) = A_0$. Then let x_0 be any nonzero vector in the range of A_0.

We can now define a transformation that implements ϕ. For each $B \in \mathcal{B}(\mathcal{V})$, define $S(Bx_0)$ to be $\phi(B)x_0$. We must first show that S is well-defined. If $B_1x_0 = B_2x_0$, then, since $A_0x_0 = x_0$, it follows that $(B_1 - B_2)A_0x_0 = 0$. Since A_0 has rank one, this means that $(B_1 - B_2)A_0 = 0$, so

$$(\phi(B_1) - \phi(B_2))\,\phi(A_0) = (\phi(B_1) - \phi(B_2))\,A_0 = 0,$$

and

$$(\phi(B_1) - \phi(B_2))\,x_0 = 0.$$

Thus S is well-defined.

Now, S is obviously linear. To show that S is injective, simply note that $\phi(B)x_0 = 0$ implies $\phi(B)\phi(A_0) = \phi(BA_0) = 0$, so $BA_0 = 0$ and thus $Bx_0 = 0$. Since $\mathcal{V}$ is finite-dimensional, S is also surjective, and hence is invertible.

It remains to be shown that S implements ϕ. Fix any $A \in \mathcal{B}(\mathcal{V})$. Then, for each $B \in \mathcal{B}(\mathcal{V})$,

$$S(AB)x_0 = \phi(AB)x_0 = \phi(A)\phi(B)x_0,$$

and

$$SBx_0 = \phi(B)x_0,$$

so

$$SABx_0 = \phi(A)SBx_0.$$

Thus $SAy = \phi(A)Sy$ for all y of the form Bx_0 for some B; i.e., for all y. Hence $SA = \phi(A)S$, or $SAS^{-1} = \phi(A)$. $\qquad\square$

1.3 Triangularizability of Algebras of Matrices

Note that any collection of quotients (Definition 1.1.3) of an algebra is itself an algebra. Thus, by Burnside's Theorem (Theorem 1.2.2) and the Triangularization Lemma (Lemma 1.1.4), an algebra of linear transformations is triangularizable if it satisfies any property that is inherited by quotients and is not satisfied by $\mathcal{B}(\mathcal{V})$ when the dimension of $\mathcal{V}$ is greater than 1.

We begin with a simple illustration.

Theorem 1.3.1. *An algebra of nilpotent linear transformations is triangularizable.*

Proof. If $\mathcal{M}$ and $\mathcal{N}$ are invariant subspaces of the algebra and $\mathcal{N}$ is properly contained in $\mathcal{M}$, then the quotient algebra on $\mathcal{M}/\mathcal{N}$ also consists of nilpotent transformations: $A^k = 0$ implies $\tilde{A}^k = 0$. Thus each quotient algebra is proper, since there are non-nilpotent transformations on every vector space of nonzero dimension. Burnside's Theorem (Theorem 1.2.2) therefore establishes the reducibility of all quotient algebras on quotient spaces of dimension greater than 1, so the Triangularization Lemma (Lemma 1.1.4) applies. □

This theorem can be generalized to Lie algebras (Corollary 1.7.5) and to semigroups (Theorem 2.1.7). Also, as we now show, it can be strengthened substantially in the case of algebras. It will be further strengthened in Chapter 4.

Theorem 1.3.2. *If $\mathcal{A}$ is an algebra of linear transformations, then $\mathcal{A}$ is triangularizable if and only if each commutator of the form $BC - CB$ with B and C in $\mathcal{A}$ is nilpotent.*

Proof. If $\mathcal{A}$ is triangularizable, then by the Spectral Mapping Theorem (Theorem 1.1.8), the eigenvalues of $BC - CB$ are all of the form $\beta\gamma - \gamma\beta$ with $\beta \in \sigma(B)$ and $\gamma \in \sigma(C)$. Since multiplication in the field is commutative, $\sigma(BC - CB) = \{0\}$, and thus $BC - CB$ is nilpotent.

For the converse, we employ Burnside's Theorem and the Triangularization Lemma. Note that on a space of dimension greater than 1 there are pairs of linear transformations with non-nilpotent commutators: If

$$B = \begin{pmatrix} 1 & 0 \\ 0 & 0 \end{pmatrix} \text{ and } C = \begin{pmatrix} 0 & 1 \\ 1 & 0 \end{pmatrix},$$

then

$$BC - CB = \begin{pmatrix} 0 & 1 \\ -1 & 0 \end{pmatrix}, \text{ and } \begin{pmatrix} 0 & 1 \\ -1 & 0 \end{pmatrix}^2 = \begin{pmatrix} -1 & 0 \\ 0 & -1 \end{pmatrix}.$$

(On spaces of dimension greater than 2, B and C can be constructed by adding 0 as a direct summand.) Also, the property of having nilpotent commutators is inherited by quotients. Thus every quotient of $\mathcal{A}$ where the dimension of $\mathcal{M}/\mathcal{N}$ is greater than 1 is a proper subalgebra of $\mathcal{B}(\mathcal{M}/\mathcal{N})$, and is therefore reducible by Burnside's Theorem (Theorem 1.2.2). Hence the Triangularization Lemma (Lemma 1.1.4) gives the result. □

Note that this theorem illustrates the fact that triangularizability is a generalization of commutativity: Commutators being nilpotent is weaker than commutators being 0. Further generalizations are given below (Theorem 1.5.5 (iv) and Corollary 1.7.8).

The following corollary seems surprising.

Corollary 1.3.3. *An algebra of linear transformations is triangularizable if and only if every pair of transformations in the algebra is triangularizable.*

Proof. If every pair is triangularizable, then, for B and C in the algebra, $\sigma(BC - CB) = \{0\}$ by the Spectral Mapping Theorem (Theorem 1.1.8). Hence the result follows from Theorem 1.3.2. $\qquad\square$

A generalization of the above corollary to semigroups is given in Corollary 4.2.14 below.

Triangularizability of a pair of linear transformations can be characterized a little differently.

Theorem 1.3.4. (McCoy's Theorem) *The pair $\{A, B\}$ is triangularizable if and only if $p(A, B)(AB - BA)$ is nilpotent for every noncommutative polynomial p.*

Proof. If $\{A, B\}$ is triangularizable, then so is the algebra they generate, so the Spectral Mapping Theorem (Theorem 1.1.8) implies that $\sigma(p(A, B)(AB - BA)) = \{0\}$.

For the converse it suffices (by the Triangularization Lemma (1.1.4)) to show that $p(A, B)(AB - BA)$ always nilpotent implies that the algebra $\mathcal{A}$ generated by $\{A, B\}$ is reducible (on spaces of dimension greater than 1). If $AB = BA$, then $\mathcal{A}$ is reducible by Theorem 1.1.5. If $AB - BA \neq 0$, let $(AB - BA)x$ be different from 0. Choose any linear transformation C such that $C(AB - BA)x = x$. If $\mathcal{A}$ were not reducible, Burnside's Theorem (Theorem 1.2.2) would imply that C is in $\mathcal{A}$. But $\mathcal{A}$ consists of the noncommutative polynomials in $\{A, B\}$, and $C(AB - BA)$ is not nilpotent, so this is impossible. $\qquad\square$

Corollary 1.3.5. *If $AB = 0$ then the pair $\{A, B\}$ is triangularizable.*

Proof. Since $AB = 0$, $(p(A, B)(AB - BA))^2 = 0$ for all noncommutative polynomials p. $\qquad\square$

There is a curious sufficient condition that a pair of transformations be triangularizable.

Theorem 1.3.6. (Laffey's Theorem) *If $AB - BA$ has rank 1, then $\{A, B\}$ is triangularizable.*

Proof. Since quotients cannot increase rank, and since commuting transformations are triangularizable (Theorem 1.1.5), it suffices to show that there is a common nontrivial invariant subspace when A and B operate on a space of dimension greater than 1 (by the Triangularization Lemma (1.1.4)).

Choose λ in $\sigma(B)$; then the kernel and range of $(B - \lambda)$ are each nontrivial invariant subspaces of B. We show that at least one of them is invariant under A.

Every vector in the range of $AB - BA$ is a multiple of some fixed vector y. If the kernel of $B - \lambda$ is not invariant under A, then there is an x such that $(B-\lambda)x = 0$ but $(B-\lambda)Ax \neq 0$. Then $A(B-\lambda)x - (B-\lambda)Ax = ABx - BAx$ is a multiple of y. Thus $(B - \lambda)Ax$ is a multiple of y, and y is in the range of $(B - \lambda)$. For any z in $\mathcal{V}$, then,

$$A(B - \lambda)z = (B - \lambda)Az + \gamma y \quad \text{for some } \gamma,$$

so $A(B - \lambda)\mathcal{V} \subseteq (B - \lambda)\mathcal{V}$, and the range of $B - \lambda$ is a common invariant subspace of A and B. $\qquad\Box$

Corollary 1.3.7. *If $\mathcal{A}$ is an algebra of linear transformations with the property that the rank of $AB - BA$ is at most 1 when A and B are in $\mathcal{A}$, then $\mathcal{A}$ is triangularizable.*

Proof. By Corollary 1.3.3, it suffices to show that each pair of transformations in $\mathcal{A}$ is triangularizable. But if $\{A, B\} \subseteq \mathcal{A}$, then $\{A, B\}$ is triangularizable either by Laffey's Theorem (1.3.6) or by the commutative case (1.1.5). $\qquad\Box$

The following result will be needed in Chapter 4.

Theorem 1.3.8. *A collection $\mathcal{E}$ of operators is triangularizable if and only if it has the property that, for every choice of integer m and members $R_1, R_2, \ldots, R_m, S, T$ of $\mathcal{E}$, the operator $R_1 R_2 \cdots R_m(ST - TS)$ is nilpotent.*

Proof. The necessity of the condition is clear from Theorem 1.1.8. To prove sufficiency, note that the property is certainly inherited by quotients, so, by the Triangularization Lemma, it suffices to show reducibility. We can assume, in view of Theorem 1.1.5, that $\mathcal{E}$ is not commutative. Thus fix a noncommutative pair S and T in $\mathcal{E}$ and let $\mathcal{A}$ be the algebra generated by $\mathcal{E}$. Since $\mathcal{A}$ is the linear span of products of the form $R_1 R_2 \cdots R_m$ with R_i in $\mathcal{E}$, it follows from the hypothesis that $A(ST - TS)$ has trace zero for all A in $\mathcal{A}$. If $\mathcal{E}$ were irreducible, then it would follow from Burnside's Theorem that the trace of $E(ST - TS)$ is 0 for every operator E. But this would yield $ST - TS = 0$, for if D is any nonzero linear transformation, there is a linear transformation C such that the trace of CD is not 0. $\qquad\Box$

The following sharpens McCoy's Theorem (1.3.4).

Corollary 1.3.9. *The pair $\{A, B\}$ is triangularizable if and only if $w(A, B)(AB - BA)$ is nilpotent for every word $w(A, B)$ in A and B.*

Proof. The proof is almost identical to that of the previous theorem. To show reducibility, simply note that the trace of $p(A, B)(AB - BA)$ is 0 for every noncommutative polynomial p (since the trace is linear), so the algebra generated by $\{A, B\}$ is proper, and Burnside's Theorem (1.2.2) implies that the algebra is reducible.

$\square$

1.4 Triangularization and the Radical

We show (Theorem 1.4.6) that a subalgebra of $\mathcal{B}(\mathcal{V})$ is triangularizable if and only if it is commutative modulo its radical.

For the following basic definitions and properties of the radical of an algebra, $\mathcal{A}$ can be any unital algebra over any algebraically closed field (not necessarily a subalgebra of $\mathcal{B}(\mathcal{V})$).

Definition 1.4.1. For A any element of the unital algebra $\mathcal{A}$, A is *quasinilpotent* if $\sigma(A) \subseteq \{0\}$.

(The terminology comes from the case where $\mathcal{A}$ is a Banach algebra; then $\sigma(A) = \{0\}$ if and only if $\lim_{n \to \infty} \|A^n\|^{\frac{1}{n}} = 0$ (cf. [Theorem 6.1.10])). If A is a quasinilpotent linear transformation on a finite-dimensional space, then A is nilpotent.

Lemma 1.4.2. *For A and B in the unital algebra $\mathcal{A}$, $\sigma(AB) \cup \{0\} = \sigma(BA) \cup \{0\}$. In particular, $\sigma(AB) \subseteq \{0\}$ if and only if $\sigma(BA) \subseteq \{0\}$. If A and B are in $\mathcal{B}(\mathcal{V})$, then $\sigma(AB) = \sigma(BA)$.*

Proof. We must show that $\lambda - AB$ invertible implies $\lambda - BA$ invertible when $\lambda \neq 0$. If $(\lambda - AB)^{-1}$ exists, then

$$(\lambda - BA) \left(\frac{1}{\lambda} B (\lambda - AB)^{-1} A + \frac{1}{\lambda} \right)$$

$$= B(\lambda - AB)^{-1} A - \frac{1}{\lambda} BAB(\lambda - AB)^{-1} A + 1 - \frac{1}{\lambda} BA$$

$$= B \left((1 - \frac{1}{\lambda} AB)(\lambda - AB)^{-1} A \right) + 1 - \frac{1}{\lambda} BA$$

$$= \frac{1}{\lambda} B \left((\lambda - AB)(\lambda - AB)^{-1} A \right) + 1 - \frac{1}{\lambda} BA$$

$$= 1.$$

Similarly, $\left(\frac{1}{\lambda} B(\lambda - AB)^{-1} A + \frac{1}{\lambda} \right) (\lambda - BA) = 1$, so $\lambda - BA$ is invertible. For A and B in $\mathcal{B}(\mathcal{V})$, AB is invertible if and only if A and B are both invertible, so $\sigma(AB) = \sigma(BA)$.

$\square$

Definition 1.4.3. The *radical* of the algebra $\mathcal{A}$ is

$$\{A \in \mathcal{A} : \sigma(AB) \subseteq \{0\} \quad \text{for all} \quad B \in \mathcal{A}\}.$$

The radical of $\mathcal{A}$ is denoted by Rad $\mathcal{A}$.

By Lemma 1.4.2, the following is an equivalent definition.

$$\text{Rad}\mathcal{A} = \{A \in \mathcal{A} : \sigma(BA) \subseteq \{0\} \quad \text{for all} \quad B \in \mathcal{A}\}.$$

The radical of a unital algebra is thus the set of "permanent" quasinilpotents; i.e., the elements whose products with all other elements are quasinilpotent.

There are many equivalent definitions of Rad $\mathcal{A}$, including the following.

Theorem 1.4.4. *For any unital algebra $\mathcal{A}$, Rad $\mathcal{A}$ is the intersection of all maximal right ideals of $\mathcal{A}$, and is also the intersection of all maximal left ideals of $\mathcal{A}$.*

Proof. Suppose that $A \in$ Rad $\mathcal{A}$ and $\mathcal{R}$ is a maximal right ideal of $\mathcal{A}$. If A were not in $\mathcal{R}$, then the right ideal $\mathcal{R} + A\mathcal{A}$ would properly contain $\mathcal{R}$ and would therefore have to be $\mathcal{A}$. Then $R + AB = I$ for some R in $\mathcal{R}$ and B in $\mathcal{A}$. But then $R = I - AB$ would be an invertible element of $\mathcal{R}$, contradicting the fact that $\mathcal{R}$ is proper. Therefore, A is in $\mathcal{R}$. The same proof shows that A is in every maximal left ideal (using Lemma 1.4.2).

Conversely, suppose that A is in the intersection of all maximal right ideals of $\mathcal{A}$. If $\lambda - AB$ were not invertible for some B in $\mathcal{A}$ and nonzero λ, then, after multiplication by $\frac{1}{\lambda}$, it would follow that $I - AB_0$ is not invertible for $B_0 = \frac{1}{\lambda}B$. We claim that then there is a B_1 in $\mathcal{A}$ such that $I - AB_1$ has no right inverse in $\mathcal{A}$. To see this, assume that for every B_1 there is a C in $\mathcal{A}$ with $(I - AB_1)C = I$. For the given B_0, then, let $D = I - C$, so that $(I - AB_0)(I - D) = I$. Then

$$I - AB_0 - D + AB_0 D = I,$$

so

$$D = AB_0 D - AB_0 = A(B_0 D - B_0).$$

Since $D = AB_2$ with $B_2 = B_0 D - B_0$ in $\mathcal{A}$, the current assumption implies that $I - D$ has a right inverse in $\mathcal{A}$. But $I - D$ has a left inverse, $I - AB_0$, so $I - D$ is invertible. Thus

$$I - AB_0 = (I - D)^{-1}$$

is also invertible, which is a contradiction. We conclude that $I - AB_1$ has no right inverse for some B_1 in $\mathcal{A}$.

It now follows that $(I - AB_1)\mathcal{A}$ is a proper right ideal, and hence is contained in some maximal right ideal $\mathcal{R}_0$. In particular, $I - AB_1 \in \mathcal{R}_0$. Since A is also in $\mathcal{R}_0$, by hypothesis, this implies that the identity is in $\mathcal{R}_0$, which is a contradiction. Thus, the intersection of the maximal right ideals is contained in Rad $\mathcal{A}$. The same proof (but multiplying on the other side) shows that the intersection of the maximal left ideals is contained in Rad $\mathcal{A}$.   □

Corollary 1.4.5. *For any unital algebra* $\mathcal{A}$, Rad $\mathcal{A}$ *is a two-sided ideal of* $\mathcal{A}$.

Proof. By Theorem 1.4.4, Rad $\mathcal{A}$ is a right ideal since it is an intersection of right ideals, and Rad $\mathcal{A}$ is a left ideal since it is an intersection of left ideals.   □

Since Rad $\mathcal{A}$ is a two-sided ideal, the quotient $\mathcal{A}/\mathrm{Rad}\,\mathcal{A}$ is an algebra under the standard operations on cosets (e.g., $(A + \mathrm{Rad}\,\mathcal{A})(B + \mathrm{Rad}\,\mathcal{A}) = AB + \mathrm{Rad}\,\mathcal{A}$).

Triangularizability can be neatly characterized in terms of the radical.

Theorem 1.4.6. *If* $\mathcal{A}$ *is a unital subalgebra of* $\mathcal{B}(\mathcal{V})$, *then* $\mathcal{A}$ *is triangularizable if and only if* $\mathcal{A}/\mathrm{Rad}\,\mathcal{A}$ *is commutative.*

Proof. The proof is like that of McCoy's Theorem (Theorem 1.3.4). For if $\mathcal{A}$ is triangularizable, and if B and C are in $\mathcal{A}$, then for every A in $\mathcal{A}$, $\sigma((BC - CB)A) = \{0\}$ by the Spectral Mapping Theorem (Theorem 1.1.8). Hence every commutator $BC - CB$ is in Rad $\mathcal{A}$, and $\mathcal{A}/\mathrm{Rad}\,\mathcal{A}$ is commutative.

For the converse, note that $\mathcal{A}/\mathrm{Rad}\,\mathcal{A}$ commutative implies $(BC - CB) \in$ Rad $\mathcal{A}$ for all B and C in $\mathcal{A}$, so $\mathcal{A}$ is triangularizable by Theorem 1.3.2.   □

Definition 1.4.7. *A unital algebra is* semisimple *if its radical is* $\{0\}$.

Corollary 1.4.8. *A semisimple subalgebra of* $\mathcal{B}(\mathcal{V})$ *is triangularizable if and only if it is commutative.*

Proof. This follows immediately from Theorem 1.4.6.   □

1.5 Block Triangularization and Characterizations of Triangularizability

Given any subalgebra $\mathcal{A}$ of $\mathcal{B}(\mathcal{V})$, we can "triangularize $\mathcal{A}$ as much as possible" by finding maximal chains of invariant subspaces for $\mathcal{A}$. The following theorem on block triangularization is cumbersome to state and to prove but has several interesting consequences.

Theorem 1.5.1. *If $\mathcal{A}$ is a unital subalgebra of $\mathcal{B}(\mathcal{V})$, there exists a basis for $\mathcal{V}$ and a direct-sum decomposition $\mathcal{V} = \mathcal{N}_1 + \mathcal{N}_2 + \cdots + \mathcal{N}_k$ with respect to which every transformation A in $\mathcal{A}$ has the block upper triangular form*

$$
A = \begin{pmatrix}
A_{11} & A_{12} & \cdots & & A_{1k} \\
0 & A_{22} & \cdots & & A_{2k} \\
0 & 0 & A_{33} & & \vdots \\
\vdots & \vdots & & \ddots & \vdots \\
0 & 0 & \cdots & & A_{kk}
\end{pmatrix},
$$

where the set $\{1, 2, \ldots, k\}$ is the disjoint union of subsets $J_1, J_2, \ldots, J_\ell$ such that

(i) $\{A_{ii} : A \in \mathcal{A}\} = \mathcal{B}(\mathcal{N}_i)$ *for* $i = 1, 2, \ldots, k$;
(ii) *if i and j are both in the same J_s, then*

$$A_{ii} = A_{jj} \quad \text{for all} \quad A \text{ in } \mathcal{A};$$

(iii) *if i and j are in different subsets J_i and J_j, then the set of pairs (A_{ii}, A_{jj}) that arise as A ranges over $\mathcal{A}$ is $\mathcal{B}(\mathcal{N}_i) \times \mathcal{B}(\mathcal{N}_j)$; and*
(iv) *if i is in J_s, there is an A in $\mathcal{A}$ such that $A_{ii} = I$ and $A_{jj} = 0$ when j is not in J_s.*

Proof. If $\mathcal{A}$ has no nontrivial invariant subspaces, then Burnside's Theorem (Theorem 1.2.2) gives the result, with $k = 1$ and $\mathcal{N}_1 = \mathcal{V}$.

If $\mathcal{A}$ is reducible, let

$$\mathcal{M}_0 = \{0\} \subset \mathcal{M}_1 \subset \mathcal{M}_2 \subset \cdots \subset \mathcal{M}_{m-1} \subset \mathcal{M}_m = \mathcal{V}$$

be a maximal chain of invariant subspaces of $\mathcal{A}$. For each $i = 1, \ldots, m$, choose any complementary subspace $\mathcal{N}_i$ to $\mathcal{M}_{i-1}$ in $\mathcal{M}_i$, so that $\mathcal{M}_i = \mathcal{M}_{i-1} + \mathcal{N}_i$. Then $\mathcal{V}$ is the direct sum $\mathcal{N}_1 + \mathcal{N}_2 + \cdots + \mathcal{N}_m$ and, since the $\{\mathcal{M}_i\}$ are invariant under $\mathcal{A}$, the transformations in $\mathcal{A}$ are all block upper triangular with respect to this decomposition of $\mathcal{V}$. For each i, let P_i denote the projection of $\mathcal{V}$ onto $\mathcal{N}_i$ along $\sum_{j \neq i} \mathcal{N}_j$; then $P_1 + P_2 + \cdots + P_m = I$.

For each i, let $\mathcal{A}_i$ denote $\{P_i A |_{\mathcal{N}_i} : A \in \mathcal{A}\}$, where $P_i A |_{\mathcal{N}_i}$ is the restriction of $P_i A$ to $\mathcal{N}_i$. That is, $\mathcal{A}_i$ is the "compression" of $\mathcal{A}$ to $\mathcal{N}_i$. Then each $\mathcal{A}_i$ is an algebra, since $\mathcal{M}_i$ and $\mathcal{M}_{i-1}$ are both invariant under $\mathcal{A}$. Moreover, each $\mathcal{A}_i$ is irreducible, for if $\mathcal{L}_i$ were a nontrivial invariant subspace of $\mathcal{A}_i$ then $\mathcal{M}_{i-1} + \mathcal{L}_i$ would be invariant under $\mathcal{A}$ and properly between $\mathcal{M}_{i-1}$ and $\mathcal{M}_i$, contradicting the maximality of the chain $\{\mathcal{M}_0, \mathcal{M}_1, \ldots, \mathcal{M}_m\}$.

By Burnside's Theorem (Theorem 1.2.2), $\mathcal{A}_i = \mathcal{B}(\mathcal{N}_i)$ for each i.

We now show that the $\{\mathcal{A}_i\}$ are pairwise either independent or linked, in the following senses. Fix distinct i and j. If there is an A in $\mathcal{A}$ such that

$P_i A \big|_{\mathcal{N}_i}$ is the identity on $\mathcal{N}_i$ and $P_j A \big|_{\mathcal{N}_j}$ is 0 on $\mathcal{N}_j$, then we say that $\mathcal{A}_i$ is independent of $\mathcal{A}_j$. This relation is symmetric, for if A shows that $\mathcal{A}_i$ is independent of $\mathcal{A}_j$, then $(I - A)$ shows that $\mathcal{A}_j$ is independent of $\mathcal{A}_i$.

If $\mathcal{A}_i$ and $\mathcal{A}_j$ are not independent, we say that they are linked; this terminology is justified by the following. Suppose that there is some A in $\mathcal{A}$ with $P_i A \big|_{\mathcal{N}_i} \neq 0$ and $P_j A \big|_{\mathcal{N}_j} = 0$. Then $\{ P_i A \big|_{\mathcal{N}_i} : A \in \mathcal{A} \text{ and } P_j A \big|_{\mathcal{N}_j} = 0 \}$ is a two-sided ideal of $\mathcal{A}_i$ that is not $\{0\}$. But $\mathcal{A}_i = \mathcal{B}(\mathcal{N}_i)$, and the only two-sided ideals of $\mathcal{B}(\mathcal{N}_i)$ are $\{0\}$ and $\mathcal{B}(\mathcal{N}_i)$ (Theorem 1.2.3). Thus the identity is in this ideal, and it follows that $\mathcal{A}_i$ and $\mathcal{A}_j$ are independent. Thus if $\mathcal{A}_i$ and $\mathcal{A}_j$ are not independent and A is in $\mathcal{A}$, $P_i A \big|_{\mathcal{N}_i} = 0$ if and only if $P_j A \big|_{\mathcal{N}_j} = 0$.

Now assume that $\mathcal{A}_i$ and $\mathcal{A}_j$ are linked. Then we can define a mapping ϕ from $\mathcal{A}_i$ to $\mathcal{A}_j$ by

$$\phi\left(P_i A \big|_{\mathcal{N}_i} \right) = P_j A \big|_{\mathcal{N}_j},$$

and the failure of independence implies, by the above, that ϕ is well-defined and injective. The fact that $\mathcal{A}$ is an algebra makes ϕ a homomorphism. Moreover, ϕ is clearly surjective. Thus ϕ is an isomorphism between the algebras $\mathcal{A}_i$ and $\mathcal{A}_j$.

Now, $\mathcal{A}_i = \mathcal{B}(\mathcal{N}_i)$ and $\mathcal{A}_j = \mathcal{B}(\mathcal{N}_j)$. Thus $\mathcal{N}_i$ and $\mathcal{N}_j$ have the same dimension and can be identified with each other, so that ϕ becomes an automorphism of $\mathcal{B}(\mathcal{N}_i)$. Since every automorphism of $\mathcal{B}(\mathcal{N}_i)$ is spatial (Theorem 1.2.4), there is an invertible linear transformation S from $\mathcal{N}_i$ onto $\mathcal{N}_j$ that implements ϕ; i.e.,

$$P_i A \big|_{\mathcal{N}_i} = S^{-1} \left(P_j A \big|_{\mathcal{N}_j} \right) S$$

for all A in $\mathcal{A}$. Hence the terminology "linked."

The proof of the theorem can now be completed by simply identifying linked $\mathcal{A}_i$ with each other. If some $\mathcal{A}_j$ is linked to $\mathcal{A}_i$, choose $S : \mathcal{N}_i \to \mathcal{N}_j$ satisfying

$$P_i A \big|_{\mathcal{N}_i} = S^{-1} \left(P_j A \big|_{\mathcal{N}_j} \right) S \quad \text{for all} \quad A \text{ in } \mathcal{A}.$$

Then let T be the transformation on $\mathcal{V}$ defined, with respect to the decomposition

$$\mathcal{V} = \mathcal{N}_1 + \mathcal{N}_2 + \cdots + \mathcal{N}_m,$$

by the block-diagonal matrix having suitable-size identities everywhere on the main diagonal, except that S occurs in position (j, j). Then every element of $T^{-1} A T$ has the same entry in position (j, j) as it has in position (i, i).

Divide the set $\{1, 2, \ldots, k\}$ into disjoint subsets $\{\mathcal{J}_s\}$ where i and j are in the same subset if and only if $\mathcal{A}_i$ and $\mathcal{A}_j$ are linked.

For each index set $\mathcal{J}_s$, let i_s be the smallest integer in $\mathcal{J}_s$ and identify every $\mathcal{A}_j$ for j in $\mathcal{J}_s$ with $\mathcal{A}_{i_s}$ by creating a T as above. If R is the product of all such T's, then $R^{-1}AR$ has the required properties.   □

A number of characterizations of triangularizability follow from this theorem. We begin with an observation that has several applications.

Lemma 1.5.2. *If a collection $\mathcal{S}$ is triangularizable, then every chain of invariant subspaces of $\mathcal{S}$ is contained in a triangularizing chain.*

Proof. Given a chain of invariant subspaces of $\mathcal{S}$, consider the collection of all chains of invariant subspaces of $\mathcal{S}$ that contain the given chain. There is a maximal such chain, say

$$\{0\} = \mathcal{M}_0 \subset \mathcal{M}_1 \subset \cdots \subset \mathcal{M}_m = \mathcal{V}.$$

We show that any such maximal chain is triangularizing. If not, there would be an i such that the dimension of $\mathcal{M}_i/\mathcal{M}_{i-1}$ is greater than 1. By the maximality of the chain the set $\tilde{\mathcal{S}}$ of quotients of $\mathcal{S}$ on $\mathcal{M}_i/\mathcal{M}_{i-1}$ would be irreducible. Hence the algebra $\tilde{\mathcal{A}}$ generated by $\tilde{\mathcal{S}}$ would be $\mathcal{B}(\mathcal{M}_i/\mathcal{M}_{i-1})$ (Theorem 1.2.2) and there would be elements $\tilde{A}$ and $\tilde{B}$ in $\tilde{\mathcal{A}}$ such that $\sigma(\tilde{A}\tilde{B} - \tilde{B}\tilde{A})$ contains a nonzero scalar. Hence there would be transformations A and B in the algebra generated by $\mathcal{S}$ such that $\sigma(AB - BA) \neq \{0\}$. But the algebra generated by $\mathcal{S}$ is triangularizable, so this would contradict the Spectral Mapping Theorem (Theorem 1.1.8).   □

Corollary 1.5.3. *An algebra $\mathcal{A}$ is triangularizable if and only if the spaces $\{\mathcal{N}_i\}$ occurring in its block triangularization of Theorem 1.5.1 all have dimension 1.*

Proof. The spaces $\{\mathcal{N}_i\}$ are constructed in the proof of Theorem 1.5.1 as complements of $\mathcal{M}_{i-1}$ in $\mathcal{M}_i$, where $\{\mathcal{M}_i\}_{i=0}^m$ is a maximal chain of invariant subspaces of $\mathcal{A}$. If $\mathcal{A}$ is triangularizable, then $\{\mathcal{M}_i\}_{i=0}^m$ is triangularizing, by Lemma 1.5.2.   □

There are many conditions equivalent to triangularizability in addition to those given above.

Theorem 1.5.4. *For unital subalgebras $\mathcal{A}$ of $\mathcal{B}(\mathcal{V})$, the following are equivalent:*

   (i) *$\mathcal{A}$ is triangularizable,*
   (ii) *$A + B$ is nilpotent whenever A and B are nilpotent elements of $\mathcal{A}$,*

(iii) *AB is nilpotent whenever A and B are elements of $\mathcal{A}$ one of which is nilpotent.*

If the characteristic of the field is zero, then each of the above is equivalent to the following assertion:

(iv) *The trace of $(AB - BA)^2$ is 0 for all A and B in $\mathcal{A}$.*

Proof. First, (i) implies (ii) and (iii) by the Spectral Mapping Theorem (Theorem 1.1.8). Similarly, if $\{A, B\}$ is triangularized, then the diagonal entries of $AB - BA$, and thus also of $(AB - BA)^2$, are all 0, so (i) implies (iv) as well.

To show that each of the other conditions implies triangularizability, we use Theorem 1.5.1. If $\mathcal{A}$ is not triangularizable, then at least one of the $\{\mathcal{N}_i\}$ occurring in Theorem 1.5.1 has dimension greater than 1. We show that this implies that none of (ii) through (iv) holds.

Choose any $\mathcal{N}_i$ of dimension greater than 1. The corresponding $\mathcal{A}_i$ may or may not be linked to other $\mathcal{A}_j$'s. In either case, however, there exist in $\mathcal{A}_i$ transformations A_i and B_i such that

$$A_i = \begin{pmatrix} 0 & 1 \\ 0 & 0 \end{pmatrix} \oplus 0 \quad \text{and} \quad B_i = \begin{pmatrix} 0 & 0 \\ 1 & 0 \end{pmatrix} \oplus 0,$$

where the 0 direct summand is absent if the dimension of $\mathcal{N}_i$ is 2, and is of size $(k - 2) \times (k - 2)$ if the dimension of $\mathcal{N}_i$ is $k > 2$. Then in $\mathcal{A}$ there are transformations A and B whose compressions to $\mathcal{N}_i$ are A_i and B_i, respectively, which have the further property that their block diagonal entries are each either 0 or are A_i and B_i, respectively.

Thus A^2 and B^2 both have all of their diagonal blocks equal to 0, so A and B are nilpotent. However, the compression of $A + B$ to $\mathcal{N}_i$ is $\begin{pmatrix} 0 & 1 \\ 1 & 0 \end{pmatrix} \oplus 0$,

so $(A + B)^2$ has diagonal block $\begin{pmatrix} 1 & 0 \\ 0 & 1 \end{pmatrix} \oplus 0$ and is therefore not nilpotent; thus (ii) fails.

Similarly, AB has compression $\begin{pmatrix} 1 & 0 \\ 0 & 0 \end{pmatrix} \oplus 0$ to $\mathcal{N}_i$ and is not nilpotent, so (iii) does not hold.

For (iv), note that

$$A_i B_i - B_i A_i = \begin{pmatrix} 1 & 0 \\ 0 & -1 \end{pmatrix} \oplus 0,$$

so the trace of $(AB - BA)^2$ is at least 2 (since the characteristic of the field is 0).

$\square$

If the field of scalars is $\mathbb{C}$, we can state other equivalent conditions. We use $\rho(A)$ to denote the spectral radius of the linear transformation A defined by $\rho(A) = \sup\{|z| : z \in \sigma(A)\}$. The trace of the matrix A is denoted by $tr\,A$.

Theorem 1.5.5. *If A is a unital subalgebra of $\mathcal{B}(V)$ and V is a finite-dimensional vector space over $\mathbb{C}$, then the following are equivalent:*

(i) *A is triangularizable,*

(ii) *there is an $L > 0$ such that $\rho(A + B) \leq L(\rho(A) + \rho(B))$ for all A and B in A,*

(iii) *there is an $M > 0$ such that $\rho(AB) \leq M((\rho(A)\rho(B))$ for all A and B in A,*

(iv) *there is an $N > 0$ such that if A and B are in A, and if $\sigma(A)$ and $\sigma(B)$ are both contained in $\{t : t \geq 0\}$, then $tr\,(AB) \leq N(tr\,A)(tr\,B)$.*

Proof. If A is in triangular form, then it is clear that conditions (ii), (iii), and (iv) hold with $L = M = N = 1$.

Conversely, if A is not triangularizable, then conditions (ii) and (iii) of Theorem 1.5.4 show that conditions (ii) and (iii) of this theorem do not hold.

To show that (iv) fails if A is not triangularizable we proceed as in the proof of Theorem 1.5.4. If A and B are as defined in that proof, then the traces of A and B are both 0, but the trace of AB is a positive integer. $\qquad\square$

1.6 Approximate Commutativity

In this section we restrict our attention to transformations on finite-dimensional vector spaces over $\mathbb{C}$. There are several concepts of approximate commutativity that are equivalent to triangularizability. In the following, $\|\cdot\|$ is the usual operator norm

$$\|T\| = \sup\{\|Tx\| : \|x\| = 1\}.$$

Definition 1.6.1. If A is a matrix, then the *diagonal of A* is the diagonal matrix with the same entries as A on the main diagonal and all other entries 0. We use $\mathcal{D}(A)$ to denote the diagonal of A.

Similarities can bring an upper triangular matrix close to its diagonal.

Theorem 1.6.2. *If $\mathcal{A}$ is the algebra of upper triangular matrices relative to a given basis, then for every $\varepsilon > 0$ there is an invertible matrix S_ε such that*

$$\|S_\varepsilon^{-1} A S_\varepsilon - \mathcal{D}(A)\| \le \varepsilon \|A\|$$

for all A in $\mathcal{A}$.

Proof. Since all norms on a finite-dimensional space are equivalent (see Conway [1], p. 69), we can assume that the space is an inner-product space. Fix $\varepsilon > 0$, let $\eta \in (0,1)$ be a constant to be determined, and let S be the diagonal matrix with diagonal $\{\eta, \eta^2, \ldots, \eta^n\}$, where n is the dimension of the underlying space.

For each A in $\mathcal{A}$, the matrix $A - \mathcal{D}(A)$ is strictly upper triangular. If $Q = ((q_{ij}))$ is any strictly upper triangular matrix, then $S^{-1}QS$ is

$$
\begin{pmatrix}
\eta^{-1} & 0 & \cdots & 0 \\
0 & \eta^{-2} & & \vdots \\
\vdots & & \ddots & 0 \\
0 & & \cdots & \eta^{-n}
\end{pmatrix}
\begin{pmatrix}
0 & q_{12} & \cdots & q_{1n} \\
 & 0 & & \\
 & & \ddots & \vdots \\
 & & \ddots & q_{n-1,n} \\
0 & & & 0
\end{pmatrix}
\begin{pmatrix}
\eta & 0 & \cdots & 0 \\
0 & \eta^2 & & \vdots \\
\vdots & & \ddots & 0 \\
0 & & \cdots & \eta^n
\end{pmatrix}
$$

$$
=
\begin{pmatrix}
0 & q_{12}\eta & q_{13}\eta^2 & & & q_{1n}\eta^{n-1} \\
 & 0 & q_{23}\eta & q_{24}\eta^2 & & \\
 & & 0 & q_{34}\eta & & \vdots \\
\vdots & & & 0 & \ddots & \\
 & & & & \ddots & q_{n-2,n}\eta^2 \\
 & & & & \ddots & q_{n-1,n}\eta \\
0 & & \cdots & & & 0
\end{pmatrix}.
$$

That is, the entries of $S^{-1}QS$ are obtained from those of Q by multiplying the ith superdiagonal by η^i. If we factor η from $S^{-1}QS$ and estimate very roughly (e.g., each superdiagonal of $\frac{1}{\eta}S^{-1}QS$ has norm at most $\|Q\|$), we see that $\|S^{-1}QS\| \le n\eta\|Q\|$ (since there are $n-1$ superdiagonals). Now in the case where $Q = A - \mathcal{D}(A)$, $\|Q\|$ is certainly at most $2\|A\|$ (since $\|\mathcal{D}(A)\| \le \|A\|$), which yields $\|S^{-1}(A - \mathcal{D}(A))S\| \le 2n\eta\|A\|$ for all A in $\mathcal{A}$. For given $\varepsilon > 0$, let $\eta = \frac{\varepsilon}{2n}$, and let S_ε be the corresponding S. Then, for each A in $\mathcal{A}$,

$$\|S_\varepsilon^{-1} A S_\varepsilon - \mathcal{D}(A)\| = \|S_\varepsilon^{-1}(A - \mathcal{D}(A))S_\varepsilon\| \le \varepsilon \|A\|. \qquad \square$$

Corollary 1.6.3. *If $\mathcal{C} = \{A_\alpha : \alpha \in I\}$ is a norm-bounded triangularizable family of linear transformations on a finite-dimensional inner-product* space, then there is a commutative family $\{D_\alpha : \alpha \in I\}$ of normal transformations such that for every $\varepsilon > 0$ there is an invertible transformation S_ε satisfying

$$\|S_\varepsilon^{-1} A_\alpha S_\varepsilon - D_\alpha\| < \varepsilon \quad \text{for all} \quad \alpha \quad \text{in } I.$$

Proof. Triangularize $\mathcal{C}$ by an appropriate similarity T and then let $D_\alpha = \mathcal{D}(T^{-1} A_\alpha T)$ for each α. Theorem 1.6.2 then gives the result. $\qquad \square$

Theorem 1.6.2 and Corollary 1.6.3 show that triangularizability implies approximate commutativity in a certain sense. The converses of these results also hold. In fact, even much weaker notions of approximate commutativity imply triangularizability. In particular, the commuting operators can depend on ε and not be normal.

Theorem 1.6.4. *Suppose that $\mathcal{C}$ is a family of linear transformations with the property that, for each finite subfamily $\{A_1, A_2, \dots, A_m\}$, there is a constant $K > 0$ such that for every $\varepsilon > 0$ there exist a commutative family $\{D_1, D_2, \dots, D_m\}$ and an invertible S satisfying $\|S^{-1} A_j S - D_j\| < \varepsilon$ and $\|D_j\| < K$ for every j. Then $\mathcal{C}$ is triangularizable.*

Proof. Let $\mathcal{A}$ be the unital algebra generated by $\mathcal{C}$. By Theorem 1.3.2, it suffices to show that every commutator $(BC - CB)$ of elements of $\mathcal{A}$ is nilpotent.

Fix B and C in $\mathcal{A}$. There are a set of transformations $\{A_1, \dots, A_m\}$ in $\mathcal{C}$ and noncommutative polynomials p and q such that $B = p(A_1, \dots, A_m)$ and $C = q(A_1, \dots, A_m)$. Choose an appropriate K for the set $\{A_1, \dots, A_m\}$. Let $h(A_1, \dots, A_m)$ be the polynomial

$$p(A_1, \dots, A_m) q(A_1, \dots, A_m) - q(A_1, \dots, A_m) p(A_1, \dots, A_m).$$

We must show that $\sigma(h(A_1, \dots, A_m)) = \{0\}$.

Now, h is a uniformly continuous function of its arguments on any bounded set in $\mathcal{B}(\mathcal{V})$. In particular, for every $\eta > 0$ there is a $\delta > 0$ such that

$$\|h(X_1, \dots, X_m) - h(Y_1, \dots, Y_m)\| < \eta$$

whenever

$$\|X_i - Y_i\| < \delta, \ \|X_i\| \leq K + 1 \quad \text{and} \quad \|Y_i\| \leq K + 1$$

for all i.

Fix any $\eta > 0$ and determine the corresponding $\delta > 0$; we can, and do, assume that $\delta < 1$. Then, by the hypothesis, choose a commutative family $\{D_1, \dots, D_m\}$ and an invertible S such that $\|S^{-1} A_i S - D_i\| < \delta$ for all i. In particular, $\|S^{-1} A_i S\| < \delta + \|D_i\|$, so $\|S^{-1} A_i S\| < K + 1$ for all i.

Thus $\|h(S^{-1}A_1S, S^{-1}A_2S, \ldots, S^{-1}A_mS) - h(D_1, D_2, \ldots, D_m)\| < \eta$. But $h(D_1, D_2, \ldots, D_m) = 0$, since the $\{D_i\}$ commute, and

$$h(S^{-1}A_1S, S^{-1}A_2S, \ldots, S^{-1}A_mS) = S^{-1}(h(A_1, A_2, \ldots, A_m))S.$$

Therefore,

$$\|S^{-1}(h(A_1, A_2, \ldots, A_m))S\| < \eta, \text{ so } \rho(S^{-1}(h(A_1, A_2, \ldots, A_m))S) < \eta.$$

Since similarities do not change the spectrum, $\rho(h(A_1, A_2, \ldots, A_m)) < \eta$. This holds for all $\eta > 0$, so $\rho(h(A_1, A_2, \ldots, A_m)) = 0$. $\square$

If the transformations $\{D_j\}$ in the hypothesis of Theorem 1.6.4 are assumed to be normal, then we need not assume that they are bounded. To see this we require a lemma.

Lemma 1.6.5. *If A is a linear transformation, $\{D_n\}$ is a sequence of normal transformations, and $\{S_n\}$ is a sequence of invertible transformations such that*

$$\lim_{n \to \infty} \|S_n^{-1}AS_n - D_n\| = 0,$$

then $\{\|D_n\|\}$ is bounded.

Proof. If $\{\|D_n\|\}$ is not bounded, we can discard some D_n's and re-index so that $\|D_n\| > n$ and $\|S_n^{-1}AS_n - D_n\| < \frac{1}{n}$ for all n. Let $\rho_n = \rho(D_n)$. Since each D_n is normal, $\rho(D_n) = \|D_n\|$.
    Then

$$\|S_n^{-1}(\frac{1}{\rho_n}A)S_n - \frac{1}{\rho_n}D_n\| < \frac{1}{n^2}$$

for all n. But $\rho(\frac{1}{\rho_n}D_n) = 1$ for all n, and $\rho(S_n^{-1}\frac{1}{\rho_n}AS_n) = \frac{1}{\rho_n}\rho(A)$ approaches 0 as n approaches ∞. This contradicts the fact that the spectral radius is continuous (and therefore uniformly continuous on compact sets of operators). The continuity of the spectral radius is a special case of Lemma 3.1.2 below. $\square$

These results can be combined in several ways to get characterizations of triangularizability in terms of approximate commutativity. We specify two such characterizations.

Corollary 1.6.6. *For any collection $\mathcal{C}$ of linear transformations on a finite-dimensional inner-product space, the following are equivalent:*

(i) $\mathcal{C}$ *is triangularizable;*

(ii) *for each finite subset $\{A_1, \ldots, A_m\}$ of $\mathcal{C}$ there exists a commutative set of transformations $\{D_1, \ldots, D_m\}$ such that for every $\varepsilon > 0$ there is an invertible transformation S with*

$$\|S^{-1}A_iS - D_i\| < \varepsilon \quad \text{for all } i;$$

(iii) *for each finite subset $\{A_1, \ldots, A_m\}$ of $\mathcal{C}$ and every $\varepsilon > 0$ there exists a commutative set $\{D_1, \ldots, D_m\}$ of normal transformations and an invertible transformation S such that $\|S^{-1}A_iS - D_i\| < \varepsilon$ for all i.*

Proof. Corollary 1.6.3 shows that (i) implies (ii) and (iii). Theorem 1.6.4 is a stronger assertion than the statement that (ii) implies (i) (since here we have fixed $\{D_1, \ldots, D_n\}$ for all ε), and Theorem 1.6.4 and Lemma 1.6.5 together yield the assertion that (iii) implies (i). $\qquad\qquad\square$

1.7 Nonassociative Algebras

There are certain much-studied linear subspaces of $\mathcal{B}(\mathcal{V})$ that are closed under nonassociative multiplications.

Definition 1.7.1. A *Lie algebra of linear transformations* is a linear subspace $\mathcal{L}$ of $\mathcal{B}(\mathcal{V})$ such that the commutator $[A, B] = AB - BA$ is in $\mathcal{L}$ whenever A and B are in $\mathcal{L}$.

Definition 1.7.2. A *Jordan algebra of linear transformations* is a linear subspace $\mathcal{J}$ of $\mathcal{B}(\mathcal{V})$ such that the Jordan product $AB + BA$ is in $\mathcal{J}$ whenever A and B are in $\mathcal{J}$.

The best-known result on triangularization of nonassociative algebras is Engel's Theorem (Corollary 1.7.6 below) about Lie algebras of nilpotent transformations. This can be obtained as a consequence of a much more general theorem.

Theorem 1.7.3. *A set $\mathcal{N}$ of nilpotent transformations is triangularizable if it has the property that whenever A and B are in $\mathcal{N}$, there is a noncommutative polynomial p such that $AB + p(A, B)A$ is in $\mathcal{N}$.*

Proof. We proceed by induction on the dimension of the vector space $\mathcal{V}$. The result is trivially true for $n = 1$; assume it for dimensions up to n and let $\mathcal{V}$ have dimension $n + 1$.

Consider the collection $\mathcal{F}$ of all subspaces that are intersections of kernels of members of $\mathcal{N}$; i.e., subspaces $\mathcal{M}$ such that there is a subset $\mathcal{S}$ of $\mathcal{N}$ with

$$\mathcal{M} = \{x : Ax = 0 \quad \text{for} \quad A \in \mathcal{S}\}.$$

Let $\mathcal{K}$ be an element of $\mathcal{F}$ of minimal positive dimension and let

$$\mathcal{N}_0 = \{A \in \mathcal{N} : Ax = 0 \quad \text{for all} \quad x \in \mathcal{K}\}.$$

Note that the set of quotients of $\mathcal{N}_0$ with respect to $\mathcal{V}/\mathcal{K}$ satisfies the hypothesis of the theorem (this requires that $p(A, B)$ be a polynomial in A and B, as opposed to an arbitrary linear transformation) and is therefore triangularizable by the inductive hypothesis. Since the restrictions of the elements of $\mathcal{N}_0$ to $\mathcal{K}$ are all 0, it follows that $\mathcal{N}_0$ itself is triangularizable.

We will finish the proof by showing that $\mathcal{N}_0 = \mathcal{N}$. If not, there would be a B in $\mathcal{N}$ such that $Bx \neq 0$ for some $x \in \mathcal{K}$. Then $\mathcal{K}$ would not be invariant under B, for if it were, there would be an $x_0 \in \mathcal{K}$ with $x_0 \neq 0$ and $Bx_0 = 0$ (since B is nilpotent), which would imply that

$$\{x : Bx = 0 \quad \text{and} \quad Ax = 0 \quad \text{for all} \quad A \in \mathcal{N}_0\}$$

is an element of $\mathcal{F}$ of smaller positive dimension than $\mathcal{K}$.

Thus $\mathcal{K}$ is not invariant under B; since $\mathcal{K}$ is the intersection of the kernels of the members of $\mathcal{N}_0$, there is an x_1 in $\mathcal{K}$ and an A_1 in $\mathcal{N}_0$ such that $A_1 B x_1 \neq 0$. Now, $A_1 B + p_1(A_1, B)A_1$ is in $\mathcal{N}$ for some p_1. Let $B_1 = A_1 B + p_1(A_1, B)A_1$, and note that $B_1 x_1 \neq 0$. As shown in the case of B in the preceding paragraph, it follows that $\mathcal{K}$ is not invariant under B_1 and there is an A_2 in $\mathcal{N}_0$ such that $A_2 B_1 x_2 \neq 0$ for some $x_2 \in \mathcal{K}$. Choose p_2 such that $A_2 B_1 + p_2(A_2, B_1)A_2$ is in $\mathcal{N}$, and let $B_2 = A_2 B_1 + p_2(A_2, B_1)A_2$. Then $A_3 B_2 x_3 \neq 0$ for some $A_3 \in \mathcal{N}_0$ and $x_3 \in \mathcal{K}$.

Continue in this manner to obtain

$$\{A_1, A_2, \ldots, A_{n+1}\} \subseteq \mathcal{N}_0, \quad \{B_1, B_2, \ldots, B_n\} \subseteq \mathcal{N},$$

and vectors $\{x_1, x_2, \ldots, x_{n+1}\}$. Then

$$A_{n+1} A_n \cdots A_2 A_1 B x_{n+1} = A_{n+1} B_n x_{n+1} \neq 0.$$

However, $\mathcal{N}_0$ is a triangularizable collection of nilpotent transformations on a space of dimension $(n + 1)$, so the product of $(n + 1)$ members of $\mathcal{N}_0$ is 0. Thus $A_{n+1} A_n \cdots A_2 A_1 = 0$. This contradiction shows that $\mathcal{N} = \mathcal{N}_0$ and is therefore triangularizable. $\square$

Corollary 1.7.4. (Jacobson's Theorem) *A set $\mathcal{N}$ of nilpotent transformations is triangularizable if whenever A and B are in $\mathcal{N}$, there is a scalar c such that $AB - cBA$ is in $\mathcal{N}$.*

Proof. This is a special case of Theorem 1.7.3. $\square$

Corollary 1.7.5. *A set $\mathcal{L}$ of nilpotent transformations is triangularizable if it is closed under Lie products (i.e., if $A, B \in \mathcal{L}$ implies $(AB - BA) \in \mathcal{L}$).*

Proof. This is the case $c = 1$ of Corollary 1.7.4.   □

Corollary 1.7.6. (Engel's Theorem) *A Lie algebra of nilpotent transformations is triangularizable.*

Proof. This is a special case of the previous corollary.   □

Corollary 1.7.7. *A set $\mathcal{J}$ of nilpotent transformations is triangularizable if it is closed under Jordan products (i.e., if $A, B \in \mathcal{J}$ implies $(AB + BA) \in \mathcal{J}$).*

Proof. This is the case $c = -1$ of Corollary 1.7.4.   □

Engel's Theorem (Corollary 1.7.6) can be strengthened.

Corollary 1.7.8. *Let $\mathcal{L}$ be any set of linear transformations that is closed under commutation (i.e., $(AB - BA)$ is in $\mathcal{L}$ whenever A and B are in $\mathcal{L}$). Then $\mathcal{L}$ is triangularizable if and only if every commutator of elements of $\mathcal{L}$ is nilpotent.*

Proof. If $\mathcal{L}$ is triangularizable, then every commutator is nilpotent, by the Spectral Mapping Theorem (1.1.8).

For the converse, the Triangularization Lemma (1.1.4) implies that it suffices to show that $\mathcal{L}$ has a nontrivial invariant subspace. Let $\mathcal{N}$ denote the set of commutators of members of $\mathcal{L}$; if $\mathcal{N} = \{0\}$, then $\mathcal{L}$ is commutative and therefore triangularizable (Theorem 1.1.5). If $\mathcal{N} \neq \{0\}$, then Jacobson's Theorem (1.7.4) with $c = 1$ shows that $\mathcal{N}$ is triangularizable. Since the elements of $\mathcal{N}$ are nilpotent, the intersection of the kernels of the transformations in $\mathcal{N}$ is some nontrivial subspace $\mathcal{K}$.

We claim that $\mathcal{K}$ is invariant under $\mathcal{L}$. For suppose that $x \in \mathcal{K}$ and $B \in \mathcal{L}$. To show that $Bx \in \mathcal{K}$ we must show that $ABx = 0$ for all $A \in \mathcal{N}$. But if $A \in \mathcal{N}$, then $Ax = 0$ and $(AB - BA)x = 0$, so $ABx = 0$.   □

Corollary 1.7.9. *If $\mathcal{L}$ is a set of linear transformations that is closed under commutation and that has the property that the rank of every commutator is at most 1, then $\mathcal{L}$ is triangularizable.*

Proof. Since the trace of $(AB - BA)$ is always 0, and a transformation of rank 1 has trace λ if λ is a nonzero eigenvalue, it follows that $AB - BA$ is always nilpotent, so Corollary 1.7.8 applies.   □

A collection of linear transformations that are nilpotent of index two is triangularizable if certain very mild conditions are satisfied, such as in the following.

Theorem 1.7.10. *A collection of linear transformations each of which has square 0 is triangularizable if it is closed under addition.*

Proof. By the Triangularization Lemma (1.1.4), it suffices to show that such a collection, acting on a space of dimension greater than 1, has a non-trivial invariant subspace. For this, let A be any member of the collection other than 0 (the theorem is certainly true if the collection is simply $\{0\}$). Then $A\mathcal{V}$ is a nontrivial subspace, and $A\mathcal{V}$ is invariant under the collection, for if B is in the collection, then $(A + B)^2 = 0$ yields $AB = -BA$, so $BA\mathcal{V} \subseteq AB\mathcal{V} \subseteq A\mathcal{V}$. ☐

In spite of the above, linear subspaces consisting of nilpotent transformations are generally not triangularizable.

Example 1.7.11. *The set*

$$\left\{ \begin{pmatrix} 0 & \alpha & 0 \\ \beta & 0 & -\alpha \\ 0 & \beta & 0 \end{pmatrix} : \alpha, \beta \in \mathbb{C} \right\}$$

consists of nilpotent matrices but is irreducible.

Proof. Direct computation shows that

$$\begin{pmatrix} 0 & \alpha & 0 \\ \beta & 0 & -\alpha \\ 0 & \beta & 0 \end{pmatrix}^3 = 0 \qquad \text{for all } \alpha \text{ and } \beta.$$

Now, the algebra generated by this set contains

$$\begin{pmatrix} 0 & 0 & 0 \\ 1 & 0 & 0 \\ 0 & 1 & 0 \end{pmatrix} \begin{pmatrix} 0 & 1 & 0 \\ 0 & 0 & -1 \\ 0 & 0 & 0 \end{pmatrix} = \begin{pmatrix} 0 & 0 & 0 \\ 0 & 1 & 0 \\ 0 & 0 & -1 \end{pmatrix}$$

and its square, and therefore also contains

$$\begin{pmatrix} 0 & 0 & 0 \\ 0 & 1 & 0 \\ 0 & 0 & 0 \end{pmatrix} \qquad \text{and} \qquad \begin{pmatrix} 0 & 0 & 0 \\ 0 & 0 & 0 \\ 0 & 0 & 1 \end{pmatrix}.$$

Similarly, the algebra contains

$$\begin{pmatrix} 0 & 1 & 0 \\ 0 & 0 & -1 \\ 0 & 0 & 0 \end{pmatrix} \begin{pmatrix} 0 & 0 & 0 \\ 1 & 0 & 0 \\ 0 & 1 & 0 \end{pmatrix} = \begin{pmatrix} 1 & 0 & 0 \\ 0 & -1 & 0 \\ 0 & 0 & 0 \end{pmatrix}$$

and thus also $\begin{pmatrix} 1 & 0 & 0 \\ 0 & 0 & 0 \\ 0 & 0 & 0 \end{pmatrix}$. Hence the algebra contains all diagonal matrices. The only subspaces invariant under all diagonal matrices are the spans of basis vectors, and the only two of those that are invariant under both

$$\begin{pmatrix} 0 & 1 & 0 \\ 0 & 0 & -1 \\ 0 & 0 & 0 \end{pmatrix} \quad \text{and} \quad \begin{pmatrix} 0 & 0 & 0 \\ 1 & 0 & 0 \\ 0 & 1 & 0 \end{pmatrix}$$

are $\{0\}$ and $\mathcal{V}$. □

1.8 Notes and Remarks

The subject of triangularization began with the proof of Schur's Theorem (Corollary 1.1.6) by I. Schur in 1909 (Schur [1]; he stated it in the following form: If A is a square matrix over $\mathbb{C}$, then there is a unitary matrix U such that $U^{-1}AU$ is upper triangular). Schur [1] applied this to the study of integral equations; it has since been used in a wide variety of contexts.

The term "simultaneously triangularizable" is often used for what we have called "triangularizable." In our view, the latter is enough of a mouthful and completely expresses the idea. The Triangularization Lemma (1.4.1) is implicit in most proofs of triangularization theorems, although it appears to have first been explicitly formulated in Radjavi-Rosenthal [5]. Burnside's Theorem (1.2.2) was established in 1905 in Burnside [1] for groups of matrices; shortly afterwards, it was extended by Frobenius-Schur [1] to its present form. It is important in group theory as a characterization of irreducible representations; see Lam [2] for a discussion of this context. The proof presented above is from Halperin-Rosenthal [1]. The standard proof can be found in Jacobson [1, page 276]; other elementary proofs can be found in E. Rosenthal [1] and Lam [1].

A family $\mathcal{L}$ of operators on a finite-dimensional vector space V is said to be *transitive* if for every $x \neq 0$ and y in V there exists a member A of F such that $Ax = y$. For an algebra, the concepts of irreducibility and transitivity are easily seen to coincide, but in general, the latter is stronger. For extensions of Burnside-type results to Jordan algebras see Grunenfelder-Omladič-Radjavi [1], which includes the result that a transitive Jordan algebra in $\mathcal{M}_n(\mathbb{F})$ (where $\mathbb{F}$ is algebraically closed) is either $\mathcal{M}_n(\mathbb{F})$ or is simultaneously similar to the Jordan algebra of all symmetric matrices.

Frobenius [1] showed that a commutative pair of matrices has the spectral mapping property. In 1935, Williamson [1] investigated the relationship between simultaneous triangularity of a pair of matrices and the spectral mapping property, and showed that they are equivalent if one of the matrices is nonderogatory. Shortly afterwards, McCoy [2] showed that they are always equivalent, and proved the stronger assertion that we have called "McCoy's Theorem" (1.3.4). McCoy [1] had earlier investigated quasi-commutative matrices (i.e., pairs of matrices that commute with their commutator) and shown that they are triangularizable — this has a proof similar to that of McCoy's Theorem (1.3.4). The fact that $\{A, B\}$ is triangularizable if $AB = 0$ is a special case of McCoy's Theorem (Corollary 1.3.5) was observed years ago in Schneider [2]. Theorem 1.3.8 is due to Guralnick [1]. The strengthening of McCoy's Theorem given in Corollary 1.3.9 was pointed out by Eric Nordgren after he read the proof of Theorem 1.3.8 in a preliminary manuscript of this book.

Laffey's Theorem (1.3.6) is established in Laffey [3]; it was independently discovered by Barth-Elencwajg [1]. The proof in this book is from Choi-Laurie-Radjavi [1]. Some results in the case where the commutator has rank 2 are presented in Laffey [4]. Theorem 1.4.6, characterizing triangularity for algebras as commutativity modulo the radical, is due to McCoy [2].

The Block Triangularization Theorem (1.5.1) was conjectured by Hans Schneider, proven by Barker-Eifler-Kezlan [1] for matrices over the complex numbers, and proven by Watters [1] over arbitrary algebraically closed fields. The proof in the text is similar to the proof of a related result in Radjavi-Rosenthal [6, Theorem 1] and Radjavi-Rosenthal [3, Lemma 3.3]. Many of the consequences of block triangularization given in Section 1.5 are taken from Radjavi-Rosenthal [5], although part (ii) of Theorem 1.5.4 was discovered much earlier by Goldhaber-Whaples [1]. Shapiro [1] discusses the size of the blocks that occur in a block triangularization.

The theorems on approximate commutativity contained in Section 1.6 are from Jafarian-Radjavi-Rosenthal-Sourour [1], although the essence of the proof of Theorem 1.6.2 (using a diagonal matrix to make a triangular matrix similar to one with small off-diagonal entries) has been known for many years: it was apparently first used by Perron [1] in 1929. An interesting differential-equations approach can be found in Bellman [1, p. 205].

Engel's Theorem (1.7.6), stated somewhat differently, was really the first theorem on triangularization, having been proven before 1890 (see Bourbaki [1, pp. 410–429] for a brief history of Lie algebras and Lie groups). It plays an important role in the development of Lie algebras, as does Lie's theorem that a Lie algebra of matrices is solvable if and only if it is triangularizable (see Jacobson [2,p. 50], Sagle-Walde [1, pp. 212–214], Varadarajan [1, p. 202].) Lie's Theorem has applications to certain systems of linear differential equations (see, e.g., Lee-Ma [1]). Jacobson's Theorem (Corollary 1.7.4) is given in Jacobson [2, p. 34]; it includes both the Lie and Jordan algebra cases. The further generalization of Engel's Theorem to Theorem 1.7.3 is due to Radjavi [3], which also contains Corollary 1.7.8. Theorem 1.7.10 is established in Hadwin-Nordgren-Radjabalipour-Radjavi-Rosenthal [2, Theorem 4.3].

CHAPTER 2

Semigroups of Matrices

In this and the next two chapters we examine various obvious necessary conditions for triangularizability of semigroups to determine whether they are also sufficient. When they are not, we consider whether they at least imply reducibility. These investigations lead to insights into the structure of irreducible collections satisfying certain spectral conditions.

We assume, as before, that the underlying field is algebraically closed.

2.1 Basic Definitions and Propositions

Definition 2.1.1. A collection of matrices or linear transformations is called a *semigroup* if it is closed under multiplication.

If S is a semigroup, then it is clear that the algebra generated by it is simply the linear span of S. It therefore follows from Burnside's Theorem (1.2.2) that a semigroup of linear transformations on a vector space $\mathcal{V}$ is irreducible if and only if it contains a basis for $\mathcal{B}(\mathcal{V})$.

Lemma 2.1.2. *Let S be a semigroup in $\mathcal{B}(\mathcal{V})$ where the dimension of $\mathcal{V}$ is greater than 1. If there exists a nonzero linear functional f on $\mathcal{B}(\mathcal{V})$ with $f\mid_S = 0$, then S is reducible.*

Proof. The hypothesis implies that f is zero on the algebra generated by S but not on all of $\mathcal{B}(\mathcal{V})$. Thus S has to be reducible by Burnside's Theorem (Theorem 1.2.2). □

A more general lemma will be given below.

Definition 2.1.3. Let ϕ be a function from $\mathcal{B}(\mathcal{V})$ into a set $\mathcal{E}$. We say ϕ is *permutable* on a collection $\mathcal{F}$ of linear transformations if for every integer n, all members $A_1, \ldots, A_n$ of $\mathcal{F}$, and all permutations τ of $\{1, 2, \cdots, n\}$, we have

$$\phi(A_1 A_2 \cdots A_n) = \phi(A_{\tau(1)} A_{\tau(2)} \cdots A_{\tau(n)}).$$

Lemma 2.1.4. *Let $\mathcal{F}$ be any collection in $\mathcal{B}(\mathcal{V})$ where the dimension of $\mathcal{V}$ is greater than 1. If there is a nonzero linear functional on $\mathcal{B}(\mathcal{V})$ whose restriction to $\mathcal{F}$ is permutable, then $\mathcal{F}$ is reducible.*

Proof. Let S and $\mathcal{A}$ be, respectively, the semigroup and the algebra generated by $\mathcal{F}$ respectively. The permutability of the functional, say f, on S

follows directly from the hypothesis. However, it is also easily verified that f is actually permutable on all of $\mathcal{A}$. To see this, first observe that if

$$\mathcal{T} = \{\alpha S : S \in \mathcal{S}, \alpha \text{ scalar}\},$$

then f is permutable on $\mathcal{T}$ by linearity. Now, if $A_1, \ldots, A_n$ are any members of $\mathcal{A}$, then there exist members $\{T_{ij}\}$ of $\mathcal{T}$ such that $A_i = \Sigma_j T_{ij}$ for each i. Then for any permutation τ,

$$
\begin{aligned}
f(A_1 \cdots A_n) &= f[(\Sigma_j T_{1j}) \cdots (\Sigma_j T_{nj})] \\
&= \Sigma_{j_1, \ldots, j_n} f(T_{1j_1} \cdots T_{nj_n}) \\
&= \Sigma_{j_1, \ldots, j_n} f(T_{\tau(1)j_{\tau(1)}} \cdots T_{\tau(n)j_{\tau(n)}}) \\
&= f[\Sigma_{j_{\tau(1)}, j_{\tau(2)}, \ldots, j_{\tau(n)}} T_{\tau(1)j_{\tau(1)}} \cdots T_{\tau(n)j_{\tau(n)}}] \\
&= f[(\Sigma_{j_{\tau(1)}} T_{\tau(1)j_{\tau(1)}}) \cdots (\Sigma_{j_{\tau(n)}} T_{\tau(n)j_{\tau(n)}})] \\
&= f(A_{\tau(1)} \cdots A_{\tau(n)}).
\end{aligned}
$$

Thus f is permutable on $\mathcal{A}$. Suppose that $\mathcal{F}$ is irreducible. Then $\mathcal{A} = \mathcal{B}(\mathcal{V})$ by Burnside's Theorem. Pick any A and B in $\mathcal{B}(\mathcal{V})$ with $C = AB - BA \neq 0$. Then the permutability of f on $\mathcal{B}(\mathcal{V})$ implies that $f(XCY) = 0$ for all X and Y in $\mathcal{B}(\mathcal{V})$. It follows that f is zero on the (two-sided) ideal of $\mathcal{B}(\mathcal{V})$ generated by C. But $\mathcal{B}(\mathcal{V})$ has no nontrivial ideals (Theorem 1.2.3), and thus f is zero on $\mathcal{B}(\mathcal{V})$. This contradiction shows that $\mathcal{F}$ is reducible. □

Corollary 2.1.5. *If there is a nonzero linear functional on $\mathcal{B}(\mathcal{V})$ whose restriction to a semigroup is multiplicative, then the semigroup is reducible (assuming that the dimension of $\mathcal{V}$ is greater than 1).*

Proof. For any function from $\mathcal{S}$ into any commutative semigroup, multiplicativity obviously implies permutability. □

Corollary 2.1.6. *If there is a nonzero linear functional on $\mathcal{B}(\mathcal{V})$ that is constant on a semigroup, then the semigroup is reducible (as long as $\dim \mathcal{V} > 1$).*

Proof. Constancy is a special case of permutability. □

The following immediate corollaries are special cases of a later result (Corollary 2.2.3.).

Theorem 2.1.7. (Levitzki's Theorem) *Every semigroup of nilpotent operators is triangularizable.*

Proof. Let $\mathcal{S}$ be such a semigroup. Then the trace is a linear functional that vanishes on all $S \in \mathcal{S}$, so it follows from Lemma 2.1.2 that $\mathcal{S}$ is reducible. Since quotients of nilpotent operators are nilpotent, the Triangularization Lemma (1.1.4) yields the result.   □

Theorem 2.1.8. (Kolchin's Theorem) *If every member of a semigroup S is unipotent, (i.e., of the form $I + N$ with N nilpotent) then S is triangularizable.*

Proof. The hypothesis implies that the trace of every member of S is n, so the result follows from Corollary 2.1.6 and the Triangularization Lemma (1.1.4).   □

For the next lemma we require a definition.

Definition 2.1.9. A subset $\mathcal{J}$ of a semigroup S is called an *ideal* if JS and SJ belong to $\mathcal{J}$ for all $J \in \mathcal{J}$ and all $S \in S$.

The following simple but very useful lemma is the semigroup analogue of the corresponding well-known fact about algebras and their ideals (which is implicitly established in the proof of Theorem 1.2.3).

Lemma 2.1.10. *A nonzero ideal of an irreducible semigroup is irreducible.*

Proof. We show, equivalently, that a semigroup is reducible if it has a reducible ideal. Let $\mathcal{M}$ be a nontrivial invariant subspace of the ideal $\mathcal{J}$. Define

$$\mathcal{M}_1 = \vee\{J\mathcal{M} : J \in \mathcal{J}\},$$
$$\mathcal{M}_2 = \cap\{\ker J : J \in \mathcal{J}\}.$$

It is easily seen that both $\mathcal{M}_1$ and $\mathcal{M}_2$ are invariant under the whole semigroup. Also, $\mathcal{M}_1 \subseteq \mathcal{M}$ implies $\mathcal{M}_1 \neq \mathcal{V}$. Thus we are done if $\mathcal{M}_1 \neq \{0\}$. On the other hand, if $\mathcal{M}_1 = \{0\}$, then $\mathcal{M}_2 \supseteq \mathcal{M}$; since $\mathcal{M}_2 \neq \mathcal{V}$ by the assumption that $\mathcal{J} \neq \{0\}$, the subspace $\mathcal{M}_2$ is nontrivial.   □

The following example shows that this lemma does not extend to triangularizability.

Example 2.1.11. *Let S_0 be the semigroup consisting of all $n \times n$ matrices that have at most one entry equal to 1 and all other entries equal to 0 (i.e., S_0 is the zero matrix together with the standard basis for $n \times n$ matrices). This is clearly an irreducible semigroup. Let S be the semigroup of $(n + 1) \times (n + 1)$ matrices*

$$\left\{ \begin{bmatrix} 1 & X \\ 0 & A \end{bmatrix} : A \in S_0,\ X \text{ a } 1 \times n \text{ matrix} \right\}.$$

*Then S has only the one exhibited nontrivial invariant subspace (by irre-
ducibility of S_0 and the arbitrariness of X). Yet the ideal of S generated
by the rank-one idempotent $E = \begin{pmatrix} 1 & 0 \\ 0 & 0 \end{pmatrix}$ is*

$$\left\{ \begin{bmatrix} 1 & X \\ 0 & 0 \end{bmatrix} : X \ arbitrary \right\},$$

which is triangular.

Lemma 2.1.12. *Let S be a semigroup in $\mathcal{B}(\mathcal{V})$ and E an idempotent of
rank at least 2, not necessarily in S. If the collection $ESE\big|_{E\mathcal{V}}$, where*

$$ESE = \{ESE : S \in \mathcal{S}\},$$

is reducible, then so is S.

Proof. Let $\mathcal{M}$ be a nontrivial invariant subspace of $ESE|_{E\mathcal{V}}$. Pick a
nonzero x in $\mathcal{M}$ and a linear functional f on $\mathcal{V}$ with $f\big|_{\mathcal{M}} = 0$ and $f\big|_{E\mathcal{V}} \neq 0$.
Let ϕ be the functional on $\mathcal{B}(\mathcal{V})$ defined by $\phi(T) = f(ETEx)$ for all T.
Then $\phi \neq 0$ and $\phi\big|_S = 0$. Thus S is reducible by Lemma 2.1.2. $\square$

The following simple lemma will be used frequently.

Lemma 2.1.13. *Let S be an irreducible semigroup and let*

$$m = \min\{\operatorname{rank} S : 0 \neq S \in \mathcal{S}\}.$$

*Then there exists an element of the form $S_0 \oplus 0$ in S, with respect to a
suitable basis, where S_0 is invertible and has rank m.*

Proof. Let $A \in S$ have rank m. Consider the ideal $\mathcal{J} = SAS$. Irreducibility of S implies that $AS \neq \{0\}$, which implies, in turn, that $\mathcal{J} \neq \{0\}$. Thus
$\mathcal{J}$ is irreducible by Lemma 2.1.10, and has non-nilpotent elements by Levitzki's Theorem (2.1.7). Replace A by a nonzero member of $\mathcal{J}$ that is not
nilpotent. By the minimality of m, all powers of A have the same rank as
A, so the Jordan form of A is of the desired type. $\square$

Determining permutability of certain functions is made easier by the
following lemma.

Lemma 2.1.14. *Let S be a semigroup in $\mathcal{B}(\mathcal{V})$ and ϕ any function with
domain S. Then ϕ is permutable on S if and only if*

(i) $\phi(ST) = \phi(TS)$ *and*
(ii) $\phi(STR) = \phi(TSR)$

for all $R, S,$ and $T \in \mathcal{S}$.

Proof. Assume that ϕ satisfies (i) and (ii). Since (i) implies that every cyclic permutation of letters in a word leaves the value of ϕ unchanged, we deduce from (ii) that

$$\phi(A_1 A_2 A_3) = \phi(A_{\tau(1)} A_{\tau(2)} A_{\tau(3)})$$

for every permutation τ on three letters. Now assume permutability of ϕ for all words with fewer than n letters. To complete the proof by induction we have to show that ϕ is equal on any pair of words W_1 and W_2 with the same n letters. By (i), any given letter can be brought to the end of a word, so we can assume that W_1 and W_2 both end in the same letter A_n. Let

$$W_1 = A_1 \cdots A_{n-1} A_n \text{ and } W_2 = A_{\tau(1)} \cdots A_{\tau(n-1)} A_n.$$

Now, by use of (ii), A_{n-1} can be assumed to be at the beginning of W_2, so W_2 has the form $W_2 = A_{n-1}(A_* A_* \cdots A_*) A_n$. Using (ii) again we can assume that $W_2 = A_* A_* \cdots A_* A_{n-1} A_n$. Since $A_{n-1} A_n$ is an element of $\mathcal{S}$, this reduces to the case of $n - 1$ letters. $\square$

We shall also need the following simple purely algebraic fact. We include a proof for completeness.

Lemma 2.1.15. *Let* $\{\alpha_1, \ldots, \alpha_n\}$ *and* $\{\beta_1, \ldots, \beta_n\}$ *be n-tuples of elements of any field whose characteristic is either zero or larger than n.*

(i) *If* $\displaystyle\sum_{i=1}^{n} \alpha_i^k = \sum_{i=1}^{n} \beta_i^k$ *for* $k = 1, \ldots, n$, *then there is a permutation τ on n letters such that* $\beta_i = \alpha_{\tau(i)}$ *for all i.*

(ii) *If* $\displaystyle\sum_{i=1}^{n} \alpha_i^k = 0$ *for* $k = 1, \ldots, n$, *then* $\alpha_i = 0$ *for all i.*

(iii) *If* $\displaystyle\sum_{i=1}^{n} \alpha_i^k = c$ *with c fixed for* $k = 1, \ldots, n+1$, *then c is an integer and each α_i is either 0 or 1.*

Proof. (i) The functions T_k defined by

$$T_k = T_k(x_1, \ldots, x_n) = \sum_{i=1}^{n} x_i^k$$

are symmetric polynomials in n variables. For each k, let S_k denote the elementary symmetric polynomial in $x_1, \ldots, x_n$ of degree k; i.e., S_k is the sum of all products of k variables:

$$S_1 = x_1 + \cdots + x_n$$
$$S_2 = x_1 x_2 + \cdots + x_{n-1} x_n$$
$$\vdots$$
$$S_n = x_1 x_2 \cdots x_n.$$

Then it can be verified that

$$T_k - T_{k-1} S_1 + T_{k-2} S_2 - \cdots + (-1)^{k-1} T_1 S_{k-1} + (-1)^k k S_k = 0$$

for $k = 1, \ldots, n$. This formula enables us to determine each S_k inductively in terms of the T_j.

Now, the hypothesis that $T_k(\alpha_1, \ldots, \alpha_n) = T_k(\beta_1, \ldots, \beta_n)$ for $1 \le k \le n$ implies that $S_k(\alpha_1, \ldots, \alpha_n) = S_k(\beta_1, \ldots, \beta_n)$ for $k = 1, \ldots, n$. This means that the monic polynomial of degree n with zeros at $\alpha_1, \ldots, \alpha_n$ (whose coefficients are the elementary symmetric functions of its roots) coincides with the corresponding polynomial for $\beta_1, \ldots, \beta_n$. Thus the two n-tuples are the same except for a permutation.

(ii) This is a corollary of (i) obtained by taking $\beta_i = 0$ for all i.

(iii) We can assume $c \ne 0$, in view of (ii). After a permutation, if necessary, assume that the first m of the $\{\alpha_i\}$ are nonzero and $\alpha_i = 0$ for $i > m$. We must show that $c = m$. By disregarding α_i for $i > m$, we can assume $m = n$.

Now, an easy calculation shows that, in general, in the case of n variables,

$$T_{n+1} = T_n S_1 - T_{n-1} S_2 + \cdots + (-1)^{n-1} T_1 S_n.$$

Denoting $S_k(\alpha_1, \ldots, \alpha_n)$ by s_k and using the hypothesis $T_k(\alpha_1, \ldots, \alpha_n) = c$ for $k = 1, \ldots, n + 1$ yields

$$1 = s_1 - s_2 + \cdots + (-1)^{n-1} s_n.$$

But the recursive equation in the proof of (i) above, for $k = n$, also gives

$$c(1 - s_1 + s_2 - \cdots + (-1)^{n-1} s_{n-1}) + (-1)^n n s_n = 0.$$

Since $s_n \ne 0$, the last two equations imply $c = n$. Now (i) applies with $\beta_i = 1$ for all i, proving that $\alpha_i = 1$. $\square$

Theorem 2.1.16. *Let A and B be $n \times n$ matrices over a field whose characteristic is either zero or larger than n. If $\mathrm{tr}(A^k) = \mathrm{tr}(B^k)$ for $k = 1, \ldots, n$ then A and B have the same eigenvalues, counting multiplicity. In particular, if $\mathrm{tr}(A^k) = 0$ for $k = 1, \ldots, n$, then A is nilpotent.*

Proof. This follows immediately from the preceding lemma and the facts that $\sigma(A^k) = \{\alpha^k : \alpha \in \sigma(A)\}$ and $\mathrm{tr}\, C$ is the sum of the eigenvalues of C, counting multiplicity. $\square$

2.2 Permutable Trace

If a collection $\mathcal{F}$ is triangularizable, then the trace function is obviously permutable on $\mathcal{F}$. We shall prove the converse.

It follows from Lemma 2.1.4 that a semigroup on a space of dimension greater than 1 with permutable trace is reducible, but it is not so obvious that this property is inherited by quotients.

Theorem 2.2.1. *Let $\mathcal{F}$ be any collection of operators on an n-dimensional space over a field whose characteristic is either zero or larger than $n/2$. Then $\mathcal{F}$ is triangularizable if and only if trace is permutable on $\mathcal{F}$.*

Proof. As in the proof of Lemma 2.1.4, the hypothesis extends to the algebra $\mathcal{A}$ generated by $\mathcal{F}$. Let $\{A_{ij}\}$ be the block triangularization of $\mathcal{A}$ obtained in Theorem 1.5.1. We must show that the diagonal blocks A_{ii} all act on one-dimensional spaces $\mathcal{N}_i$. Assume otherwise. Then one of the subsets J_k of Theorem 1.5.1, say J_1, corresponds to equal diagonal blocks $\{A_{ii} : i \in J_1\}$ acting on spaces of dimension at least 2. Thus the number m of elements in J_1 is no greater than $n/2$. It follows from the permutability hypothesis together with parts (i) and (iv) of Theorem 1.5.1 that the function f defined by $f(A) = m\,\mathrm{tr}(A)$ is permutable on $\mathcal{B}(\mathcal{N}_i)$ for any fixed i in J_1. Since m is different from the characteristic of the field underlying $\mathcal{V}$ by hypothesis, trace is permutable on $\mathcal{B}(\mathcal{N}_i)$, which is a contradiction to Lemma 2.1.4. □

It should be noted that a shorter proof of the above theorem can be given if the field has characteristic 0. Just observe that the permutability of trace implies $\mathrm{tr}((AB - BA)C) = 0$ for all A, B, C in the algebra generated by $\mathcal{F}$. In particular, $\mathrm{tr}((AB - BA)^k) = 0$ for all natural numbers k. The characteristic of the field then implies that $AB - BA$ is nilpotent for all pairs A and B in the algebra (by Theorem 2.1.16), so the algebra is triangularizable by Theorem 1.3.2.

The assumption on the characteristic of the field is necessary in Theorem 2.2.1. For example, let $\mathcal{V}$ be a vector space of dimension $2p$ over a field of characteristic p. Express $\mathcal{V}$ as a direct sum of p two-dimensional subspaces (identified with) $\mathcal{V}_0$ and let

$$\mathcal{S} = \{A \oplus \cdots \oplus A : A \in \mathcal{B}(\mathcal{V}_0)\}.$$

Since $\mathcal{B}(\mathcal{V}_0)$ is irreducible, $\mathcal{S}$ cannot be triangularizable (as is seen by Lemma 1.5.2). Yet $\mathrm{tr}\,\mathcal{S} = 0$ for all $S \in \mathcal{S}$.

Corollary 2.2.2. *Let $\mathcal{S}$ be a semigroup of operators on an n-dimensional space over a field whose characteristic is either zero or larger than $n/2$. Then $\mathcal{S}$ is triangularizable if and only if*

$$\mathrm{tr}(ABC) = \mathrm{tr}(BAC)$$

for all A, B, C in $\mathcal{S}$.

Proof. Since $\text{tr}(AB) = \text{tr}(BA)$ for all operators A and B, Lemma 2.1.14 implies that the condition is equivalent to permutability of trace.   □

In the remaining corollaries of this section we assume, for simplicity, that the characteristic of the underlying field is either zero or larger than n. (The necessary adjustments for other characteristics between $n/2$ and n are not hard to formulate.)

The next result generalizes Theorems 2.1.7 and 2.1.8.

Corollary 2.2.3. (Kaplansky's Theorem) *A semigroup with constant trace is triangularizable. Moreover, every diagonal entry in a triangularization of such a semigroup is either constantly zero or constantly one.*

Proof. The triangularizability follows from Theorem 2.2.1. The fact that the spectrum of each member S of the semigroup is contained in $\{0, 1\}$ is a consequence of the constancy of $\text{tr}(S^k)$ for all k together with Lemma 2.1.15. Now, if the diagonal zeros of any two members S and T did not occur at the same positions, then we would have

$$\text{tr}(ST) \neq \text{tr } S,$$

contradicting the constancy of the trace.   □

Corollary 2.2.4. *A semigroup with multiplicative trace is triangularizable. Moreover, in every triangularization of such a semigroup there is an integer j such that all diagonal entries other than those in the (j, j) position are zero.*

Proof. The triangularizability follows from Theorem 2.2.1. We need only show that every S in the semigroup has at most one nonzero eigenvalue. Let $a_1, \ldots, a_n$ be the eigenvalues of S. By hypothesis $\Sigma a_i^k = (\Sigma a_i)^k$ for every k. If $\Sigma a_i = 0$, we have $a_i = 0$ for all i by Lemma 2.1.15. If $\Sigma a_i = b \neq 0$, then $\Sigma(a_i/b)^k = 1$ for every k, and it follows from Lemma 2.1.15 that $a_j = b$ for some j and $a_i = 0$ for $i \neq j$. The multiplicativity implies that the integer j obtained is the same for every member of the semigroup.   □

Corollary 2.2.5. *Let $\mathcal{G}$ be a group of invertible operators and denote its commutator subgroup by $\mathcal{H}$. (Recall that $\mathcal{H}$ is the normal subgroup generated by all elements of the form $A^{-1}B^{-1}AB$ with A and B in $\mathcal{G}$.) Then the following conditions are mutually equivalent:*

(i) *$\mathcal{G}$ is triangularizable.*
(ii) *The trace is constant on each coset of $\mathcal{G}$ relative to $\mathcal{H}$.*
(iii) *The trace is constant on $\mathcal{H}$.*

(iv) $\mathcal{H}$ consists of unipotent operators (i.e., $\sigma(A) = \{1\}$ for all A in $\mathcal{H}$).

Proof. Since (i) implies permutability of trace on $\mathcal{G}$, for $G \in \mathcal{G}$ and $H \in \mathcal{H}$ we have $\mathrm{tr}(GH) = \mathrm{tr}(GI) = \mathrm{tr}\, G$. Thus (i) implies (ii). The implication (ii) $\Rightarrow$ (iii) is immediate. The equivalence of (iii) and (iv) is a consequence of Corollary 2.2.3 and the invertibility of members of $\mathcal{H}$. Hence we need only show (iv) $\Rightarrow$ (i).

Assume (iv). Then, by Kolchin's Theorem (2.1.8), $\mathcal{H}$ is triangularizable: Let $\{0\} = \mathcal{V}_0 \subset \cdots \subset \mathcal{V}_{n-1} \subset \mathcal{V}_n = \mathcal{V}$ be a triangularizing chain. We first show that $\mathcal{G}$ is reducible. We can assume $\mathcal{H} \neq \{I\}$, by Corollary 1.1.5. Thus the subspace

$$\mathcal{M} = \vee\{(H - I)\mathcal{V} : H \in \mathcal{H}\}$$

is nonzero. Also, since $(H - I)\mathcal{V} \subseteq \mathcal{V}_{n-1}$ for all $H \in \mathcal{H}$, we have $\mathcal{M} \neq \mathcal{V}$. We show that $\mathcal{M}$ is invariant under $\mathcal{G}$.

Fix $G \in \mathcal{G}$. For any $H \in \mathcal{H}$

$$G(H - I) = ((GHG^{-1}) - I)G,$$

so

$$G(H - I)\mathcal{V} = (GHG^{-1} - I)\mathcal{V},$$

and, since $GHG^{-1} \in \mathcal{H}$, we conclude that $G\mathcal{M} \subseteq \mathcal{M}$.

We have proven that $\mathcal{G}$ is reducible if its commutator subgroup consists of unipotent operators. Since the property of having such a commutator subgroup is inherited by quotients, the Triangularization Lemma (Lemma 1.1.4) implies that $\mathcal{G}$ is triangularizable. □

Before stating the next corollary we need a general lemma (which does not require any hypotheses on the field).

Lemma 2.2.6. Let $\mathcal{S}$ be a semigroup of linear transformations and k a natural number. Denote by $\mathcal{S}^k$ the ideal of $\mathcal{S}$ consisting of words in $\mathcal{S}$ of length at least k:

$$\mathcal{S}^k = \{S_1 S_2 \cdots S_m : m \geq k, S_i \in \mathcal{S}, i = 1, \ldots, m\}.$$

If $\mathcal{S}^k$ is triangularizable, then so is $\mathcal{S}$.

Proof. Assume that $\mathcal{S}^k$ is triangularizable. We first show that $\mathcal{S}$ is reducible. If $\mathcal{S}^k \neq \{0\}$, this follows from Lemma 2.1.10, since $\mathcal{S}^k$ is an ideal of $\mathcal{S}$. Otherwise, every member of $\mathcal{S}$ is nilpotent (of order at most k), so the result follows from Levitzki's Theorem (2.1.7).

Now, triangularizability is a property that quotients inherit, by Lemma 1.5.2. Moreover, we have shown that any semigroup $\mathcal{S}$ such that $\mathcal{S}^k$ is triangularizable is reducible. Thus $\mathcal{S}$ is triangularizable by the Triangularization Lemma (1.1.4). □

Corollary 2.2.7. *Let $\mathcal{F}$ be a family of linear transformations. If there exists an integer k such that trace is permutable on $\mathcal{F}^k$, then trace is permutable on all words in $\mathcal{F}$.*

Proof. This follows directly from the preceding lemma and Theorem 2.2.1.
□

Recall that a family $\mathcal{F}$ of complex matrices is called *self-adjoint* if $F \in \mathcal{F}$ implies $F^* \in \mathcal{F}$, where F^* is the adjoint (i.e., conjugate transpose) of F.

Corollary 2.2.8. *A self-adjoint family $\mathcal{F}$ of complex matrices is commutative if and only if trace is permutable on $\mathcal{F}$.*

Proof. For a self-adjoint set, triangularizability is equivalent to diagonalizability. (To see this, suppose that $\{0\} = \mathcal{M}_0 \subset \mathcal{M}_1 \subset \cdots \subset \mathcal{M}_n = \mathcal{V}$ is a triangularizing chain of subspaces for $\mathcal{F}$. Let $\mathcal{V}_j$ denote the orthogonal complement of $\mathcal{M}_{j-1}$ in $\mathcal{M}_j$, for $j = 1, \ldots, n$. Pick a unit vector e_j in each $\mathcal{V}_j$ to form an orthonormal basis $\{e_j : j = 1, \ldots, n\}$ for $\mathcal{V}$. Then every member of $\mathcal{F}$ is diagonal relative to this basis.) Also, a diagonalizable family is obviously commutative, so the result follows.
□

We conclude this section with a slight extension of Kolchin's Theorem.

Corollary 2.2.9. *Let $\mathcal{S}$ be a semigroup of $n \times n$ matrices such that, for every S in $\mathcal{S}$, the algebraic multiplicity of 1 in $\sigma(S)$ is at least $n - 1$. Then $\mathcal{S}$ is triangularizable.*

Proof. The characteristic polynomial of each S in $\mathcal{S}$ is clearly of the form $(x-1)^{n-1}(x-\alpha)$, where α depends on S. Thus

$$\operatorname{tr} S = n - 1 + \alpha = n - 1 + \det S.$$

Since the determinant is permutable, so is the trace, and the corollary follows from Theorem 2.2.1.
□

2.3 Zero-One Spectra

In this section we consider semigroups whose members all have spectra contained in $\{0, 1\}$. Examples of these are semigroups of nilpotents, unipotents, and idempotents, the first two of which were treated in Section 2.1. Operators with zero-one spectrum are those with characteristic polynomial $(x-1)^r x^{n-r}$ for some r.

The mere assumption that $\sigma(S) \subseteq \{0, 1\}$ for every member of a semigroup is too weak to yield reducibility, as the semigroup $\mathcal{S}_0$ of Example 2.1.11 demonstrates. However, reducibility theorems can be obtained with various additional hypotheses.

Theorem 2.3.1. *Let S be a semigroup of operators with spectra contained in $\{0,1\}$. Assume that whenever 1 is in $\sigma(AB)$ for A and B in S, it follows that 1 is in $\sigma(A) \cap \sigma(B)$. Then S is reducible (if S acts on a space of dimension greater than 1).*

Proof. The hypotheses imply that the set of nilpotents in S is an ideal. If this ideal is nonzero, we are done by Lemma 2.1.10. Otherwise, $1 \in \sigma(S)$ for all S in S. In this case, let m be the minimum of $\{\text{rank}(S) : 0 \neq S \in S\}$. Then every member A of rank m has the property that the (algebraic) multiplicity of 1 in $\sigma(A)$ is exactly m. (If it were less, say $k < m$, then some power of A would have rank k, contradicting the minimality of m.)

Now the members of S with rank m form a nonzero ideal on which the trace is always m. The result then follows from Corollary 2.1.6 together with Lemma 2.1.10. (If the underlying field was restricted as in Section 2.2, we could use Corollary 2.2.3 instead.) $\qquad\square$

Corollary 2.3.2. *Let S be a semigroup of operators with spectra contained in $\{0,1\}$. If $1 \in \sigma(S)$ for all $S \in S$, then S is reducible (assuming $\dim V > 1$).*

Proof. The hypothesis of Theorem 2.3.1 is satisfied. $\qquad\square$

The next result is a special case of Corollary 2.2.3 if the characteristic of the field is 0 or greater than n.

Theorem 2.3.3. *Let S be a semigroup whose members all have characteristic polynomial $(x-1)^r x^{n-r}$ for a fixed r. Then S is triangularizable.*

Proof. By Theorem 2.3.1, S is reducible. The proof will be completed by showing that the Triangularization Lemma is applicable. For this, it suffices to prove the following assertion: If S is any semigroup of matrices with spectra in $\{0,1\}$ and with constant multiplicity r for the eigenvalue 1, then the restriction of S to an invariant subspace also has this constant multiplicity property. In other words, if the members of S all have the form

$$\begin{pmatrix} R & X \\ 0 & T \end{pmatrix},$$

then there exists an integer k such that the multiplicity of 1 in $\sigma(R)$ is always k (so that the corresponding multiplicity for $\sigma(T)$ is forced to be $r - k$).

Assume that the restrictions R_1 and R_2 of two members S_1 and S_2 have different multiplicities for 1, say k_1 and k_2 respectively, with $k_1 < k_2$. Then for a sufficiently large integer m, the matrices R_1^m, R_2^m, T_1^m, and T_2^m have ranks $k_1, k_2, r - k_1$, and $r - k_2$ respectively. Thus

$$\text{rank } (R_1^m R_2^m) + \text{rank } (T_1^m T_2^m) \leq k_1 + r - k_2 < r.$$

This shows that the multiplicity of 1 in $\sigma(S_1^m S_2^m)$, which is the sum of the corresponding multiplicities for $R_1^m R_2^m$ and $T_1^m T_2^m$, is less than r. This is a contradiction.

$\square$

The semigroup in Example 2.1.11 shows that if r is allowed to vary from member to member, then the conclusion of Theorem 2.3.3 need not hold.

Definition 2.3.4. A *band* is a semigroup whose members are idempotents (i.e., $S^2 = S$ for every S in the semigroup).

Theorem 2.3.5. *Every band is triangularizable.*

Proof. If we make the hypotheses of Theorem 2.2.1 on the field, then this result can be established by verifying that trace (or, equivalently, rank) is permutable on a band. The following is a different proof that applies over all fields.

By the Triangularization Lemma (1.1.4), it suffices to show reducibility. For this we use Theorem 2.3.1: We must prove that $1 \in \sigma(AB)$ for A, B in the band implies $1 \in \sigma(A) \cap \sigma(B)$. But if $1 \in \sigma(AB)$, then A and B are both nonzero idempotents. Thus $\sigma(A)$ and $\sigma(B)$ both contain 1. $\square$

Corollary 2.3.6. *Rank is permutable on a band.*

Proof. This is a consequence of triangularizability together with the fact that if A, B, and C are any three matrices in triangular form, then the diagonals of ABC and BAC coincide. $\square$

A commutative band is diagonalizable (as is verified by a simple induction on the dimension n of the underlying space). This makes a commutative band quite "thin," in the sense that the algebra generated by it is at most n-dimensional. A noncommutative band, on the other hand, can generate as large an algebra as triangularizability allows, as in the following example.

Example 2.3.7. *For each k with $1 \leq k \leq n$ let*

$$\mathcal{S}_k = \left\{ \begin{bmatrix} I_k & X \\ 0 & 0 \end{bmatrix} : X \text{ a } k \times (n-k) \text{ matrix} \right\},$$

where I_k is the $k \times k$ identity. Then $\mathcal{S}_k$ is clearly a band for each k. Moreover, if $S \in \mathcal{S}_k$ and $T \in \mathcal{S}_j$ with $k \leq j$, then straightforward calculations with 3×3 blocks show that ST and TS both belong to $\mathcal{S}_k$. Thus

$$\mathcal{S} = \bigcup_{k=1}^{n} \mathcal{S}_k$$

is a band. It is easy to see that the algebra generated by $\mathcal{S}$ has dimension $n(n+1)/2$ and thus is the algebra of all upper triangular matrices.

Bands of constant rank have a particularly simple triangularization.

Lemma 2.3.8. *Let S be a band on an n-dimensional space $\mathcal{V}$ and assume that the rank is constant on S, and the constant is neither 0 nor n. Then at least one of the two subspaces*

$$\mathcal{N} = \cap\{\ker S : S \in \mathcal{S}\} \text{ and } \mathcal{M} = \vee\{S\mathcal{V} : S \in \mathcal{S}\}$$

is a nontrivial invariant subspace for S.

Proof. The subspaces are clearly invariant, and $\mathcal{N} \neq \mathcal{V}$ and $\mathcal{M} \neq \{0\}$. Assume $\mathcal{M} = \mathcal{V}$. We must show that this implies $\mathcal{N} \neq \{0\}$. Note that constancy of the rank implies $TST = T$ for all S and T in $\mathcal{S}$ (since TST is a subprojection of T). Thus for any R, S, T in $\mathcal{S}$ we have

$$TR(I - T)S = TR(I - T)STS = (TRST - TRTST)S$$
$$= (T - T)S = 0.$$

Since $\vee\{S\mathcal{V} : S \in \mathcal{S}\} = \mathcal{V}$, we get $TR(I - T) = 0$ for all R and T. Fixing $T \in \mathcal{S}$, we note that

$$R(I - T) = RTR(I - T) = 0,$$

so that the range of $I - T$ is in the kernel of every $R \in \mathcal{S}$, and consequently $\mathcal{N} \neq 0$.

$\square$

Corollary 2.3.9. *Every band of constant rank k has a triangularization of the form*

$$\left\{ \begin{bmatrix} 0 & X & XY \\ 0 & I & Y \\ 0 & 0 & 0 \end{bmatrix} : X \in \mathcal{X}, Y \in \mathcal{Y} \right\},$$

where $\mathcal{X}$ and $\mathcal{Y}$ are two given sets of matrices of appropriate size and I is the identity of size k. Moreover, the triangularizing chain can be chosen so as to contain the two subspaces $\mathcal{N}$ and $\mathcal{M}$ given in the preceding lemma.

Proof. If k is the dimension of $\mathcal{V}$, the band is $\{I\}$. Otherwise, let $\mathcal{N}'$ be a complement of $\mathcal{N}$ in $\mathcal{N} + \mathcal{M}$, and $\mathcal{N}''$ a complement of $\mathcal{N} + \mathcal{M}$ in $\mathcal{V}$. Then we get a block triangularization, relative to the decomposition $\mathcal{N} \oplus \mathcal{N}' \oplus \mathcal{N}''$, in which every member of the band is of the form

$$
S = \begin{bmatrix} 0 & X & Z \\ 0 & E & Y \\ 0 & 0 & 0 \end{bmatrix},
$$

where E is an idempotent of rank k. It follows from $S = S^2$ that $X = XE, Y = EY$, and $Z = XY$. The collection of all blocks E occurring in the elements S of $\mathcal{S}$ is a band of constant rank k. We claim that this collection is the singleton $\{I\}$ with I the $k \times k$ identity. Otherwise, Lemma 2.3.8 would be applicable to this band: Either the intersection of kernels of all E or the span of their ranges would be nontrivial. In the the former case, $X = XE$ for all S would contradict the choice of $\mathcal{N}$; in the latter, $Y = EY$ would contradict the choice of $\mathcal{M}$. Thus $E = I$ for every $S \in \mathcal{S}$.

Now, let

$$
\mathcal{X} = \left\{ X : \begin{bmatrix} 0 & X & XY \\ 0 & I & Y \\ 0 & 0 & 0 \end{bmatrix} \in \mathcal{S} \right\},
$$

$$
\mathcal{Y} = \left\{ Y : \begin{bmatrix} 0 & X & XY \\ 0 & I & Y \\ 0 & 0 & 0 \end{bmatrix} \in \mathcal{S} \right\}.
$$

To complete the proof we must show that the band contains all matrices of the given form. For this, observe that if X and Y are arbitrary members of $\mathcal{X}$ and $\mathcal{Y}$ respectively, then there are some $X_1 \in \mathcal{X}$ and $Y_1 \in \mathcal{Y}$ such that

$$
S = \begin{pmatrix} 0 & X & XY_1 \\ 0 & I & Y_1 \\ 0 & 0 & 0 \end{pmatrix} \text{ and } T = \begin{pmatrix} 0 & X_1 & X_1Y \\ 0 & I & Y \\ 0 & 0 & 0 \end{pmatrix}
$$

are in $\mathcal{S}$. It follows that

$$
ST = \begin{bmatrix} 0 & X & XY \\ 0 & I & Y \\ 0 & 0 & 0 \end{bmatrix}
$$

belongs to $\mathcal{S}$. $\square$

A pair of idempotents is triangularizable if it satisfies the simplest necessary condition.

Theorem 2.3.10. *If P and Q are idempotents, then $\{P, Q\}$ is triangularizable if and only if $PQ - QP$ is nilpotent.*

Proof. If $\{P,Q\}$ is triangularizable, then $PQ - QP$ is nilpotent by the Spectral Mapping Theorem (1.1.8).

For the converse, it suffices to prove reducibility (by the Triangularization Lemma (1.1.4)). We consider two different cases. First, if P and Q operate on a space of dimension 2, we can choose a similarity such that $P = \begin{pmatrix} 1 & 0 \\ 0 & 0 \end{pmatrix}$. With respect to the same basis, let $Q = \begin{pmatrix} a & b \\ c & d \end{pmatrix}$. Then $PQ - QP = \begin{pmatrix} 0 & b \\ -c & 0 \end{pmatrix}$, so $(PQ - QP)^2 = 0$ implies that at least one of b and c is 0. Thus Q is upper or lower triangular, and $\{P,Q\}$ is triangularizable.

If P and Q act on a space of dimension greater than 2, note that the operator

$$R = PQP + (I - P)(I - Q)(I - P)$$

commutes with both P and Q (as seen by an easy direct computation). Thus each eigenspace of R is invariant under both P and Q, so we are done unless R is a multiple of I. If R is a multiple of I, let x be any vector other than 0 in the range of $I - P$, and let $\mathcal{M}$ be the linear span of x and PQx. Then $\mathcal{M}$ is nontrivial (recall that the dimension of the space is assumed to be greater than 2 for this part of the proof), so we need only show that $\mathcal{M}$ is invariant under both P and Q. Note that R a multiple of I implies that PQP is a multiple of P and $(I - P)(I - Q)(I - P)$ is a multiple of $I - P$. Thus

$$
\begin{aligned}
(I - Q)x &= (I - Q)(I - P)x \\
&= (I - P)(I - Q)(I - P)x + P(I - Q)(I - P)x \\
&= \lambda(I - P)x + P(I - Q)x \qquad \text{(for some } \lambda) \\
&= \lambda x - PQx \in \mathcal{M}.
\end{aligned}
$$

Hence $Qx \in \mathcal{M}$. Applying Q to both sides of the above equation shows that $QPQx$ is a multiple of Qx, and so is also in $\mathcal{M}$. Thus $\mathcal{M}$ is invariant under Q. Also, $Px = 0$ and $P(PQx) = PQ$, so $\mathcal{M}$ is invariant under P as well.   □

2.4 Notes and Remarks

Levitzki's Theorem (Theorem 2.1.7) was one of the earliest in which conditions imposed on spectra were shown to imply reducibility (Levitzki [1]). Kolchin's Theorem (Theorem 2.1.8) came a little later, in Kolchin [1]. Kaplansky [2] unified these two results in several directions and proved, in particular, Corollaries 2.2.3 and 2.3.3; this paper also essentially contains Lemma 2.1.15 and Theorem 2.1.16. The elementary results of Section 2.1

have been used, explicitly or implicitly, in proofs of various results by many authors, and it is hard to determine who first discovered them.

The permutability condition on traces was used in Radjavi [2] to prove Theorem 2.2.1; this paper also contains all the results of Section 2.2 except for Kaplansky's Theorem. Some proofs given here are simpler than the original ones. Theorem 2.3.1 and Corollary 2.3.2 on zero-one spectra were proven in Radjavi [1]. Permutability of trace in the general case of nonzero characteristic is treated in Grunenfelder-Guralnick-Košir-Radjavi [1].

Abstract bands have been extensively studied by semigroup theorists; a good reference on the subject is the book by Petrich [1]. Triangulariz-ability of a matrix band (Theorem 2.3.5) was proven in Radjavi [1]. The examples and results on the structure of matrix bands given in this section come from Fillmore-MacDonald-Radjabalipour-Radjavi [1]. The classification of matrix bands is a topic far from trivial. Even unicellular bands (i.e., those with a unique triangularizing chain) remain unclassified. The paper mentioned above studies the classification problem for maximal unicellular bands; it is perhaps surprising that in $\mathcal{M}_4(\mathbb{F})$ there are fifty maximal unicellular bands up to similarity. Theorem 2.3.10 is due to Szep [1], who gives a more computational proof.

A matrix band S is a principal-ideal band (that is, $I \in S$ and every ideal of S is of the form STS for some T in S) if and only if, in a tri-angularization of S, the diagonals form a totally ordered (abelian) band. Example 2.3.7 illustrates this type, but there are many others (Fillmore-MacDonald-Radjabalipour-Radjavi [1]). It is not known whether a maxi-mal principal-ideal band in $\mathcal{M}_n(\mathbb{F})$ contains members of all ranks between 0 and n. (Example 2.3.7 does.) These and other classification problems are discussed in Fillmore-MacDonald-Radjabalipour-Radjavi [2].

There is a vast literature on abstract semigroups. For semigroups of matrices, a recent reference is the book by Okniński [2], which has a small intersection with the present book. Okniński [1] gives necessary and suf-ficient conditions for triangularizability of semigroups in terms of their structure as semigroups.

CHAPTER 3
Spectral Conditions on Semigroups

In this chapter we show that properties such as submultiplicativity of spectrum or of spectral radius imply reducibility or triangularizability, under various additional hypotheses.

From now on, we assume not only that the underlying field is algebraically closed but also that it has characteristic zero. Although not needed for some of the results (e.g., the first theorem of Section 3.2), this assumption allows us to use certain techniques to great advantage. Once this assumption is made, we can further reduce all our arguments to the case of complex numbers, as the following lemma shows.

3.1 Reduction to the Field of Complex Numbers

Lemma 3.1.1. *Let $\mathbb{F}$ be an algebraically closed field of characteristic zero and let S be a semigroup of matrices over $\mathbb{F}$. Pick a maximal linearly independent set $\{S_1, \dots, S_m\}$ in S (i.e., a basis for the linear span of S). Denote by S_0 the subsemigroup of S generated by this finite set. Then*

(i) S_0 has the same invariant subspaces as S, and

(ii) S_0 is, up to a field isomorphism, a semigroup of complex matrices.

Proof. The first assertion follows trivially from the fact that S and S_0 have the same span over $\mathbb{F}$. To prove (ii), form a subfield $\mathbb{K}$ of $\mathbb{F}$ by adjoining all the entries of all the S_j to the set $\mathbb{Q}$ of rational numbers. The field $\mathbb{K}$ is obtainable from $\mathbb{Q}$ by a finite sequence of extensions

$$\mathbb{Q} = \mathcal{F}_0 \subset \mathcal{F}_1 \subset \mathcal{F}_2 \subset \cdots \subset \mathcal{F}_k = \mathbb{K},$$

where, for each $j = 1, 2, \dots, k$, there is an r_j in $\mathcal{F}_j$ such that $\mathcal{F}_j = \mathcal{F}_{j-1}(r_j)$. To proceed by induction, suppose that $\mathcal{F}_{j-1}$ is isomorphic to a subfield of $\mathbb{C}$. If r_j is algebraic over $\mathcal{F}_{j-1}$, then a subfield of $\mathbb{C}$ isomorphic to $\mathcal{F}_j$ can be constructed by adjoining a complex root of the corresponding polynomial to the isomorphic copy of $\mathcal{F}_{j-1}$. If r_j is transcendental over $\mathcal{F}_{j-1}$, then any subfield of $\mathbb{C}$ generated by the isomorphic copy of $\mathcal{F}_{j-1}$ and any complex transcendental number algebraically independent of that copy is isomorphic to $\mathcal{F}_j$. In either case, there is a subfield of $\mathbb{C}$ isomorphic to $\mathcal{F}_j$, so we can embed $\mathcal{F}_j$ in $\mathbb{C}$. It follows that the algebraic closure $\mathbb{F}_0$ of $\mathbb{K}$ can also be embedded in $\mathbb{C}$, so (ii) holds. □

Since we are dealing with questions of reducibility and triangularizability, the above lemma implies that we lose no generality by assuming $\mathbb{F} = \mathbb{C}$. We do so without further explanation.

We start with several very useful lemmas. Note that all norms on $\mathcal{B}(\mathcal{V})$ are equivalent, since $\mathcal{V}$ is finite-dimensional. Thus statements like "A_n

converges to A in $\mathcal{B}(V)$" make unambiguous sense. Also, we can use an inner-product norm on V and speak of unitary operators, normal operators, and so on. Every subalgebra of $\mathcal{B}(V)$ is automatically closed; this, of course, does not hold for semigroups, but the closure of a semigroup is clearly a semigroup.

Lemma 3.1.2. *If $\{A_n\}$ is a sequence of complex matrices with $A = \lim_{n\to\infty} A_n$, then the following hold:*

> (i) $\sigma(A) = \lim_{n\to\infty}\{\sigma(A_n)\}$ *(which means $\{\lambda : \lambda = \lim_{n\to\infty}\lambda_n, \lambda_n \in \sigma(A_n)\}$).*
>
> (ii) *If m_n is the multiplicity of $\lambda_n \in \sigma(A_n)$ as a zero of the characteristic polynomial of A_n, and if $\lim_{n\to\infty}\lambda_n = \lambda$, then the multiplicity of λ in $\sigma(A)$ is at least $\limsup_n m_n$.*

Proof. (i) Let $\lambda_n \in \sigma(A_n)$ and $\lambda = \lim_{n\to\infty}\lambda_n$; then $\det(A_n - \lambda_n) = 0$, and, since the determinant is a continuous function, we get $\det(A - \lambda) = 0$; so $\lambda \in \sigma(A)$. For the converse, assume that λ lies in $\sigma(A)$. We must show that every neighborhood of λ contains an eigenvalue of A_n for sufficiently large n. Suppose otherwise. Then there is a closed disk $\mathcal{D}$ centered at λ intersecting $\sigma(A)$ only in $\{\lambda\}$, and a subsequence $n(j)$ of integers, such that $\det(A_{n(j)} - z) \neq 0$ for every $z \in \mathcal{D}$. Define polynomials f_j by

$$f_j(z) = \det(A_{n(j)} - z).$$

Observe that $f(z) = \det(A - z)$ is the uniform limit of $f_j(z)$ on $\mathcal{D}$. Since $|f(z)|$ is bounded away from zero on the boundary $\partial\mathcal{D}$ of $\mathcal{D}$, there is an $\epsilon > 0$ such that $|f_j(z)| \geq \epsilon > 0$ for all j and all z in $\partial\mathcal{D}$. Now, the maximum modulus principle, applied to the functions $1/f_j$ analytic on $\mathcal{D}$, implies that $|f_j(z)| \geq \epsilon$ on the whole disk and for all j. This yields $|f(\lambda)| = \lim_{n\to\infty}|f_j(\lambda)| \geq \epsilon$, which is a contradiction.

(ii) Note that m_n equals the nullity of $(A_n - \lambda)^d$, where the matrices have size $d \times d$. Thus the assertion to be proven is equivalent to

$$\operatorname{rank}((A - \lambda)^d) \leq \liminf_n \operatorname{rank}((A_n - \lambda_n)^d).$$

Denote $(A_n - \lambda_n)^d$ by B_n, and $(A - \lambda)^d$ by B. We need only show that

$$\operatorname{rank}(B) \leq \min\{\operatorname{rank}(B_n) : n \geq n_0\}$$

for some n_0. Let $r = \operatorname{rank}(B)$. There is an $r \times r$ submatrix B_0 of B with nonzero determinant. The corresponding submatrices of B_n tend to B_0, so that their determinant is nonzero for all n larger than some n_0. Thus $\operatorname{rank}(B_n) \geq r$ for $n > n_0$. $\qquad\square$

Lemma 3.1.3. *Let S be a semigroup of linear transformations and let Λ be a (multiplicative) semigroup of complex numbers. Then each of the following semigroups has the same invariant subspaces as S:*

(i) *The semigroup $\Lambda S = \{\lambda S : \lambda \in \Lambda, S \in S\}$.*

(ii) *The norm closure $\overline{S}$ of S.*

Proof. The algebra generated by either semigroup is the same as the one generated by S. $\square$

Recall that $\rho(A)$ denotes the spectral radius of A. Note that continuity of ρ follows from Lemma 3.1.2.

Lemma 3.1.4. *An irreducible semigroup is bounded if and only if the spectral radius is bounded on the semigroup.*

Proof. Since $\rho(A) \leq \|A\|$ for every operator, boundedness of ρ on the semigroup S follows from that of the norm. For the converse, we can assume, by continuity of ρ, that S is closed. Also, we can enlarge S to ΛS, where Λ is the closed interval $[0, 1]$. Now suppose that S is unbounded. Pick a sequence $\{S_n\} \subseteq S$ with $\lim_{n\to\infty} \{\|S_n\|\} = \infty$. Replacing $\{S_n\}$ by a subsequence, we can assume that the bounded sequence $\{S_n/\|S_n\|\}$ converges, say to S. Then $S \in S$ and $\|S\| = 1$. Since $\{\rho(S_n)\}$ is bounded,

$$\rho(S) = \lim_{n\to\infty} (\rho(S_n)/\|S_n\|) = 0,$$

so that S is nilpotent. For any member T of S, the boundedness of $\{\rho(S_nT)\}$ implies that

$$\rho(ST) = \lim_{n\to\infty} (\rho(S_nT)/\|S_n\|) = 0,$$

and, similarly, that $\rho(TS) = 0$. Thus the ideal $\mathcal{J}$ of S generated by S consists of nilpotents. Since $\mathcal{J} \neq \{0\}$, this, together with Lemma 2.1.10, contradicts irreducibility (by Theorem 2.1.7). $\square$

The following result is well known.

Theorem 3.1.5. *Every bounded group in $\mathcal{M}_n(\mathbb{C})$ is simultaneously similar to a group of unitary matrices.*

Proof. By taking its closure, we can assume that $\mathcal{G}$ is a compact subgroup of $\mathcal{M}_n(\mathbb{C})$. We present two proofs. The first is the standard one, based on Haar measure; the second is more elementary and geometric.

Let μ be Haar measure on $\mathcal{G}$ (i.e., μ is a positive regular Borel measure on $\mathcal{G}$ such that $\mu(\mathcal{G}) = 1$ and $\int_{\mathcal{G}} f(GG_0)d\mu(G) = \int_{\mathcal{G}} f(G)d\mu(G)$ for all measurable f and all $G_0 \in \mathcal{G}$. Proofs of the existence of Haar measure and discussions of its properties may be found in many texts including Rudin ([2], p. 128); Halmos ([1], p. 251).

We define a new inner product $\langle \cdot, \cdot \rangle$ on $\mathbb{C}^n$ as follows:

$$\langle x, y \rangle = \int_{\mathcal{G}} (Gx, Gy) d\mu(G),$$

where $(\cdot, \cdot)$ is the standard inner product on $\mathbb{C}^n$. It is easily verified that $\langle \cdot, \cdot \rangle$ is an inner product. For any $G_0 \in \mathcal{G}$ and $x, y \in \mathbb{C}^n$,

$$\langle G_0 x, G_0 y \rangle = \int_{\mathcal{G}} (GG_0 x, GG_0 y) d\mu(G)$$

$$= \int_{\mathcal{G}} (Gx, Gy) d\mu(G) = \langle x, y \rangle.$$

Thus every $G_0 \in \mathcal{G}$ is unitary with respect to the new inner product. This finishes the first proof.

We now present the second proof.

First observe that if $\|x\|$ denotes the given norm of $x \in \mathbb{C}^n$, then the relation

$$|x| = \sup\{\|Tx\| : T \in \mathcal{G}\}$$

defines a new norm on $\mathbb{C}^n$, which is equivalent to the given norm by boundedness. Then every S in $\mathcal{G}$ is an isometry relative to the new norm, because for $x \in \mathbb{C}^n$,

$$|Sx| = \sup\{\|TSx\| : T \in \mathcal{G}\} = \sup\{\|Rx\| : R \in \mathcal{G}\} = |x|.$$

We shall change the norm once more to an equivalent inner-product norm rendering every member of $\mathcal{G}$ unitary. The combined norm change will then give the claimed simultaneous similarity.

Denote the unit ball $\{x : |x| \leq 1\}$ by $\mathcal{B}$ and let $\mathcal{E}$ be the set of all $n \times n$ positive-definite matrices A such that

$$x^* A x \leq |x|^2 \quad \text{for all} \quad x \in \mathbb{C}^n$$

(where $x \in \mathbb{C}^n$ is represented by a column vector). We next prove that $\mathcal{E}$ has a member of maximal determinant. Note that

$$p = \sup\{|z| : z \in \sigma(A), A \in \mathcal{E}\} < \infty.$$

For otherwise there would be an A in $\mathcal{E}$ with an arbitrarily large eigenvalue λ, and then a corresponding eigenvector x could be chosen with $|x| < 1$ and $\lambda^2 x^* x = x^* A^2 x \geq 1$, which is a contradiction. This shows that

$$\|A\| \leq p \quad \text{and} \quad \det A \leq p^n$$

for all A in $\mathcal{E}$. Let

$$q = \max\{\det A : A \in \mathcal{E}\},$$

and choose a sequence $\{A_k\} \subseteq \mathcal{E}$ with $\lim_{k \to \infty} (\det A_k) = q$. Since $\{A_k\}$ is bounded, we can assume, by passing to a subsequence, that

$$A = \lim_{k \to \infty} A_k \quad \text{and} \quad \det A = q.$$

Then A is positive-definite and is easily seen to be in $\mathcal{E}$. Thus we have shown the existence of a member A with maximal determinant q.

We now show that A is unique. To this end, suppose $B \neq A$, $B \in \mathcal{E}$, and $\det B = q$. Note that $A^{-1}B^2A^{-1}$ has determinant 1 and is different from the identity matrix. Define the positive-definite matrix C by

$$C^2 = \frac{1}{2}(A^2 + B^2);$$

i.e., let C be the positive-definite square root of the right side. Clearly, $C \in \mathcal{E}$. We claim that $\det C > q$; equivalently, $\det(A^{-1}C^2A^{-1}) > 1$. Note that

$$\det(I + A^{-1}B^2A^{-1}) = 2^n \det(A^{-1}C^2A^{-1}).$$

So we need to show that $\det(I + A^{-1}B^2A^{-1}) > 2^n$. Denote the eigenvalues of $A^{-1}B^2A^{-1}$ by $\{d_1, \ldots, d_n\}$, counting multiplicities, so that

$$\det(I + A^{-1}B^2A^{-1}) = \prod_{j=1}^{n}(1 + d_j),$$

and note that $\prod_{j=1}^{n} d_j = 1$, but not all the d_j are equal to 1. Now, $(1 + d_j)^2 \geq 4d_j$ for every j, with inequality strict at least for one d_j. It follows that

$$\left[\prod_{j=1}^{n}(1 + d_j)\right]^2 > 4^n \prod_{j=1}^{n} d_j = 4^n,$$

proving the claim, and thus the uniqueness of A by maximality of q.

We next show that the set

$$\mathcal{A} = \{x : x^*A^2x \leq 1\}$$

is invariant under every member of $\mathcal{G}$. Observe that if $S \in \mathcal{G}$, then $|\det S| = 1$ (because the eigenvalues of an isometry are all of modulus 1). Also,

$$S\mathcal{A} = \{Sx : x^*A^2x \leq 1\} = \{x : x^*(S^*)^{-1}A^2S^{-1}x \leq 1\}.$$

Thus if B is the (unique) positive-definite square root of the matrix $(AS^{-1})^*(AS^{-1})$, then

$$S\mathcal{A} = \{x : x^*B^2x \leq 1\},$$

and $B \in \mathcal{E}$. Since $\det B = \det A$, the uniqueness of A in $\mathcal{E}$ implies $A = B$. Hence $S\mathcal{A} = \mathcal{A}$ for every $S \in \mathcal{G}$.

The final step is easy: The desired inner product is defined by

$$(x, y) = x^* A^2 y$$

for column vectors x and y in $\mathbb{C}^n$. If $S \in \mathcal{G}$, and if $\|x\|$ denotes the norm of x induced by this inner product, then

$$\|Sx\|^2 = (Sx)^* A^2 (Sx) = x^* A^2 x = \|x\|^2$$

for all x.   □

The following lemma will be used many times in various contexts.

Lemma 3.1.6. *Let S be a semigroup of matrices satisfying $S = \overline{\mathbb{R}^+ S}$, where $\mathbb{R}^+$ is the set of positive real numbers. Let m be the minimal rank of nonzero members of S.*

(i) *If E is any idempotent in S of rank m, then the restriction of $ESE \setminus \{0\}$ to the range of E is a group $\mathcal{G}$.*

(ii) *Up to a simultaneous similarity, each such group $\mathcal{G}$ is contained in $\mathbb{R}^+ \mathcal{U}$, where $\mathcal{U}$ is the group of unitary $m \times m$ matrices.*

(iii) *If S is irreducible, then it contains idempotents of rank m, and, for each such idempotent, the corresponding group $\mathcal{G}$ is irreducible.*

Proof. To prove (i) and (ii), note that ESE is closed and $ESE = \overline{\mathbb{R}^+ ESE}$, so we can assume with no loss of generality that E is the identity matrix and $ESE = S$.

By minimality of m, every nonzero member of S is invertible. We first claim that every nonzero S in S is similar to a scalar multiple of a unitary operator. Fix such an S. Since $S = \overline{\mathbb{R}^+ S}$, we can assume $\rho(S) = 1$. Then S is, up to similarity, of the form

$$\begin{bmatrix} B & 0 \\ 0 & C \end{bmatrix},$$

where $\sigma(B)$ is on the unit circle, and $\sigma(C)$ is inside it. Furthermore, B can be assumed to be in its Jordan form:

$$B = U + N, \text{ with } U \text{ unitary}, N \text{ nilpotent, and } UN = NU.$$

Note that $\rho(C) < 1$ implies $\lim_{n \to \infty} \|C^n\| = 0$.

Our claim amounts to the assertion that $N = 0$ and C acts on the zero-dimensional space. Let $k \geq 0$ be such that $N^k \neq 0$ and $N^{k+1} = 0$. Then, for every $n \geq k$, the binomial expansion yields

$$(U + N)^n = U^n + \binom{n}{1} U^{n-1} N + \cdots + \binom{n}{k} U^{n-k} N^k.$$

Since U is unitary, by taking successive subsequences of integers, we can get a subsequence of $\{U^n\}$ that converges to I. Choose $\{n_j\}$ such that

$$\lim_{j \to \infty} U^{n_j - k} = I.$$

If $k \neq 0$, divide both sides of the expansion equation by $\binom{n}{k}$, which is the dominant coefficient on the right, to get

$$\lim_{j \to \infty} \frac{(U + N)^{n_j}}{\binom{n_j}{k}} = \lim_j (U^{n_j - k} N^k) = N^k.$$

Since $\lim_{n \to \infty} \|C^n\| = 0$, this implies

$$\lim_{j \to \infty} \frac{S^{n_j}}{\binom{n_j}{k}} = \begin{bmatrix} N^k & 0 \\ 0 & 0 \end{bmatrix} \in \mathcal{S},$$

which is a contradiction, because N^k is nonzero and has rank less than m. Thus $k = 0$, and $N = 0$. Now, $N = 0$ implies that

$$\lim_{j \to \infty} B^{n_j} = I \quad \text{and} \quad \lim_{j \to \infty} S^{n_j} = \begin{bmatrix} I & 0 \\ 0 & 0 \end{bmatrix} \in \mathcal{S},$$

which shows that C acts on the zero-dimensional space (by the minimality of m). Thus we have proven that every nonzero member of $\mathcal{S}$ is similar to a multiple of a unitary operator.

To show that $\mathcal{G} = \mathcal{S} \setminus \{0\}$ is a group we must verify that $\mathcal{S}$ contains the inverse of each of its elements other than 0. This follows by setting $N = 0$ in the expansion above and observing that

$$\lim_{j \to \infty} U^{n_j - k - 1} = \lim_{j \to \infty} U^{-1}(U^{n_j - k}) = U^{-1} I = U^{-1}.$$

We have shown that $G/\rho(G)$ is similar to a unitary matrix for every G in $\mathcal{G}$, but it remains to be established that there is a simultaneous similarity. Since $\rho(G)^m = |\det G|$ for every G, it follows that $\rho(G_1 G_2) = \rho(G_1)\rho(G_2)$ for all G_1 and G_2 in $\mathcal{G}$. This implies that $\mathcal{G}_0 = \{G/\rho(G) : G \in \mathcal{G}\}$ is a group. Since $\mathcal{G}_0$ is closed (by continuity of ρ), the proof of (ii) will be completed by Lemma 3.1.5 if we show that $\mathcal{G}_0$ is bounded. If not, choose (G_n) in $\mathcal{G}_0$ with $\lim_{n \to \infty} \|G_n\| = \infty$. Pick a subsequence of the bounded sequence $\{G_n / \|G_n\|\}$ that converges to, say, A in $\mathcal{S}$. Since $\|A\| = 1$ and $\rho(A) = 0$ (as $\rho(G_n) = 1$ and ρ is continuous), we conclude that the rank of A is less than m. This is a contradiction.

To prove (iii), assume that $\mathcal{S}$ is irreducible. Then, by Lemma 2.1.13, it contains a member of the form $A = A_0 \oplus 0$, where A_0 is an invertible $m \times m$ matrix. By multiplying by the reciprocal of its spectral radius, assume that $\rho(A_0) = 1$. Apply the argument used on the matrix $B \oplus C$ in the proof of

(i) above to conclude that A_0 is similar to a unitary matrix, by minimality of m. Thus $A_0^{-1} \oplus 0$ belongs to $\mathcal{S}$, because $\mathcal{S}$ is closed. Hence

$$E = (A_0 \oplus 0)(A_0^{-1} \oplus 0) = I \oplus 0$$

is an idempotent of rank m in $\mathcal{S}$. The irreducibility of $\mathcal{G}$ now follows from Lemma 2.1.12. $\square$

Definition 3.1.7. A *minimal idempotent* in a semigroup $\mathcal{S}$ is a nonzero idempotent E in $\mathcal{S}$ such that no other nonzero idempotent in $\mathcal{S}$ has range contained in that of E and kernel containing that of E.

An equivalent form of this definition is often useful.

Lemma 3.1.8. *A nonzero idempotent E in a semigroup $\mathcal{S}$ is minimal if and only if $EF = E$ whenever F is an idempotent in $\mathcal{S}$ such that $EF = FE \neq 0$.*

Proof. Assume that E is minimal and $EF = FE \neq 0$. The commutativity implies that EF is an idempotent. Since the range of EF is contained in that of E and its kernel contains that of E, and since $EF \neq 0$, we have $EF = E$ by minimality.

For the converse, assume that E is not minimal, and represent it as

$$E = \begin{pmatrix} I & 0 \\ 0 & 0 \end{pmatrix}.$$

Since E is not minimal, there is a different nonzero idempotent F in $\mathcal{S}$ whose range is contained in that of E and whose kernel contains that of E. Thus the matrix of F has the form

$$F = \begin{pmatrix} F_0 & 0 \\ 0 & 0 \end{pmatrix},$$

where $F_0 \neq I$ (because $E \neq F$). Then $EF = FE \neq 0$, but $EF \neq E$. $\square$

Note that in any semigroup $\mathcal{S}$, an idempotent is minimal if it has minimum rank among all the nonzero idempotents in $\mathcal{S}$. There are obvious examples of semigroups containing minimal idempotents whose rank is not minimal. For example, let E be a rank-one idempotent on a space of dimension greater than 2; then $\{E, I - E\}$ is a semigroup in which both E and $I - E$ are minimal idempotents with different ranks. For many semigroups we are concerned with, however, the two concepts coincide.

Theorem 3.1.9. *Let $\mathcal{S}$ be an irreducible semigroup such that $\mathcal{S} = \overline{\mathbb{R}^+\mathcal{S}}$. Then an idempotent E in $\mathcal{S}$ is minimal if and only if it has minimal rank among nonzero idempotents of $\mathcal{S}$.*

Proof. We need only show that if E is minimal and F is a nonzero idempotent in $\mathcal{S}$, then $\operatorname{rank} F \geq \operatorname{rank} E$. Assume that this is not the case, so that the ideal $\mathcal{S}F\mathcal{S}$ (necessarily irreducible, by Lemma 2.1.10) has members of nonzero rank less than that of E. Now, $ESFSE$ is irreducible when restricted to the range of E, and thus contains a nonzero idempotent G by Lemma 3.1.6(iii). Since $G = EGE$, we have $GE = EG = G$. Since $0 \neq G \neq E$, this contradicts the minimality of E, by Lemma 3.1.8. $\qquad\square$

Example 3.1.10. *Let $\mathcal{S}_0$ be the semigroup consisting of zero and the standard basis for $n \times n$ matrices. Let $E = \operatorname{diag}(1, 0, \ldots, 0)$. Then*

$$\mathcal{S} = \{E, I - E\} \cup \{rS : S \in \mathcal{S}_0, \ 1 < r \in \mathbb{R}\}$$

is easily seen to be a semigroup with only two idempotents E and $I - E$, which are both minimal. If $n \geq 3$, then E and $I - E$ have different ranks. Also, $\mathcal{S}$ is irreducible, because $\mathcal{S}_0$ is.

This example shows that the preceding theorem does not hold if $\mathcal{S}$ is merely assumed to be irreducible.

3.2 Permutable Spectrum

In this section, we consider permutability of the spectrum on a collection of linear transformations (see Definition 2.1.3). Since $\sigma(AB) = \sigma(BA)$ for every pair A and B (cf. Lemma 1.4.2), it follows from Lemma 2.1.14 that, on a semigroup, σ is permutable if and only if $\sigma(ABC) = \sigma(BAC)$ for all $A, B,$ and C in the semigroup. Permutability of σ is clearly a necessary condition for triangularizability.

It should be noted that permutability of spectrum does not, in general, imply that of trace, because $\sigma(A) = \sigma(B)$ means that the spectra are equal as sets, not as sets with multiplicities.

It turns out that permutability of spectrum is a stronger condition on groups than it is on semigroups: On the former it implies triangularizability, on the latter merely reducibility.

Theorem 3.2.1. *Every group with permutable spectrum is triangularizable.*

Proof. Let $\mathcal{G}$ be a group on which σ is permutable, and let $\mathcal{H}$ be its commutator subgroup. For $H \in \mathcal{H}$ we have $\sigma(H) = \sigma(I) = \{1\}$, by permutability. Thus $\mathcal{H}$ is a group of unipotent operators. The result then follows from Corollary 2.2.5 (the implication (iv) $\Rightarrow$ (i)). $\qquad\square$

Theorem 3.2.2. *Every semigroup acting on a space of dimension greater than 1 that has permutable spectrum is reducible.*

Proof. By continuity of spectrum (Lemma 3.1.2), permutability of spectrum on $\mathcal{S}$ extends to its norm closure; it also clearly extends to the homogeneous semigroup $\mathbb{C}\mathcal{S}$ generated by $\mathcal{S}$. Thus (Lemma 3.1.3) we can assume $\mathcal{S} = \overline{\mathbb{C}\mathcal{S}}$. Let

$$m = \min\{\text{rank } S : 0 \neq S \in \mathcal{S}\},$$

and assume that $\mathcal{S}$ is irreducible. Then, by Lemma 3.1.6, there is an idempotent E of rank m in $\mathcal{S}$.

If $m = 1$, we consider the ideal $\mathcal{J} = \mathcal{S}E\mathcal{S}$. Since every member of $\mathcal{J}$ has at most one (counting multiplicity) nonzero element in its spectrum, the permutability of spectrum on $\mathcal{J}$ implies permutability of trace. Thus $\mathcal{J}$ is triangularizable by Theorem 2.2.1, which contradicts Lemma 2.1.10.

If $m > 1$, we consider the subsemigroup $E\mathcal{S}E$. The restriction of $E\mathcal{S}E \setminus \{0\}$ to the range of E is, by Lemma 3.1.6, a group $\mathcal{G}$. It is easily seen that σ is permutable on $\mathcal{G}$: Just observe that if G is the restriction of $E\mathcal{S}E$ to $E\mathcal{V}$, then $\sigma(G) = \sigma(E\mathcal{S}E) \setminus \{0\}$. Thus $\mathcal{G}$ is triangularizable by Theorem 3.2.1. Reducibility of $\mathcal{S}$ now follows from that of $E\mathcal{S}E$ (Lemma 2.1.12). □

There are nontriangularizable semigroups with permutable (even constant) spectrum. For instance, the semigroup $\mathcal{S}$ of Example 2.1.11 has the property that $\sigma(S) = \{0,1\}$ for all $S \in \mathcal{S}$, yet it has only one nontrivial invariant subspace. A theorem requiring less than permutability is presented in the next chapter (Theorem 4.3.5).

For further results we need the following simple algebraic lemmas.

Lemma 3.2.3. *Assume that the $2n$ numbers $\{\alpha_1, \dots, \alpha_n; \beta_1, \dots, \beta_n\}$ are algebraically independent over $\mathbb{Q}$ (that is, there is no nontrivial polynomial p in $2n$ indeterminates over $\mathbb{Q}$ with $p(\alpha_1, \dots, \beta_n) = 0$). Then there exists a field automorphism ϕ of $\mathbb{C}$ such that $\phi(\alpha_i) = \beta_i$ and $\phi(\beta_i) = \alpha_i$ for every i.*

Proof. For each rational function f in $2n$ variables with coefficients in $\mathbb{Q}$, define ϕ by

$$\phi(f(\alpha_1, \dots, \alpha_n; \beta_1, \dots, \beta_n)) = f(\beta_1, \dots, \beta_n; \alpha_1, \dots, \alpha_n).$$

The hypothesis of algebraic independence implies that the image under ϕ is well-defined. It is easy to see that ϕ is an automorphism of the extension field $\mathbb{F} = \mathbb{Q}(\alpha_1, \dots \beta_n)$. But any automorphism of $\mathbb{F}$ can be extended to one of $\mathbb{C}$, by Zorn's Lemma. □

Definition 3.2.4. Let ϕ be any field automorphism of $\mathbb{C}$. Then the ring automorphism $\Phi : \mathcal{M}_n(\mathbb{C}) \to \mathcal{M}_n(\mathbb{C})$ defined entrywise by ϕ (i.e., by the formula

$$(\Phi(A))_{ij} = \phi(A_{ij})$$

for all i and j) is called the *automorphism of $\mathcal{M}_n(\mathbb{C})$ induced by ϕ* . Its restriction to a semigroup $\mathcal{S} \subseteq \mathcal{M}_n(\mathbb{C})$ is called the *induced isomorphism* on $\mathcal{S}$.

When no confusion is likely to arise, we use the same notation, Φ, for all values of n.

Lemma 3.2.5. *If ϕ is an automorphism of $\mathbb{C}$ and Φ is the induced automorphism, then*

$$\sigma(\Phi(A)) = \phi(\sigma(A)) = \{\phi(\lambda) : \lambda \in \sigma(A)\}.$$

Proof. This is a direct consequence of the equation

$$\phi(\det(A - \lambda I)) = \det(\Phi(A) - \phi(\lambda)I).$$

$\square$

Theorem 3.2.6. *Let $\mathcal{S}$ be a semigroup with permutable spectrum on an n-dimensional $\mathcal{V}$. If $\mathcal{S}$ contains a member A with n algebraically independent eigenvalues, then $\mathcal{S}$ is triangularizable.*

Proof. Let $\mathcal{P}$ be the following set of properties for a collection $\mathcal{E}$ of operators:

(i) $\mathcal{E}$ is a semigroup,
(ii) $\mathcal{E}$ has permutable spectrum,
(iii) $\mathcal{E}$ contains an operator with distinct and algebraically independent eigenvalues.

By Theorem 3.2.2, every collection satisfying $\mathcal{P}$ (even without (iii)) is reducible. The proof will be completed by the Triangularization Lemma (Lemma 1.1.4) if we demonstrate that $\mathcal{P}$ is inherited by quotients. Thus let $\mathcal{E}$ satisfy $\mathcal{P}$ and let $\mathcal{M}$ and $\mathcal{N}$ be any two invariant subspaces of $\mathcal{E}$, with $\mathcal{M}$ properly contained in $\mathcal{N}$. Consider a block triangularization of $\mathcal{E}$ relative to the chain

$$0 \subseteq \mathcal{M} \subset \mathcal{N} \subseteq \mathcal{V}$$

in which a typical member of $\mathcal{E}$ is of the form

$$S = \begin{bmatrix} S_1 & X & Y \\ 0 & S_2 & Z \\ 0 & 0 & S_3 \end{bmatrix}.$$

Let $\mathcal{T}$ be the set of all the middle blocks S_2. We must show that $\mathcal{T}$ satisfies $\mathcal{P}$. Since $\mathcal{T}$ is a semigroup, and since every $A \in \mathcal{E}$ with distinct, algebraically independent eigenvalues gives rise to a block $A_2 \in \mathcal{T}$ having the same property, it suffices to show that $\mathcal{T}$ has permutable spectrum.

Let $\{\alpha_1, \ldots, \alpha_n\}$ be a listing of eigenvalues of such a member A of $\mathcal{S}$ with $\{\alpha_1, \ldots, \alpha_m\} = \sigma(A_2)$ and, consequently, $\{\alpha_{m+1}, \ldots, \alpha_n\} = \sigma(A_1) \cup \sigma(A_3)$. By an inductive argument, using successive field extensions, we can produce numbers $\beta_1, \ldots, \beta_n$ such that $\{\alpha_1, \ldots, \alpha_n; \beta_1, \ldots, \beta_n\}$ is an algebraically independent set and, moreover,

$$|\beta_i| = 1 \text{ for } i \le m \text{ and } |\beta_i| < 1 \text{ for } i > m.$$

By Lemma 3.2.3, there is an automorphism ϕ of $\mathbb{C}$ with $\phi(\alpha_i) = \beta_i$. Let Φ be the isomorphism on $\mathcal{S}$ induced by ϕ. Then the semigroup $\Phi(\mathcal{S})$ has permutable spectrum, by Lemma 3.2.5, and so does its closure $\overline{\Phi(\mathcal{S})}$ (by Lemma 3.1.2). The image of a typical $S \in \mathcal{S}$, as decomposed above, has the form

$$\Phi(S) = \begin{bmatrix} \Phi(S_1) & X' & Y' \\ 0 & \Phi(S_2) & Z' \\ 0 & 0 & \Phi(S_3) \end{bmatrix}.$$

Now, $\Phi(A)$ has distinct eigenvalues $\{\beta_i\}$, and is thus diagonalizable. The eigenvalues of the middle diagonal block, $\Phi(A_2)$, of $\Phi(A)$ are all on the unit circle, and those of $\Phi(A_1)$ and $\Phi(A_3)$ are inside it. Thus, for some sequence $\{n_i\}$ of integers, $(\Phi(A))^{n_i}$ converges to an idempotent of the form

$$E = \begin{bmatrix} 0 & X'' & Y'' \\ 0 & I & Z'' \\ 0 & 0 & 0 \end{bmatrix}$$

in $\overline{\Phi(\mathcal{S})}$. If R, S, and T are any members of $\mathcal{S}$, then

$$\sigma(\Phi(R)E \cdot \Phi(S)E \cdot \Phi(T)E) = \sigma(\Phi(S)E \cdot \Phi(R)E \cdot \Phi(T)E),$$

or, since the first and last diagonal blocks of $\Phi(R)E, \Phi(S)E$ and $\Phi(T)E$ are zero,

$$\sigma(\Phi(R_2) \cdot \Phi(S_2) \cdot \Phi(T_2)) \cup \{0\} = \sigma(\Phi(S_2) \cdot \Phi(R_2) \cdot \Phi(T_2)) \cup \{0\}.$$

Since $\Phi(R_2)\Phi(S_2)\Phi(T_2)$ is invertible if and only if $\Phi(S_2)\Phi(R_2)\Phi(T_2)$ is invertible, this implies permutability of spectrum on $\Phi(\mathcal{T})$. Applying Lemma 3.2.5 to ϕ^{-1} and Φ^{-1} shows that $\mathcal{T}$ has permutable spectrum.   □

Other results on semigroups with permutable spectrum are obtained below (Section 3.4) from theorems on permutability of spectral radius.

Triangularizability can also be deduced in the following very special case.

Theorem 3.2.7. *Every semigroup of operators of rank at most one with permutable spectrum is triangularizable.*

Proof. In view of Theorem 3.2.2 and the Triangularization Lemma, we need only show that if S_1, S_2, and S_3 are members of the semigroup of the form

$$S_j = \begin{pmatrix} A_j & * & * \\ 0 & B_j & * \\ 0 & 0 & C_j \end{pmatrix}$$

for $j = 1, 2, 3$, then $\sigma(B_1 B_2 B_3) = \sigma(B_2 B_1 B_3)$. If at least one of the B_j is zero, or if the subspace on which they act has dimension 1, we are done. So assume that the B_j are nonzero members of $\mathcal{M}_k(\mathbb{C})$ with $k \geq 2$. Then the A_j and C_j are all zero, because each S_j has rank one. Hence

$$\sigma(S_1 S_2 S_3) = \sigma(B_1 B_2 B_3) \cup \{0\} = \sigma(B_1 B_2 B_3).$$

Similarly, $\sigma(S_2 S_1 S_3) = \sigma(B_2 B_1 B_3)$. Thus $B_1 B_2 B_3$ and $B_2 B_1 B_3$ have the same spectrum. □

3.3 Submultiplicative Spectrum

Definition 3.3.1. The spectrum is *submultiplicative on a semigroup $\mathcal{S}$* if, for every S and $T \in \mathcal{S}$,

$$\sigma(ST) \subseteq \sigma(S)\sigma(T) = \{\lambda\mu : \lambda \in \sigma(S), \mu \in \sigma(T)\}.$$

If a semigroup is triangularizable, then the spectrum is clearly submultiplicative on it. The converse is, in general, not true: There are irreducible groups on which σ is submultiplicative. However, this cannot occur for groups that are "genuinely" infinite, as we shall see (Theorem 3.3.4).

Our first result concerns semigroups with at least one nonzero singular member.

Theorem 3.3.2. *Let $\mathcal{S}$ be a semigroup with submultiplicative spectrum. If the closure of $\mathbb{C}\mathcal{S}$ contains a nonzero singular element, then $\mathcal{S}$ is reducible.*

Proof. By continuity and homogeneity, σ is submultiplicative on $\overline{\mathbb{C}\mathcal{S}}$, so we can assume that $\mathcal{S} = \overline{\mathbb{C}\mathcal{S}}$. Suppose that $\mathcal{S}$ is irreducible. The submultiplicativity of the spectrum implies that the set of nilpotent members of $\mathcal{S}$ is an ideal; by Levitzki's Theorem and Lemma 2.1.10, this ideal is $\{0\}$. Then, since $\mathcal{S}$ contains a noninvertible matrix other than zero, the idempotent E given by Lemma 3.1.6 is nontrivial. The ideal $\mathcal{S}E\mathcal{S}$ is thus nonzero; by Lemma 2.1.10, it suffices to show that it is reducible.

We accomplish this by verifying that

$$\mathrm{tr}((I - E)T) = 0$$

for all $T \in SES$, for then Lemma 2.1.2 will yield reducibility. If $T = 0$, there is nothing to prove, so assume $T \neq 0$. Write $T = RES$ with R and S in S. Up to a simultaneous similarity, we can write

$$E = \begin{pmatrix} I & 0 \\ 0 & 0 \end{pmatrix}, \quad RE = \begin{pmatrix} A & 0 \\ B & 0 \end{pmatrix}, \quad \text{and} \quad ES = \begin{pmatrix} C & D \\ 0 & 0 \end{pmatrix}.$$

Since 0 is the only nilpotent member of S, neither A nor C is 0. The matrices

$$ERE = \begin{pmatrix} A & 0 \\ 0 & 0 \end{pmatrix} \text{ and } ESE = \begin{pmatrix} C & 0 \\ 0 & 0 \end{pmatrix}$$

are in S, so the minimality of the rank of E implies that A and C are invertible. By Lemma 3.1.6,

$$\begin{pmatrix} A^{-1} & 0 \\ 0 & 0 \end{pmatrix} \text{ and } \begin{pmatrix} C^{-1} & 0 \\ 0 & 0 \end{pmatrix}$$

belong to ESE. Hence,

$$S_1 = \begin{pmatrix} I & C^{-1}D \\ 0 & 0 \end{pmatrix} \text{ and } S_2 = \begin{pmatrix} I & 0 \\ BA^{-1} & 0 \end{pmatrix}$$

are in SES. Now, $\sigma(S_1 S_2) \subseteq \sigma(S_1)\sigma(S_2)$ implies that

$$\sigma(I + C^{-1}DBA^{-1}) \subseteq \{0, 1\},$$

and, since $I + C^{-1}DBA^{-1}$ is similar to a multiple of a unitary operator (by Lemma 3.1.6), it follows that $I + C^{-1}DBA^{-1}$ is either 0 or I. The former case cannot arise, since it would imply that the semigroup S_2SS_1 consists of nilpotent matrices and thus is $\{O\}$. Thus would contradict irreducibility. Thus $C^{-1}DBA^{-1} = 0$, which implies $DB = 0$. Consequently,

$$\begin{aligned} \operatorname{tr}((I - E)T) &= \operatorname{tr}((I - E)RES) \\ &= \operatorname{tr}(ES(I - E)RE) \\ &= \operatorname{tr}(DB) = 0, \end{aligned}$$

which is the desired equation.   □

The semigroup S given in Example 2.1.11 has constant spectrum $\{0, 1\}$, which is thus submultiplicative. That example shows that one nontrivial invariant subspace is all that is guaranteed by the hypotheses of Theorem 3.3.2.

Definition 3.3.3. A semigroup S is called *essentially finite* if $S \subseteq \mathbb{C}S_0$ for some finite semigroup S_0.

It follows from the definition that if S is a group, then S_0 can also be chosen to be a group.

Theorem 3.3.4. *If S is an irreducible semigroup with submultiplicative spectrum, then $S \setminus \{0\}$ is an essentially finite group.*

Proof. The semigroup $\overline{\mathbb{C}S}$ also has submultiplicative spectrum, by continuity and homogeneity of σ, so we can assume that $S = \overline{\mathbb{C}S}$ with no loss of generality. By Theorem 3.3.2, S has no nonzero singular elements, so the idempotent E of Lemma 3.1.6 is the identity. Thus $S \setminus \{0\}$ is a group, which can be assumed to be contained in $\mathbb{C}\mathcal{U}$, where $\mathcal{U}$ is the group of all unitary matrices.

Next consider the subgroup

$$\mathcal{G} = \{A \in S : \det A = 1\}$$

of $S \setminus \{0\}$. This is a unitary group, and the hypothesis $S = \mathbb{C}S$ implies that $S \subseteq \mathbb{C}\mathcal{G}$. We will be done if we show that $\mathcal{G}$ is finite. We do this in three steps.

(1) We first show that $\mathcal{G}$ is a torsion group; i.e., that given $G \in \mathcal{G}$, there is a natural number k with $G^k = I$. Let m be such that $\sigma(G^m)$ has the minimal number of eigenvalues among all positive powers of G. We need only show that G^m is a scalar multiple of I, for if $G^m = \lambda I$, then $\det G^m = 1$ implies $\lambda^n = 1$ so $G^{mn} = I$. Since $\mathcal{G}$ is irreducible (so, by Burnside's Theorem (1.2.2), its linear span is $\mathcal{B}(\mathcal{V})$), it suffices to verify that G^m is in the center of $\mathcal{G}$. Let $A = G^m$ and let B be an arbitrary member of $\mathcal{G}$. Let $\{\alpha_1, \dots, \alpha_r\}$ and $\{\beta_1, \dots, \beta_s\}$ be the distinct eigenvalues of A and B, respectively. Note that no positive power of A has fewer than r distinct eigenvalues, by assumption.

The finiteness of the set

$$\sigma(B)\sigma(B^{-1}) = \left\{ \frac{\beta_i}{\beta_j} : i, j = 1, \dots, s \right\}$$

together with the relation

$$\sigma(A^t B A^{-t} B^{-1}) \subseteq \sigma(A^t B A^{-t})\sigma(B^{-1}) = \sigma(B)\sigma(B^{-1})$$

for all positive integers t implies the existence of a subset Ω of $\sigma(B)\sigma(B^{-1})$ such that

$$\sigma(A^t B A^{-t} B^{-1}) = \Omega$$

for all t in some infinite subset $\mathbb{N}_0$ of positive integers. For each $t \in \mathbb{N}_0$,

$$\Omega \subseteq \sigma(A^t)\sigma(BA^{-t}B^{-1}) = \sigma(A^t)\sigma(A^{-t})$$

$$= \left\{ \left(\frac{\alpha_i}{\alpha_j} \right)^t : i, j = 1, \dots, r \right\}.$$

Thus each $w \in \Omega$ equals $(\alpha_i/\alpha_j)^t$ for some fixed pair (i, j) and for infinitely many t. For this pair, then, some power of α_i/α_j is 1. Thus $\alpha_i^p = \alpha_j^p$ for some p. Since no power of A has fewer distinct eigenvalues than A itself, we get $\alpha_i = \alpha_j$ and, hence, $w = 1$.

We have shown that $\Omega = \{1\}$. Now fix $t \in \mathbb{N}_0$ and observe that

$$\sigma(A^t B A^{-t} B^{-1}) = \{1\}$$

implies $A^t B A^{-t} B^{-1} = I$, because $\mathcal{G}$ is a unitary group. It follows that B commutes with A^t, and thus with A, since A is a polynomial in A^t. (This last fact is a consequence of the minimality of r.) Thus $\mathcal{G}$ is a torsion group.

(2) We next show that there is a fixed k such that $G^k = I$ for all $G \in \mathcal{G}$. Recall that the *order* of a root of unity λ is the smallest natural number r such that $\lambda^r = 1$. If every eigenvalue of a given G has order at most r, then $G^{(r!)} = I$ (since G is unitary). Thus if there were no k such that $G^k = I$ for all $G \in \mathcal{G}$, it would follow that there is a sequence $\{G_m\} \subseteq \mathcal{G}$ and $\lambda_m \in \sigma(G_m)$ such that λ_m has order greater than m. Then the set consisting of all powers of all the $\{\lambda_m\}$ is dense in the unit circle. Fix any λ of modulus 1 that is not a root of unity, and choose a sequence $\{\lambda_m^{r_m}\}$ that converges to λ. Thus some subsequence of the corresponding $\{G_m^{r_m}\}$ converges to some unitary operator G_0. By continuity of spectrum (Lemma 3.1.2), $\lambda \in \sigma(G_0)$. But $G_0^k = I$ for some k, which contradicts the fact that λ is not a root of unity. Thus there is a k such that $G^k = I$ for all $G \in \mathcal{G}$.

(3) It remains to be shown that $A^k = I$ for all $A \in \mathcal{G}$ implies that $\mathcal{G}$ is finite. For this, first observe that if $0 < \epsilon < |1 - e^{2\pi i/k}|$, then the neighborhood

$$\mathcal{N} = \{G \in \mathcal{G} : \|I - G\| < \epsilon\}$$

of I consists of I alone. This is because $A \in \mathcal{N}$ and $\lambda \in \sigma(A)$ imply $\lambda = 1$, since $\lambda^k = 1$ and $|1 - \lambda| < \epsilon$. Thus $A = I$. Similarly, for each $G_0 \in \mathcal{G}$, $\{G \in \mathcal{G} : \|G_0 - G\| < \epsilon\}$ consists of G_0 alone. The compactness of $\mathcal{G}$ implies that a finite number of such neighborhoods cover $\mathcal{G}$, so $\mathcal{G}$ is finite. $\qquad \square$

Theorem 3.3.4 can also be deduced from a more general lemma to be proven later (Lemma 4.1.1).

Essential finiteness is not the only condition that is implied by the hypotheses of Theorem 3.3.4. The group in $\mathcal{S}$ must be solvable and, in fact, nilpotent. (Recall that a finite group is nilpotent if and only if it is the direct product of its Sylow p-subgroups.)

Theorem 3.3.5. *An irreducible group $\mathcal{G}$ with submultiplicative spectrum is nilpotent.*

Proof. It suffices to show that the group

$$\mathcal{G}_0 = \{\lambda G : G \in \mathcal{G}, \lambda \in \mathbb{C}, \det(\lambda G) = 1\},$$

which is finite by Theorem 3.3.4, is nilpotent. For each prime p that divides the order of $\mathcal{G}$, let

$$\mathcal{G}_p = \{G \in \mathcal{G}_0 : G^{p^r} = I \text{ for some } r \in \mathbb{N}\}$$

and

$$\Gamma_p = \{\lambda \in \mathbb{C} : \lambda^{p^r} = 1 \text{ for some } r \in \mathbb{N}\}.$$

Then $G \in \mathcal{G}_p$ if and only if $\sigma(G) \subseteq \Gamma_p$ (since each G is similar to a unitary operator, by Lemma 3.1.6 (ii)). Thus $\mathcal{G}_p$ is a subgroup, by the submultiplicativity of σ. If p and q are distinct primes, then $\Gamma_p \cap \Gamma_q = \{1\}$ implies $\mathcal{G}_p \cap \mathcal{G}_q = \{I\}$. We need only show that if $G \in \mathcal{G}_p$ and $H \in \mathcal{G}_q$, then $HG = GH$. But

$$HGH^{-1} \in \mathcal{G}_p \text{ and } GH^{-1}G^{-1} \in \mathcal{G}_q$$

imply that $HGH^{-1}G^{-1} \in \mathcal{G}_p \cap \mathcal{G}_q$, and thus $HGH^{-1}G^{-1} = I$. □

Corollary 3.3.6. *Let S be a semigroup with submultiplicative spectrum. If S contains a nonzero member A satisfying any one of the following conditions, then S is reducible.*

   (i) *$\sigma(A)$ contains λ and μ with $|\lambda| \neq |\mu|$.*
   (ii) *$\sigma(A)$ contains λ and μ with $(\lambda/\mu)^m \neq 1$ for all $m \in \mathbb{N}$.*
   (iii) *$AB = 0$ for some $B \neq 0$ in S .*
   (iv) *A is nilpotent .*
   (v) *$A/\rho(A)$ is not similar to a unitary operator .*
   (vi) *$A/\rho(A)$ has unbounded powers .*
   (vii) *$\det A = 1$, and A has unbounded power.*
  (viii) *$\det A = 1$, and $A^m \neq I$ for all $m \in \mathbb{N}$.*

Proof. We can assume that $S = \overline{\mathbb{C}S}$. Suppose S is irreducible, so that Theorem 3.3.4 applies. Then we can assume, after a simultaneous similarity, that $S = \mathbb{C}\mathcal{G}$, where $\mathcal{G}$ is a finite unitary group with constant determinant 1. Now any one of the conditions (i) - (v) contradicts the unitary character of $\mathcal{G}$, while its finiteness is contradicted by each of the remaining conditions. □

There are finite irreducible groups with submultiplicative spectrum.

Example 3.3.7. *Let p be an odd prime, and $\{e_j\}$ an orthonormal basis for p-dimensional $\mathcal{V}$. Let U be the operator on $\mathcal{V}$ defined by $Ue_j = e_{j+1}$ for $j \leq p-1$ and $Ue_p = e_1$. Denote by $\mathcal{D}$ the group of all diagonal operators of determinant 1 whose eigenvalues are in the set*

$$\Gamma = \{1, \omega, \dots, \omega^{p-1}\},$$

where ω is a fixed primitive p-th root of unity. Let $\mathcal{G}$ be the group generated by U and $\mathcal{D}$. Then $\mathcal{G}$ has submultiplicative spectrum and is irreducible.

Proof. Observe that every nondiagonal member of $\mathcal{G}$ can be obtained from a power of U by replacing its nonzero entries with various powers of ω. Since $\det G = 1$ for all $G \in \mathcal{G}$, every nondiagonal operator in $\mathcal{G}$ is similar to a cyclic permutation (i.e., to U), so that its spectrum coincides with Γ. Each member of $\mathcal{D}$, of course, has spectrum contained in Γ. Now, if A and B are both diagonal members of $\mathcal{G}$, then the inclusion $\sigma(AB) \subseteq \sigma(A)\sigma(B)$ is obvious. If one of them, say B, is nondiagonal, then $\sigma(B) = \Gamma$. Choose an element ω^j in $\sigma(A)$ and note that

$$\sigma(A)\sigma(B) \supseteq \omega^j\Gamma \supseteq \sigma(AB).$$

Thus σ is submultiplicative on $\mathcal{G}$. It is easily verified that $\mathcal{G}$ is irreducible. □

The next result states that there are examples of irreducible groups with submultiplicative spectrum on every space of odd dimension. We first recall some definitions and facts.

If A and B are square matrices of size m and n, respectively, then the tensor product $A \otimes B$ can be defined as the $mn \times mn$ matrix C expressed in the block form $(C_{ij}) = (a_{ij}B)$, $i, j = 1, \ldots, m$.

If $\mathcal{S}$ and $\mathcal{T}$ are semigroups of matrices, then the set

$$\{S \otimes T : S \in \mathcal{S}, T \in \mathcal{T}\},$$

denoted by $\mathcal{S} \otimes \mathcal{T}$, is a semigroup, because

$$(S_1 \otimes T_1)(S_2 \otimes T_2) = (S_1 S_2) \otimes (T_1 T_2).$$

We also have $\sigma(A \otimes B) = \sigma(A)\sigma(B)$ for any A and B. One way to see this is to pick invertible K and L such that $K^{-1}AK$ and $L^{-1}BL$ are triangular. Then

$$(K \otimes L)^{-1}(A \otimes B)(K \otimes L) = (K^{-1}AK) \otimes (L^{-1}BL),$$

so that we can assume that both A and B are triangular. Then it follows from the definition of tensor products that $A \otimes B$ is triangular and that its diagonal entries consist of products of diagonal entries of A and B.

If $\mathcal{S}$ and $\mathcal{T}$ are irreducible semigroups, then it is easy to see that $\mathcal{S} \otimes \mathcal{T}$ is irreducible (e.g., one can use Burnside's Theorem (Theorem 1.2.2)). If $\mathcal{S}$ and $\mathcal{T}$ are semigroups with submultiplicative spectrum, then so is $\mathcal{S} \otimes \mathcal{T}$. For if S_1 and S_2 are in $\mathcal{S}$, and T_1 and T_2 in $\mathcal{T}$, then

$$\begin{aligned}
\sigma[(S_1 \otimes T_1)(S_2 \otimes T_2)] &= \sigma(S_1 S_2 \otimes T_1 T_2) = \sigma(S_1 S_2)\sigma(T_1 T_2) \\
&\subseteq \sigma(S_1)\sigma(S_2) \cdot \sigma(T_1)\sigma(T_2) \\
&= \sigma(S_1)\sigma(T_1) \cdot \sigma(S_2)\sigma(T_2) \\
&= \sigma(S_1 \otimes T_1)\sigma(S_2 \otimes T_2).
\end{aligned}$$

Theorem 3.3.8. *For every odd n there exists an irreducible group of $n \times n$ unitary matrices with submultiplicative spectrum.*

Proof. Let $n = p_1^{r_1} \cdots p_m^{r_m}$, where $\{p_i\}$ is a sequence of odd primes. For each p_i let $\mathcal{G}_i$ denote the group constructed in Example 3.3.7 with $p = p_i$. Let $\mathcal{S}_i$ be the tensor product

$$\mathcal{G}_i \otimes \cdots \otimes \mathcal{G}_i$$

with r_i equal factors, and let $\mathcal{S} = \mathcal{S}_1 \otimes \cdots \otimes \mathcal{S}_m$. Then $\mathcal{S}$ acts on an n-dimensional space, is irreducible, and has submultiplicative spectrum.

$\square$

In some special cases, submultiplicative spectrum does imply triangularizability.

Theorem 3.3.9. *A semigroup $\mathcal{S}$ of operators of rank ≤ 1 with submultiplicative spectrum is triangularizable.*

Proof. There is no harm in assuming that $\mathcal{S} = \overline{\mathbb{C}\mathcal{S}}$. Now let $\mathcal{S}_0$ be the subsemigroup of $\mathcal{S}$ generated by its nilpotent and idempotent elements, so that $\mathcal{S} = \mathbb{C}\mathcal{S}_0$ (by the hypothesis on the ranks). By submultiplicativity, $\sigma(A) \subseteq \{0, 1\}$ for every $A \in \mathcal{S}_0$.

Now the reducibility of $\mathcal{S}_0$ follows from Theorem 3.3.2 (or Theorem 2.3.1). To show triangularizability for $\mathcal{S}_0$, and hence for $\mathcal{S}$, we use the Triangularization Lemma (1.1.4). Since quotients of $\mathcal{S}_0$ are also semigroups of operators of rank ≤ 1 with spectra in $\{0, 1\}$, all we need to observe is the following fact: If

$$A = \begin{pmatrix} A_1 & X & Y \\ 0 & A_2 & Z \\ 0 & 0 & A_3 \end{pmatrix} \text{ and } A' = \begin{pmatrix} A_1' & X' & Y' \\ 0 & A_2' & Z' \\ 0 & 0 & A_3' \end{pmatrix}$$

are members of $\mathcal{S}_0$ with at least one of the middle blocks A_2 and A_2' nilpotent, then the product $A_2 A_2'$ is nilpotent. So assume that A_2 is nilpotent, and, with no loss of generality, nonzero. Since A has rank one, it follows that $A_1 = 0$ and $A_3 = 0$. Thus A is nilpotent, and so is AA' by hypothesis, yielding $\sigma(A_2 A_2') = 0$. Therefore, the quotient also has submultiplicative spectrum.

$\square$

The semigroup $\mathcal{S}$ of Example 2.1.11 shows that Theorem 3.3.9 does not hold if the ranks are allowed to be more than one.

Theorem 3.3.10. *Let $\mathcal{S}$ be a semigroup with submultiplicative spectrum on an n-dimensional space. If $\mathcal{S}$ contains a member with n algebraically independent eigenvalues, then $\mathcal{S}$ is triangularizable.*

Proof. Let $\mathcal{P}$ be as in the proof of Theorem 3.2.6, but with "permutable" replaced by "submultiplicative." A very similar proof works in the present situation. All we have to observe at the start of the proof is that every collection satisfying $\mathcal{P}$ is reducible, by part (ii) of Corollary 3.3.6.   □

3.4 Conditions on Spectral Radius

In this section we consider permutability and submultiplicativity of ρ on a semigroup, prove that these two conditions are equivalent, and study their effects on reducibility and triangularizability.

Since $\rho(AB) = \rho(BA)$ for any operators A and B, it follows from Lemma 2.1.14 that ρ is permutable on a semigroup $\mathcal{S}$ if and only if $\rho(ABC) = \rho(BAC)$ for all A, B, and C in $\mathcal{S}$. Submultiplicativity means, of course, that $\rho(AB) \leq \rho(A)\rho(B)$ for all A and B in $\mathcal{S}$. Clearly, each of these conditions on ρ is implied by the corresponding condition on the spectrum itself, studied in the preceding sections. The example of the full unitary group $\mathcal{U}$ on $\mathcal{V}$, on which ρ is constant, shows that permutability and submultiplicativity are too weak by themselves to yield any reducibility results for semigroups. (Recall that in the case of algebras, submultiplicativity of ρ implies not just reducibility but triangularizability, as shown in Theorem 1.5.5.)

Lemma 3.4.1. *If a semigroup $\mathcal{S}$ has permutable ρ, then so does $\overline{\mathbb{C}\mathcal{S}}$; if $\mathcal{S}$ has submultiplicative ρ, so does $\overline{\mathbb{C}\mathcal{S}}$.*

Proof. Both statements are direct consequences of continuity of ρ (Lemma 3.1.2) together with the fact that $\rho(\lambda A) = |\lambda|\rho(A)$ for all $\lambda \in \mathbb{C}$ and all operators A.   □

Lemma 3.4.2. *Let $\mathcal{S}$ be any semigroup.*

(i) *If $\overline{\mathbb{R}^+\mathcal{S}}$ contains no nonzero nilpotent operators, then for every S in $\mathcal{S}$ with $\rho(S) = 1$ there is a sequence of integers $\{n_i\}$ such that $\{S^{n_i}\}$ converges to an idempotent.*

(ii) *If $\mathcal{S}$ is irreducible and has submultiplicative ρ, then $\overline{\mathbb{C}\mathcal{S}}$ contains no nonzero divisors of zero (and, in particular, no nonzero nilpotent operators).*

Proof. We can assume that $\mathcal{S} = \overline{\mathbb{R}^+\mathcal{S}}$. To prove (i), fix S and employ a similarity so that S is of the form

$$\begin{bmatrix} U + N & 0 \\ 0 & C \end{bmatrix},$$

where, as in the proof Lemma 3.1.6, U is unitary, N is nilpotent, $UN = NU$, and $\rho(C) < 1$. The calculations given in that proof yield a nonzero nilpotent in S if N is not zero. Thus $N = 0$ by hypothesis. Letting $\{n_i\}$ be a sequence with $\lim_{n_i \to \infty} U^{n_i} = I$, we then get

$$\lim_{n_i \to \infty} S^{n_i} = \lim_{n_i \to \infty} U^{n_i} \oplus \lim_{n_i \to \infty} C^{n_i} = I \oplus 0.$$

For (ii), we can assume $S = \overline{\mathbb{C}S}$ by Lemma 3.4.1. The nilpotent members of S form an ideal, by submultiplicativity of ρ. This ideal is $\{0\}$ by Lemma 2.1.10. If $AB = 0$ with nonzero A and B in S, then BSA consists of nilpotents, and thus $BSA = \{0\}$. Hence S sends the range $\mathcal{R}$ of A to the kernel of B, implying that the span of $S\mathcal{R}$ is either $\{0\}$ or a proper invariant subspace of S, which contradicts the irreducibility of S.   □

The following result shows, in particular, that if S has submultiplicative ρ and the strict inequality $\rho(AB) < \rho(A)\rho(B)$ holds for at least one pair $\{A, B\}$ in S, then S is reducible.

Theorem 3.4.3. *Let S be a semigroup with submultiplicative ρ. Then ρ is multiplicative on S if $\overline{\mathbb{R}^+S}$ has no nonzero divisors of zero, and, in particular, if S is irreducible.*

Proof. By Lemmas 3.1.2 and 3.1.3 we can assume that $S = \overline{\mathbb{R}^+S}$. By Lemma 3.4.2 it suffices to treat the case where S has no nonzero divisors of zero. To show that $\rho(AB) = \rho(A)\rho(B)$ for any A and B in S, we need only consider the case $\rho(A) = \rho(B) = 1$. Use part (i) of Lemma 3.4.2 to get sequences $\{n_i\}$ and $\{m_i\}$ such that

$$\lim_{n_i \to \infty} A^{n_i} = E = E^2 \text{ and } \lim_{m_i \to \infty} B^{m_i} = F = F^2$$

with E and F in S. Then it follows from the relations

$$\rho(A^{n_i} B^{m_i}) = \rho(AB \cdot B^{m_i-1} A^{n_i-1})$$
$$\leq \rho(AB)\rho(B^{m_i-1})\rho(A^{n_i-1}) = \rho(AB)$$

that $\rho(EF) \leq \rho(AB)$. Thus to complete the proof it suffices to verify that $\rho(EF) = 1$.

Observe that $\rho(EF) \neq 0$, by part (ii) of Lemma 3.4.2. Now the relations

$$\rho(EF) = \rho(E^2 F^2) = \rho(EF \cdot FE) \leq \rho(EF)^2$$

imply that $\rho(EF) \geq 1$. Since $\rho(EF) \leq \rho(E)\rho(F) = 1$, it follows that $\rho(EF) = 1$.   □

Corollary 3.4.4. *Let S be an irreducible semigroup with submultiplicative ρ. Then there is a semigroup $\mathcal{T}$ such that $\rho(T) = 1$ for all $T \in \mathcal{T}$ and $S \subseteq \mathbb{C}\mathcal{T}$.*

Proof. By the preceding theorem, ρ is multiplicative. Since $\mathcal{S} \setminus \{0\}$ is a semigroup, by Lemma 3.4.2, we observe that

$$\mathcal{T} = \left\{ \frac{S}{\rho(S)} : 0 \neq S \in \mathcal{S} \right\}$$

is a semigroup with $\rho = 1$. □

Corollary 3.4.5. *Let $\mathcal{G}$ be a group that is not similar to a subgroup of $\mathbb{C}\mathcal{U}$, where $\mathcal{U}$ is the full unitary group. If $\mathcal{G}$ has submultiplicative ρ, then it is reducible. In particular, a group of normal operators is reducible if it contains an operator whose spectrum is not on a circle.*

Proof. If $\mathcal{G}$ is irreducible, then the semigroup $\mathcal{T}$ defined in the proof of Corollary 3.4.4 is a group. However, a group with $\rho = 1$ is, up to similarity, contained in $\mathcal{U}$, by Lemmas 3.1.4 and 3.1.5. Thus $\mathcal{G}$ must be reducible.

If a group $\mathcal{G}$ consists of normal operators, then

$$\rho(ST) = \|ST\| \leq \|S\| \cdot \|T\| = \rho(S)\rho(T)$$

for every pair S and T in $\mathcal{G}$. Hence, if $\mathcal{G}$ is not a subgroup of $\mathbb{C}\mathcal{U}$, then it is reducible.

□

Lemma 3.4.6. *If the sequence $\{\mathrm{tr}(T^m)\}$ is bounded for a given operator T, then $\rho(T) \leq 1$.*

Proof. If $\rho(T) = 0$, there is nothing to prove. Suppose $a = \rho(T) \neq 0$; we must show that $a \leq 1$. Enumerate the eigenvalues $\{\lambda_j\}$ of T, counting multiplicities, so that

$$a = |\lambda_1| = \cdots = |\lambda_r| > |\lambda_{r+1}| \geq \cdots \geq |\lambda_n|.$$

There is an increasing sequence $\{m_i\}$ of integers such that

$$\lim_{m_i \to \infty} \left(\frac{\lambda_j}{a} \right)^{m_i} = 1 \text{ for all } j \leq r.$$

Since $\lim_{m_i \to \infty} (\lambda_j/a)^{m_i} = 0$ for $j > r$, this yields

$$\lim_{m_i \to \infty} \mathrm{tr}\left(\left(\frac{T}{a} \right)^{m_i} \right) = \lim_{m_i \to \infty} \sum_{j=1}^{n} \left(\frac{\lambda_j}{a} \right)^{m_i} = r.$$

Thus $\lim_{m_i \to \infty} \mathrm{tr}(T^{m_i}) = \infty$ if $a > 1$. □

Theorem 3.4.7. *Let $\mathcal{J}$ be a nonzero ideal of an irreducible semigroup $\mathcal{S}$. If $\mathcal{J}$ has submultiplicative spectral radius, so does $\mathcal{S}$.*

Proof. By continuity and homogeneity of ρ, we can assume $\mathcal{S} = \overline{\mathbb{C}\mathcal{S}}$ and $\mathcal{J} = \overline{\mathbb{C}\mathcal{J}}$. (It is clear that $\overline{\mathbb{C}\mathcal{J}}$ is an ideal of $\overline{\mathbb{C}\mathcal{S}}$.) Since $\mathcal{J}$ is irreducible (by Lemma 2.1.10), ρ is multiplicative on it (by Theorem 3.4.3).

We first show that if $S \in \mathcal{S}$ and $J \in \mathcal{J}$, then $\rho(SJ) \leq \rho(S)\rho(J)$. There are no nonzero nilpotents in $\mathcal{S}$, by Lemma 3.4.2, so we can replace J by $J/\rho(J)$ and assume $\rho(J) = 1$ with no loss of generality. We then have

$$\rho(JSJ) = \rho(J)\rho(SJ) = \rho(SJ).$$

For $k > 1$,

$$\rho(JS^k J) = \rho(JS^{k-1})\rho(SJ) = \rho(JS^{k-1}J)\rho(SJ),$$

so, by induction, $\rho(JS^k J) = (\rho(SJ))^k$ for every natural number k. Thus

$$(\rho(SJ))^k \leq \|JS^k J\| \leq \|J\|^2 \cdot \|S^k\|,$$

which implies

$$\rho(SJ) \leq \lim_{k \to \infty} \|S^k\|^{\frac{1}{k}} = \rho(S)$$

(by the spectral radius formula, Theorem 6.1.10). To show that $\rho(ST) \leq \rho(S)\rho(T)$ for all S and T in $\mathcal{S}$, we can assume, with no loss of generality, that $\rho(S) = \rho(T) = 1$. By irreducibility, $\mathcal{J}$ contains a basis

$$\{J_i : i = 1, \dots, n^2\}$$

for $\mathcal{B}(\mathcal{V})$, where n is the dimension of $\mathcal{V}$. Replacing each J_i by $J_i/\rho(J_i)$, we can assume $\rho(J_i) = 1$ for every i. Express the identity operator as a linear combination $\Sigma \alpha_i J_i$ to get

$$(ST)^k = \sum_{i=1}^{n^2} \alpha_i (ST)^k J_i$$

for every k. Observe that, for every i and k,

$$\rho((ST)^{k+1} J_i) = \rho(J_i)\rho((ST)^{k+1} J_i) = \rho(J_i(ST)^{k+1} J_i)$$
$$= \rho(J_i ST)\rho((ST)^k J_i).$$

Since $\rho((ST)J_i) = \rho(J_i(ST)J_i) = \rho(J_i S)\rho(TJ_i)$, it follows by induction that

$$\rho((ST)^k J_i) = \rho(S^k J_i)\rho(T^k J_i).$$

This, together with the preceding paragraph, yields

$$|\operatorname{tr}(ST)^k| \leq \sum_i |\alpha_i| \cdot |\operatorname{tr}\left((ST)^k J_i\right)| \leq n \sum_i |\alpha_i| \cdot \rho\left((ST)^k J_i\right)$$

$$= n \sum_i |\alpha_i| \cdot \rho(S^k J_i)\rho(T^k J_i)$$

$$\leq n \sum_i |\alpha_i| \cdot \rho(S^k)\rho(T^k) = n \sum_i |\alpha_i|,$$

because $\rho(S^k) = \rho(T^k) = 1$. Thus the sequence $\{(ST)^k\}$ has bounded trace, and $\rho(ST) \leq 1$ by Lemma 3.4.6. $\square$

We have seen (in Theorem 3.4.3) that, for an irreducible semigroup, sub-multiplicativity of ρ implies its multiplicativity, which implies permutability. The significance of the following result is that it does not require irreducibility.

Theorem 3.4.8. *The spectral radius is submultiplicative on a semigroup if and only if it is permutable on the semigroup.*

Proof. First assume that ρ is permutable on a semigroup $\mathcal{S}$. Then, for S and T in $\mathcal{S}$ and $k \in \mathbb{N}$,

$$(\rho(ST))^k = \rho((ST)^k) = \rho(S^k T^k) \leq \|S^k T^k\| \leq \|S^k\| \cdot \|T^k\|.$$

Thus $\rho(ST) \leq \|S^k\|^{\frac{1}{k}} \cdot \|T^k\|^{\frac{1}{k}}$, so

$$\rho(ST) \leq \rho(S)\rho(T)$$

(by the spectral radius formula, Theorem 6.1.10). We must now prove the converse.

Let $\mathcal{S}$ be a semigroup in $\mathcal{B}(\mathcal{V})$ with submultiplicative ρ, and make the harmless assumption $\mathcal{S} = \overline{\mathbb{C}\mathcal{S}}$. Let $\mathcal{C}$ be a maximal chain of invariant subspaces for $\mathcal{S}$. Express the members of $\mathcal{C}$ as

$$\{0\} = \mathcal{V}_0 \subset \mathcal{V}_0 \oplus \mathcal{V}_1 \subset \cdots \subset \mathcal{V}_0 \oplus \mathcal{V}_1 \oplus \cdots \oplus \mathcal{V}_m = \mathcal{V},$$

and consider the corresponding block triangularization of $\mathcal{S}$, where a typical member of $\mathcal{S}$ has a matrix of the form

$$S = \begin{bmatrix} S_1 & S_{12} & \cdots & S_{1m} \\ 0 & S_2 & \cdots & S_{2m} \\ \vdots & & & \\ 0 & 0 & \cdots & S_m \end{bmatrix},$$

and where, for each i, the set $\mathcal{T}_i = \{S_i : S \in \mathcal{S}\}$ is an irreducible semigroup on $\mathcal{B}(\mathcal{V}_i)$.

Define a function Φ on $\mathcal{S}$ mapping matrices into their diagonals by

$$\Phi(S) = S_1 \oplus S_2 \oplus \cdots \oplus S_m.$$

It is easily seen that Φ is a spectrum-preserving semigroup homomorphism. Thus $\rho(\Phi(S)) = \rho(S)$, and we can therefore work with $\Phi(S)$ rather than S. Thus assume that every S in $\mathcal{S}$ is in block diagonal form. We then make a further reduction of the problem: If there is a natural number $j \leq m$ such that

$$\rho(S) = \rho(\oplus_{i \neq j} S_i)$$

for all $S \in \mathcal{S}$, we eliminate the j-th direct summand from every member S without altering ρ. Since such an elimination results in a semigroup homomorphism, we can, after a finite number of eliminations, make the following assumption without loss of generality: For each j, there exists an $S \in \mathcal{S}$ whose spectral radius is achieved only on its j-th direct summand; that is,

$$\rho(S_j) = \rho(S) \text{ and } \rho(S_i) < \rho(S) \text{ for } i \neq j.$$

For a given j, replacing this S by $S/\rho(S)$ and applying Lemma 3.4.2 to S_j in the irreducible semigroup $\mathcal{T}_j$ provides an idempotent $E_j \in \mathcal{T}_j$ as a limit of $\{S_j^{n_i}\}$. Then

$$\lim_{n_i \to \infty} S^{n_i} = 0 \oplus \cdots \oplus 0 \oplus E_j \oplus 0 \oplus \cdots \oplus 0$$

is in $\mathcal{S}$. Let $\mathcal{J}_j$ denote the ideal of $\mathcal{T}_j$ generated by E_j.

It is easy to see that ρ is submultiplicative on $\mathcal{J}_j$ for each j. Since $\mathcal{T}_j$ is irreducible, we deduce from Theorems 3.4.7 and 3.4.3 that ρ is multiplicative on each $\mathcal{T}_j$. This means that, for any S and T in $\mathcal{S}$, their j-th direct summands satisfy

$$\rho(S_j T_j) = \rho(S_j)\rho(T_j).$$

To complete the proof, let $R, S,$ and T be any members of $\mathcal{S}$. Then

$$\begin{aligned}
\rho(RST) &= \max_j \rho((RST)_j) = \max_j \rho(R_j S_j T_j) \\
&= \max_j (\rho(R_j)\rho(S_j)\rho(T_j)) \\
&= \max_j (\rho(S_j)\rho(R_j)\rho(T_j)) \\
&= \max_j (\rho(S_j R_j T_j)) = \max_j \rho((SRT)_j) = \rho(SRT).
\end{aligned}$$

It follows from Lemma 2.1.14 that ρ is permutable on $\mathcal{S}$. □

Lemma 3.4.9. *Let $\mathcal{J}$ be an ideal of a semigroup $\mathcal{S}$. If $\mathcal{J}$ has a unique triangularizing chain, then $\mathcal{S}$ is triangularizable.*

Proof. By Lemma 2.1.10, $\mathcal{S}$ is reducible. If $\widehat{\mathcal{S}}$ is any quotient of $\mathcal{S}$, then the corresponding quotient $\widehat{\mathcal{J}}$ of $\mathcal{J}$ is an ideal of $\widehat{\mathcal{S}}$ and has a unique triangularizing chain by Lemma 1.5.2. Thus the property of having an ideal with unique triangularizing chain is inherited by quotients, and the result follows from the Triangularization Lemma (1.1.4). □

Corollary 3.4.10. *Let $\mathcal{S}$ be a semigroup with submultiplicative ρ. If $\mathcal{S}$ contains a set of nilpotent operators with a unique triangularizing chain, then $\mathcal{S}$ is triangularizable.*

Proof. The set $\mathcal{S}_0$ of nilpotent members of $\mathcal{S}$ is an ideal, by submultiplicativity of ρ. This ideal is triangularizable by Levitzki's Theorem (2.1.7), and its triangularizing chain is unique by hypothesis. Thus Lemma 3.4.9 is applicable. □

Recall that an operator is *unicellular* if it has a unique triangularizing chain.

Theorem 3.4.11. *If a semigroup with submultiplicative ρ contains a unicellular operator, then it is triangularizable.*

Proof. A unicellular operator is, by the primary decomposition theorem, of the form $\lambda I + N$, where N is a nilpotent Jordan cell, i.e., has one-dimensional kernel. Since we can assume $\mathcal{S} = \overline{\mathbb{C}\mathcal{S}}$ and $I \in \mathcal{S}$ with no loss of generality, there are only two cases to deal with: $\lambda = 0$ and $\lambda = 1$.

If $\lambda = 0$, then the assertion follows directly from Corollary 3.4.10. Next assume $\lambda = 1$. Since

$$\rho((I + N)^k S) \le \rho(S)$$

for all $S \in \mathcal{S}$ and all natural numbers k, we have, for n the dimension of the underlying space,

$$\left| \sum_{j=0}^{n-1} \binom{k}{j} tr(N^j S) \right| = \left| \operatorname{tr}\left((1 + N)^k S\right) \right| \quad \text{(note that } N^n = 0\text{)}$$

$$\le n\rho((1 + N)^k S) \le n\rho(S).$$

The boundedness independent of $k \in \mathbb{N}$ implies that $\operatorname{tr}(N^j S) = 0$ for $j \ge 1$. In particular, $\operatorname{tr}(NS) = 0$ for all $S \in \mathcal{S}$, which implies, by linearity, that $\operatorname{tr}(NA) = 0$ for all A in the algebra $\mathcal{A}$ generated by $\mathcal{S}$. Now, N is in $\mathcal{A}$, and thus $\operatorname{tr}((NA)^k) = 0$, for all k and hence, by Theorem 2.1.16, $\sigma(NA) = \{0\}$ for every $A \in \mathcal{A}$. Thus the radical $\mathcal{R}$ of $\mathcal{A}$ contains N. Since $\mathcal{R}$ is an (algebra ideal and therefore a semigroup) ideal of $\mathcal{A}$, and since its triangularization (obtained by Levitzki's (Theorem 2.1.7)) is unique because $N \in \mathcal{R}$, we conclude, by Lemma 3.4.9, that $\mathcal{A}$ is triangularizable. □

Recall that a commutative band is diagonalizable. (As mentioned in Section 2.3, this can be verified easily by induction.) A maximal commutative band is, therefore, similar to the semigroup of all diagonal matrices with entries 0's and 1's.

Theorem 3.4.12. *Let $\mathcal{S}$ be a semigroup with submultiplicative ρ. If $\mathcal{S}$ contains a maximal commutative band, then $\mathcal{S}$ is triangularizable.*

Proof. Consider the following set $\mathcal{P}$ of properties for a collection $\mathcal{E}$ of operators:

(a) $\mathcal{E}$ is a semigroup.

(b) $\mathcal{E}$ has submultiplicative ρ.

(c) $\mathcal{E}$ contains a maximal commutative band.

Unless the underlying space has dimension 1, (c) implies the existence of nonzero idempotents A and B in $\mathcal{E}$ with $AB = 0$. Thus every $\mathcal{E}$ satisfying $\mathcal{P}$ is reducible, by Lemma 3.4.2(ii). The proof will be completed by the Triangularization Lemma (Lemma 1.1.4) if we show that $\mathcal{P}$ is inherited by quotients.

Let $\mathcal{M}$ and $\mathcal{N}$ be any two invariant subspaces of a collection $\mathcal{E}$ satisfying $\mathcal{P}$, with $\mathcal{M}$ properly contained in $\mathcal{N}$. Consider a block triangularization of $\mathcal{E}$ with respect to the chain

$$0 \subseteq \mathcal{M} \subset \mathcal{N} \subseteq \mathcal{V}.$$

A typical member of $\mathcal{E}$ is of the form

$$S = \begin{bmatrix} S_1 & X & Y \\ 0 & S_2 & Z \\ 0 & 0 & S_3 \end{bmatrix}.$$

We must show that the collection $\mathcal{T} = \{S_2 : S \in \mathcal{E}\}$ satisfies $\mathcal{P}$. That $\mathcal{T}$ is a semigroup is clear. It is also easy to verify that $\mathcal{T}$ satisfies (c) : if $\mathcal{E}_0$ is a maximal commutative band, so is every quotient of $\mathcal{E}_0$. We need only show that ρ is submultiplicative on $\mathcal{T}$. For this, first note that the maximality of the band in $\mathcal{E}$ implies that there is an idempotent of the form

$$E = \begin{bmatrix} 0 & X & Y \\ 0 & I & Z \\ 0 & 0 & 0 \end{bmatrix}.$$

in the block triangularization above. (Of course, $Y = XZ$, and, in fact, we can assume $X = Y = Z = 0$ by a simultaneous similarity, but we do not need this here.) Now, for any S and R in $\mathcal{E}$ we have, by hypothesis,

$$\rho(SE \cdot RE) \le \rho(SE)\rho(RE).$$

This implies $\rho(S_2 R_2) \le \rho(S_2)\rho(R_2)$, where S_2 and R_2 are the middle blocks of S and R, respectively. Hence ρ is submultiplicative on the quotient. $\square$

Perhaps the simplest example of an irreducible semigroup with multiplicative spectrum is the following.

Example 3.4.13. *For each n, there exists an irreducible semigroup $\mathcal{S}$ in $\mathcal{M}_n(\mathbb{C})$ such that, for each $A \in \mathcal{S}$, either A or $-A$ is an idempotent of rank one.*

Proof. Let $\mathcal{E}$ be the set of all column vectors whose coordinates are ± 1. Let $\mathcal{S}$ consist of those matrices with precisely one column from $\mathcal{E}$ and $n-1$ zero columns. It is easily seen that $\mathcal{S}$ is an irreducible semigroup contained in $\{\pm E : E^2 = E\}$. $\square$

3.5 The Dominance Condition on Spectra

Definition 3.5.1. We say that $\rho(A)$ is *dominant* in $\sigma(A)$ if $\rho(A) \in \sigma(A)$ and $|\lambda| < \rho(A)$ for every other member λ of $\sigma(A)$.

Examples of semigroups all of whose members have dominant spectral radii include bands and semigroups of positive-entried matrices (by the Perron-Frobenius Theorem; cf. Chapter 5).

Theorem 3.5.2. *If every member of a semigroup $\mathcal{S}$ is a nonnegative scalar multiple of an idempotent, and if ρ is submultiplicative on $\mathcal{S}$, then $\mathcal{S}$ is triangularizable.*

Proof. We first show that $\mathcal{S}$ is reducible. In view of Theorem 3.4.3, we can assume that ρ is multiplicative. But this implies that the set

$$\mathcal{S}_1 = \{S/\rho(S) : S \in \mathcal{S}, \ S \neq 0\} \cup \{0\}$$

is a band and thus triangularizable by Theorem 2.3.5. In particular, $\mathcal{S}$ is reducible. To complete the proof by the Triangularization Lemma (1.1.4), we need only verify that the properties given in the hypotheses of the theorem are inherited by quotients. Let $\mathcal{M}$ and $\mathcal{N}$ be subspaces invariant under $\mathcal{S}$ with $\mathcal{M} \subset \mathcal{N}$. For each S in $\mathcal{S}$, let S_0 denote the corresponding quotient operator. It is clear that S_0 is a nonnegative multiple of an idempotent. We must show that, for S and T in $\mathcal{S}$,

$$\rho(S_0 T_0) \leq \rho(S_0)\rho(T_0).$$

If $S_0 = 0$ or $T_0 = 0$, there is nothing to prove. So assume S_0 and T_0 are both nonzero. Then $\rho(S_0) = \rho(S)$, $\rho(T_0) = \rho(T)$, and

$$\rho(S_0 T_0) \leq \rho(ST) \leq \rho(S)\rho(T) = \rho(S_0)\rho(T_0).$$

Thus ρ is submultiplicative on the quotient. $\square$

The semigroup of all matrices of rank one whose entries are all positive is easily seen to be irreducible. Since the trace of any such matrix is nonzero, such matrices are positive multiples of idempotents of rank one. This example shows that the assumption of submultiplicativity is needed in the preceding theorem. Also, Example 3.4.13 shows that the theorem cannot be extended to the case where the members of $\mathcal{S}$ are arbitrary scalar multiples of idempotents.

The following corollary shows that a semigroup satisfying the hypotheses of the preceding theorem is essentially a band.

Corollary 3.5.3. *Let S be a semigroup satisfying the hypotheses of Theorem 3.5.2. Then there is a band S_1 such that $S \subseteq \mathbb{R}^+ S_1$.*

Proof. Let S_1 be the set of idempotents that the members of S are multiples of. We shall show that S_1 is a band. The triangularizability of S_1 (by Theorem 3.5.2) implies that if S and T are in S_1, then $\sigma(ST) \subseteq \{0, 1\}$. Since ST, as a member of the semigroup S, is a multiple of an idempotent, it follows that ST is an idempotent. It is then clear that ST belongs to S_1. $\qquad\square$

The reducibility (but not triangularizability) conclusion of Theorem 3.5.2 follows from a much weaker assumption on the spectra.

Theorem 3.5.4. *Let S be a semigroup with submultiplicative ρ. If $\rho(S)$ is dominant in $\sigma(S)$ for all S in S, then S is reducible.*

Proof. We can, without loss of generality, replace S with $\mathbb{R}^+ S$. In view of Theorem 3.4.3, we need only treat the case of multiplicative ρ. We are done if S contains a nonzero nilpotent operator, so assume otherwise. We can then assume, by scaling as in Corollary 3.4.4, that $\rho(S) = 1$ for all $S \in S$.

Let A be a member of S with minimal multiplicity, say m, for 1 in its spectrum. By Lemma 3.4.2, there is an idempotent E in the closure $\bar{S}$ of S (which is a limit of powers of A, and thus has rank m). The dominance condition may fail in $\bar{S}$. However, by Lemma 3.1.2 and the minimality of m, every member of $\bar{S}$ has 1 in its spectrum with multiplicity at least m. Now consider the subsemigroup $E\bar{S}E$ of $\bar{S}$, and note that the restriction S_0 of it to the range of E consists of unipotent operators (for if any operator in S_0 had a number other than 1 in its spectrum, m would not be minimal). Thus S_0 is triangularizable by Kolchin's Theorem (2.1.8), so there is a nonzero vector $x = Ex$ such that $ESEx = x$ for all S in $\bar{S}$. If f is a linear functional on V with $f(x) = 1$, then the functional ϕ on $\mathcal{B}(V)$ defined by

$$\phi(T) = f(ETx)$$

is constant on S. Corollary 2.1.6 then implies that S is reducible. $\qquad\square$

Corollary 3.5.5. *If 1 is dominant in $\sigma(S)$ for every S in a semigroup S, then S is reducible.*

Proof. Since $\rho(S) = 1$ for every S in S by hypothesis, ρ is multiplicative on S, so Theorem 3.5.4 applies. $\qquad\square$

The hypothesis of the proceeding corollary is not sufficient to give a common one-dimensional invariant subspace for the semigroup. For example, consider the semigroup of all block triangular matrices

$$\begin{bmatrix} R & X & Y \\ 0 & I & Z \\ 0 & 0 & T \end{bmatrix}$$

where $\|R\| < 1$, $\|T\| < 1$, and the matrices X, Y, and Z are arbitrary.

The following result shows that the number m in the proof of Theorem 3.5.4 can be used to get a lower bound for the number of distinct invariant subspaces.

Theorem 3.5.6. *Let S be a semigroup and assume that 1 is the dominant element in $\sigma(S)$ for every $S \in \mathcal{S}$. For each S, let $m(S)$ denote the algebraic multiplicity of 1 in $\sigma(S)$, and let $r = \min\{m(S) : S \in \mathcal{S}\}$. Then $\mathcal{S}$ has a block triangularization with at least r diagonal 1×1 blocks constantly equal to 1.*

Proof. Let

$$\{0\} = \mathcal{M}_0 \subset \mathcal{M}_1 \subset \cdots \subset \mathcal{M}_k = \mathcal{V}$$

be a maximal chain of invariant subspaces for $\mathcal{S}$. In the corresponding block triangularization of $\mathcal{S}$, every member S has k diagonal blocks $S_1, \ldots, S_k$. For each j, let $\mathcal{S}_j$ denote the irreducible compression semigroup $\{S_j : S \in \mathcal{S}\}$.

Observe that if $1 \in \sigma(S_j)$, then 1 is dominant in $\sigma(S_j)$, because $\sigma(S_j) \subseteq \sigma(S)$ for every j. It follows from Corollary 3.5.5, together with the irreducibility of $\mathcal{S}_j$, that if $1 \in \sigma(S_j)$ for all S_j in $\mathcal{S}_j$, then $\mathcal{S}_j$ is the 1×1 semigroup $\{1\}$.

Pick a member A of $\mathcal{S}$ such that the set

$$J = \{j : 1 \in \sigma(A_j)\}$$

has minimal size. In view of the preceding paragraph, it suffices to show that $1 \in \sigma(S_j)$ for all $j \in J$ and all $S \in \mathcal{S}$. Suppose that $\mathcal{S}$ contains a member B such that $\sigma(B_j)$ does not contain 1 for some $j \in J$. Choose a number t satisfying

$$\max\{|z| : z \in \sigma(A) \cup \sigma(B), z \neq 1\} < t < 1.$$

(Such a number exists by the dominance of 1 in all spectra.) Now, $\rho(B_i) \leq \rho(B) \leq 1$ implies that, for sufficiently large n,

$$\|B_j^n\| < t^n \quad \text{and} \quad \|B_i^n\| < \left(\frac{1}{t} - 1\right)^n \quad \text{if } i \neq j.$$

Similarly, we obtain the following relations for the compressions of A:

$$\|A_i^n\| < \left(\frac{1}{t} - 1\right)^n \text{ if } i \in J, \text{ and } \|A_i^n\| < t^n \text{ otherwise.}$$

Thus, the compressions of $A^n B^n$ satisfy the relation

$$\|(AB)_i^n\| = \|A_i^n B_i^n\| < t^n \left(\frac{1}{t} - 1\right)^n = (1-t)^n$$

for all i outside $J \setminus \{j\}$. It follows that, for each such i, $\rho((AB)_i^n) < 1$, which contradicts the minimality of J. $\square$

Note that the case $r \geq n - 1$ in the result above is a special case of Corollary 2.2.9.

3.6 Notes and Remarks

The reduction to the complex field (Lemma 3.3.1) and analogous ideas are implicit in works by many authors. The continuity of spectra in finite dimensions has been known for some time; for a treatment of this and similar topics see the book by Bhatia [1]. Theorem 3.1.5 and the measure-theoretic proof of it are in many books on group representation theory. For general groups, this result is due to Auerbach [1]. The second, more elementary and geometric, proof presented here is based on a theorem of Fritz John (see Bollobás [1, p. 68]) and is similar to the treatment in Deutsch-Schneider[1]; our presentation was arranged in collaboration with Tony Thompson.

Permutability of spectrum was studied in Lambrou-Longstaff-Radjavi [1], where Theorems 3.2.1 and 3.2.2 were proved. Theorems 3.2.6 and 3.2.7 are new. Submultiplicativity of spectrum was also studied in Lambrou-Longstaff-Radjavi [1]; this paper includes Theorem 3.3.2, Example 3.3.7, and Theorem 3.3.8. It is not known whether Theorem 3.3.8 holds if n is even; more specifically, if S is a semigroup in $\mathcal{M}_n(\mathbb{C})$ with $n = 2^k$ for some k, and if σ is submultiplicative on S, we do not know whether S can be irreducible. (It cannot if $n = 2$, a fact observed in the same paper.) Theorem 3.3.4 is from Radjabalipour-Radjavi [1]; it strengthens a result of Lambrou-Longstaff-Radjavi [1] by concluding that the groups in question are not just torsion, but finite. Theorems 3.3.9 and 3.3.10 are from Radjavi [6]. R. Guralnick (in work not yet published) has proven Theorem 3.3.4 in the case of finite characteristic.

Submultiplicativity of spectral radius was studied for general Banach algebras by Aupetit and Zemanek (see Aupetit [2]). For semigroups of operators this condition was considered in Lambrou-Longstaff-Radjavi [1], Longstaff-Radjavi [2], and Radjabalipour-Radjavi [1]. Theorems 3.4.3, 3.4.7, and 3.4.8 are from Longstaff-Radjavi [2], as are their corollaries and

Lemma 3.4.6. Theorem 3.4.11 was proved in Lambrou-Longstaff-Radjavi [1]. Theorem 3.4.12 is new. Example 3.4.13 is from Omladič-Radjavi [1].

The structure of irreducible semigroups on which spectrum is submultiplicative (and thus multiplicative, by Theorem 3.4.3) was studied in Omladič-Radjavi [1]. It was proved that, relative to a suitable norm on the underlying space, every member of such a semigroup is of the form $\lambda T \oplus N$ (not simultaneously), where T is an isometry on its domain, $\lambda \in \mathbb{C}$, and N is nilpotent. (This is an extension of Theorem 3.1.5.) The proof is complicated and technical. Another result of this paper concerns the so-called "Rota condition" for a family $\mathcal{S}$ of operators: Given $\varepsilon > 0$, there exists a norm on the underlying space relative to which

$$\|S\| \leq (1 + \varepsilon)\rho(S)$$

for all S and $\mathcal{S}$. If $\mathcal{S}$ is an irreducible semigroup satisfying this condition, then there is a norm relative to which $\|S\| = \rho(S)$ for all S in $\mathcal{S}$ (Omladič-Radjavi [1]). (See also Section 1.6.)

The results of Section 3.5 are mostly from Radjavi [2] and Radjavi [4].

CHAPTER 4
Finiteness Lemmas and Further Spectral Conditions

In this chapter we first establish a lemma reducing many questions of interest about matrix semigroups to the case of finite groups. We use the lemma to obtain reducibility and triangularizability results from assumptions of various "partial" spectral mapping properties, including sublinearity of spectra and the nilpotence of certain polynomials in semigroup elements.

4.1 Reductions to Finite Semigroups

Many of the conditions on semigroups we have considered so far or will consider in the present chapter, including permutability and submultiplicativity of spectra, are stable under the following alterations of semigroups: If a semigroup S satisfies the condition, then so do

(i) the norm closure $\overline{S}$ of S,
(ii) $\mathbb{C}S$, and
(iii) the semigroup $\Phi(S)$ for every ring automorphism Φ of matrices induced by a field automorphism of $\mathbb{C}$ (as discussed in Definition 3.2.4).

Another example of a stable condition is the hypothesis that every pair $\{A, B\}$ in the semigroup satisfies a fixed homogeneous polynomial equation in two noncommutative variables with rational coefficients, such as $(XY - YX)^n = 0$.

We do not quite need all of the above stability properties for our purposes. A weaker combination of them suffices for the following technical lemma, which has wide applicability. (In particular, this weakened stability condition will allow us to consider semigroups satisfying polynomial equations whose coefficients are not necessarily rational.)

Lemma 4.1.1. (The Finiteness Lemma) *Let $\mathcal{P}$ be a property defined for semigroups of matrices such that whenever S has the property, then so does the semigroup $\Phi^{-1}(\overline{\mathbb{C}\Phi(S)})$ for every choice of ring automorphism Φ induced by a field automorphism of $\mathbb{C}$. Let S be a maximal semigroup in $\mathcal{M}_n(\mathbb{C})$ with property $\mathcal{P}$. Denote the minimal nonzero rank in S by m. If E is an idempotent of rank m in S, then the restriction of ESE to the range of E is of the form $\mathbb{C}\mathcal{G}$, where $\mathcal{G}$ is a finite group (similar to an $m \times m$ unitary group).*

Proof. Since Φ can be taken to be the identity automorphism, the maximality hypothesis implies that $S = \overline{\mathbb{C}S}$. By Lemma 3.1.6, the restriction S_0 of ESE to the range of E is, after a similarity, contained in the set of multiples of $m \times m$ unitary matrices. Essential finiteness remains to be proven;

to establish this we show that the unitary group $\mathcal{G} = \{S \in \mathcal{S}_0 : \det S = 1\}$ is finite.

We first establish that $\mathcal{G}$ is a torsion group (i.e., that every element of it has finite order). Suppose that $A \in \mathcal{G}$ and $A^r \neq I$ for every positive integer r. This implies, since $\det A = 1$, that no positive power of A is scalar. Choose an eigenvalue α_0 of A and consider the member $B = A/\alpha_0$ of $\mathcal{S}$. Now, $1 \in \sigma(B)$ and no power of B is scalar. Thus $\sigma(B)$ contains a number α on the unit circle $\mathbb{T}$ such that $\alpha^r \neq 1$ for every r, implying that $\{\alpha^r : r > 0\}$ is dense in $\mathbb{T}$. Pick a transcendental number $\lambda \in \mathbb{T}$. Choose a subsequence of $\{\alpha^r\}$ converging to λ and a further subsequence of it to ensure that $\{B^{r_i}\}$ also converges, say to B_0. Note that $\{1, \lambda\} \subseteq \sigma(B_0)$. We now show that this contradicts the maximality of S relative to $\mathcal{P}$.

Pick a number μ, algebraically independent of λ, such that $|\mu| < 1$. By Lemma 3.2.3, there is a field automorphism ϕ of $\mathbb{C}$ such that $\phi(\lambda) = \mu$ (and $\phi(\mu) = \lambda$). Let Φ be the ring automorphism on matrices induced by ϕ. Now, the semigroup $\Phi(\mathcal{S})$ contains $\Phi(EB_0E)$, whose rank is m. Observe that

$$\sigma(\Phi(EB_0E)) = \phi(\sigma(EB_0E)) \supseteq \{1, \mu\}.$$

Also, since EB_0E is diagonalizable, so is $\Phi(EB_0E)$. Since $|\mu| < 1$, we conclude that, for some sequence n_i of integers,

$$F = \lim_{i \to \infty} \left(\frac{\Phi(EB_0E)}{\rho(\Phi(EB_0E))} \right)^{n_i}$$

is an idempotent of rank less than m. Since $F \in \overline{\mathbb{C}\Phi(\mathcal{S})}$, and since $\Phi^{-1}(F)$ is again an idempotent whose rank equals that of F, we have arrived at a semigroup $\hat{\mathcal{S}} = \Phi^{-1}(\overline{\mathbb{C}\Phi(\mathcal{S})})$ whose minimal positive rank is less than m. This, together with the relations

$$\mathcal{S} = \mathbb{C}\mathcal{S} = \Phi^{-1}(\mathbb{C}\Phi(\mathcal{S})) \subseteq \Phi^{-1}(\overline{\mathbb{C}\Phi(\mathcal{S})}) = \hat{\mathcal{S}},$$

implies that $\hat{\mathcal{S}}$ is a proper extension of $\mathcal{S}$ with property $\mathcal{P}$, which is a contradiction. Hence $\mathcal{G}$ is a torsion group.

To complete the proof, we use the fact that a compact torsion group of matrices is actually finite. This was shown explicitly in steps (2) and (3) of the proof of Theorem 3.3.4. (The irreducibility hypothesis of that theorem was not used to prove this fact.)   □

The Finiteness Lemma will be applied mainly to irreducible semigroups, in which case there are idempotents of minimal rank, by Lemma 3.1.6 (iii). Of course, much less is needed to guarantee the existence of an idempotent in the ideal

$$\mathcal{J} = \{S \in \mathcal{S} : \text{rank } S \leq m\}$$

of a semigroup where m is the minimal rank of nonzero members of $\mathcal{S}$. For example, it is easily seen (using Levitzki's Theorem (2.1.7)) that if the common kernel of $\mathcal{J}$ is trivial and if $\mathcal{S} = \overline{\mathbb{R}^+\mathcal{S}}$, then $\mathcal{J}$ contains a nonzero idempotent.

The following restates the Finiteness Lemma in a form that will be applied to reducibility questions.

Lemma 4.1.2. *Let* $\mathcal{P}$ *be a property satisfying the hypotheses of the Finiteness Lemma. Assume, furthermore, that*

(i) *if* $\mathcal{S}$ *is a semigroup with property* $\mathcal{P}$, $\mathcal{J}$ *is an ideal in* $\mathcal{S}$, *and* E *a minimal (nonzero) idempotent in* $\mathcal{S}$, *then both* $\mathcal{J}$ *and* $E\mathcal{S}E|_{EV}$ *have property* $\mathcal{P}$,

(ii) *every finite group with property* $\mathcal{P}$ *is reducible, and*

(iii) *every semigroup of operators of rank* ≤ 1 *with property* $\mathcal{P}$ *is reducible.*

Then every semigroup $\mathcal{S}$ *with property* $\mathcal{P}$ *is reducible.*

Proof. Assume, without loss of generality, that $\mathcal{S}$ is maximal with property $\mathcal{P}$. Let m be the minimal positive rank in $\mathcal{S}$. Suppose that $\mathcal{S}$ is irreducible, so that it has an idempotent of rank m (by Lemma 3.1.6 (iii)).

If $m \geq 2$, apply the Finiteness Lemma to get $E\mathcal{S}E = \mathbb{C}\mathcal{G}$, where $\mathcal{G}$ is a finite group. Since $\mathcal{G}|_{EV}$ has property $\mathcal{P}$ by (i) and is hence reducible by (ii), we obtain a contradiction to irreducibility of $\mathcal{S}$ (by Lemma 2.1.12).

If $m = 1$, consider the nonzero ideal $\mathcal{J}$ of $\mathcal{S}$ consisting of members of rank ≤ 1. Then $\mathcal{J}$ has property $\mathcal{P}$ by (i), and is reducible by (iii). The desired contradiction is now obtained from Lemma 2.1.10. $\qquad\square$

There is another kind of finiteness lemma; in certain situations it reduces a problem to one concerning matrices over a finite field. For the rest of this section, we consider matrices over any field $\mathbb{F}$, not even assumed to be algebraically closed.

Lemma 4.1.3. *Let* $\mathbb{F}$ *be a field and* $\mathcal{E}$ *a finite subset of* $\mathbb{F} \setminus \{0\}$. *Then there exist a subring* $\mathcal{R}$ *of* $\mathbb{F}$ *containing* $\mathcal{E}$ *and a ring homomorphism* ϕ *from* $\mathcal{R}$ *onto a finite field* $\mathbb{K}$ *such that* $\phi(\alpha) \neq 0$ *for all* α *in* $\mathcal{E}$. *Moreover,* $\mathbb{K}$ *can be chosen to have arbitrarily large size if* $\mathbb{F}$ *is infinite.*

Proof. We shall use the well-known fact that a field that is finitely generated as a ring is finite. (This can be derived from Hilbert's Nullstellensatz. It can also be proved directly by first showing inductively that each of the generators is algebraic over the prime subfield, and then verifying that the prime subfield cannot be $\mathbb{Q}$.) Let $\mathcal{R}$ be the unital subring of $\mathbb{F}$ generated by all the members of $\mathcal{E}$ and all the reciprocals of members of $\mathcal{E}$. Pick a maximal ideal $\mathcal{M}$ in $\mathcal{R}$ and note that $\mathcal{R}/\mathcal{M}$ is a field that is finitely generated as a ring. Thus $\mathbb{K} = \mathcal{R}/\mathcal{M}$ is a finite field. Let ϕ be the canonical map of $\mathcal{R}$ onto $\mathbb{K}$. If $\alpha \in \mathcal{E}$, then $\phi(\alpha)\phi(\alpha^{-1}) = \phi(1)$, and hence $\phi(\alpha) \neq 0$.

To achieve arbitrarily large size for $\mathbb{K}$, enlarge $\mathcal{E}$, if necessary, so that it includes distinct elements $\alpha_1, \ldots, \alpha_m$ together with all the differences $\alpha_i - \alpha_j$ for $i \neq j$. Then the elements $\phi(\alpha_i)$ of $\mathbb{K}$, as constructed above, are all distinct, so that $\mathbb{K}$ has at least m elements.   □

Definition 4.1.4. A collection in $\mathcal{M}_n(\mathbb{F})$ (the $n \times n$ matrices over the field $\mathbb{F}$) is said to be *absolutely irreducible* if it is irreducible as a subset of $\mathcal{M}_n(\overline{\mathbb{F}})$, where $\overline{\mathbb{F}}$ denotes the algebraic closure of $\mathbb{F}$. The term *absolutely nontriangularizable* is defined analogously.

Lemma 4.1.5. *Let S be an absolutely irreducible semigroup in $\mathcal{M}_n(\mathbb{F})$. Then there exists a subring $\mathcal{R}$ of $\mathbb{F}$, an absolutely irreducible subsemigroup S_0 of S contained in $\mathcal{M}_n(\mathcal{R})$, and a ring homomorphism ϕ of $\mathcal{R}$ onto a finite field $\mathbb{K}$ such that $\Phi(S_0)$ is absolutely irreducible, where Φ denotes the homomorphism from $\mathcal{M}_n(\mathcal{R})$ into $\mathcal{M}_n(\mathbb{K})$ induced by ϕ.*

Proof. By Burnside's Theorem (1.2.2), S contains a basis $\{A_i : 1 \leq i \leq n^2\}$ for $\mathcal{M}_n(\mathbb{F})$. Let S_0 be the subsemigroup of S generated by the A_i; then S_0 is absolutely irreducible. Let $\mathcal{E}_0$ be the set of nonzero entries of all the A_i.

To apply the preceding lemma, let $\mathcal{E} = \mathcal{E}_0 \cup \{\alpha, \alpha^{-1}\}$, where α is the determinant of the $n^2 \times n^2$ matrix B constructed as follows: The i-th row of B is composed of the entries of A_i written in a fixed order for all i, say the entries of the first row followed by those of the second row, and so on. Since the A_i are linearly independent, α is nonzero. Next let $\mathcal{R}$ be the subring generated by $\mathcal{E}$ and let ϕ and $\mathbb{K}$ be as in the preceding lemma. Now $S_0 \subseteq \mathcal{M}_n(\mathcal{R})$. To verify that $\Phi(S_0)$ is absolutely irreducible just observe that

$$\det \Phi(B) = \phi(\det B) = \phi(\alpha) \neq 0,$$

so that $\{\phi(A_i)\}$ is linearly independent in $\mathcal{M}_n(\overline{\mathbb{K}})$.   □

Lemma 4.1.6. *The assertion in Lemma 4.1.5 is also true if the word "irreducible" is replaced with "nontriangularizable" throughout.*

Proof. By Theorem 1.3.8, if S is an absolutely nontriangularizable semigroup, then it contains members A, B, C such that $(A(BC - CB))^n \neq 0$, where n is the dimension of the underlying vector space. Let S_0 be the subsemigroup of S generated by A, B, and C. Thus S_0 is also absolutely nontriangularizable. Let $\mathcal{E}_0$ be the set of all nonzero entries in A, B, and C. Pick a fixed nonzero entry α of $(A(BC - CB))^n$ and let $\mathcal{E} = \mathcal{E}_0 \cup \{\alpha, \alpha^{-1}\}$. Apply Lemma 4.1.3 to get $\mathbb{K}$ and ϕ as in the lemma. Since $\phi(\alpha) \neq 0$, the image of $A(BC - CB)$ under Φ is not nilpotent, and thus $\Phi(S_0)$ is a nontriangularizable semigroup in $\mathcal{M}_n(\mathbb{K})$.   □

In both of the corollaries above, if $\mathbb{F}$ is infinite, the field $\mathbb{K}$ can be chosen to be arbitrarily large by increasing the size of $\mathcal{E}$, as in the proof of the lemma.

Many of the results to which we shall apply the Finiteness Lemma could also be proven using Lemma 4.1.5 or Lemma 4.1.6. Before we resume our assumption $\mathbb{F} = \mathbb{C}$, we conclude this section with a sample application: We prove Kaplansky's unification of the theorems of Kolchin and Levitzki.

Corollary 4.1.7. *Let S be a semigroup in $\mathcal{M}_n(\mathbb{F})$. If every member of S is of the form $\alpha I + N$ with $\alpha \in \mathbb{F}$ and N nilpotent, then S is triangularizable.*

Proof. We can assume with no loss of generality that $\mathbb{F} = \overline{\mathbb{F}}$. Also, since the property of having singleton spectrum is inherited by quotients, we need only prove reducibility.

Suppose that S is irreducible. Apply Lemma 4.1.5 to get an absolutely irreducible semigroup S_1 of matrices over a finite field $\mathbb{K}$. Note that S_1 is a finite semigroup whose members have singleton spectra. The ideal $\mathcal{J}$ of singular members of S_1 coincides with the set of nilpotents in it, so that $\mathcal{J} = \{0\}$ by Levitzki's Theorem (Theorem 2.1.7) and Lemma 2.1.10. Hence the nonzero members of S_1 are all nonsingular and, by the finiteness of S_1, form a group $\mathcal{G}$. Adjoin all the n-th roots of eigenvalues of all the members of $\mathcal{G}$ to $\mathbb{K}$ to get $\mathbb{K}_1$. Assume, with no loss of generality, that $\mathcal{G} = \mathbb{K}_1\mathcal{G}$, and then let $\mathcal{G}_0 = \{A \in \mathcal{G} : \det A = 1\}$; it follows that $\mathcal{G}_0$ is irreducible.

Let p be the characteristic of $\mathbb{K}$. Pick m such that $p^m \geq n$. For each $A \in \mathcal{G}_0$, we have $A = \alpha I + N$ for some N with $N^n = 0$, so

$$A^{p^m} = (\alpha I + N)^{p^m} = \alpha^{p^m} I.$$

Hence A^{p^m} belongs to the center $\mathcal{Z}$ of $\mathcal{G}_0$ for all A, and $\mathcal{G}_0/\mathcal{Z}$ is a p-group. Thus $\mathcal{G}_0$ is a nilpotent group, i.e., the direct product of its p-subgroups (Kurosh [2], p. 216). In particular, those members of $\mathcal{G}_0$ whose order is a power of p form a subgroup to which the unipotent members of $\mathcal{G}_0$ belong. If $I + N$ and $I + M$ are any such members, then the order of their product is a power of p, so that for large enough k,

$$I = (I + MN + M + N)^{p^k} = I + (MN + M + N)^{p^k}.$$

This means that $MN + M + N$ is nilpotent. We have shown that the unipotent members of $\mathcal{G}_0$ form a group. This group is triangularizable by Kolchin's Theorem (Theorem 2.1.8). Since every member of $\mathcal{G}_0$ is a scalar multiple of a unipotent member, this contradicts the irreducibility of $\mathcal{G}_0$. $\square$

4.2 Subadditive and Sublinear Spectra

For the rest of the book we revert to the assumption that the field of scalars is $\mathbb{C}$.

Definition 4.2.1. The spectrum is said to be *sublinear* on a pair of operators S and T if, for every $\lambda \in \mathbb{C}$,

$$\sigma(S + \lambda T) \subseteq \sigma(S) + \lambda\sigma(T).$$

A *linear combination* $X + \lambda Y$ of two subsets X and Y of $\mathbb{C}$ means, of course, the set $\{x + \lambda y : x \in X \text{ and } y \in Y\}$. We say that spectrum is *subadditive* on S and T if the inclusion is merely assumed to hold for $\lambda = 1$.

A collection $\mathcal{E}$ of operators is said to have *sublinear* or *subadditive* spectrum if every pair in $\mathcal{E}$ has the property. Observe that $\mathcal{E}$ has sublinear spectrum if and only if $\mathbb{C}\mathcal{E}$ has subadditive spectrum.

It should be emphasized that these definitions are given for two summands only; the definition of sublinearity does not require that $\sigma(\Sigma\lambda_i S_i) \subseteq \Sigma\lambda_i\sigma(S_i)$ for more than two summands. Interestingly, this more general relation will follow from the restricted one, by the results below. Sublinearity is a weakening of the well-known condition called "*property L*"; the "*L*" stands for "linear" and is defined as follows for a pair S and T of operators: The eigenvalues $\{\alpha_i\}$ of S and $\{\beta_i\}$ of T, counted according to their multiplicities, can be put in fixed sequences $\{\alpha_1, \dots, \alpha_n\}$ and $\{\beta_1, \dots, \beta_n\}$ in such a way that, for every λ, the eigenvalues of $S + \lambda T$ are exactly $\{\alpha_i + \lambda\beta_i\}$. Property L is necessary for simultaneous triangularization, of course, as is the sublinearity of spectrum. The latter condition does not require a fixed pairing of $\{\alpha_i\}$ and $\{\beta_i\}$, independent of λ. Also, it does not involve multiplicities of eigenvalues. Nonetheless, we show that it is equivalent to property L and is thus sufficient for triangularizability of a semigroup.

Lemma 4.2.2. *Let A and B be operators with a common invariant subspace $\mathcal{M}$ and denote their restrictions to $\mathcal{M}$ by A_0 and B_0, respectively. If spectrum is sublinear on A and B, then it is sublinear on A_0 and B_0. The same assertion holds for subadditivity in the case where A and B have rank one.*

Proof. Denote the dimension of $\mathcal{M}$ by m. By the sublinearity hypothesis on A and B, for each λ the equation

$$\det(A_0 + \lambda B_0 - z) = 0$$

has m solutions (counting multiplicity) of the form $\alpha_i + \lambda\beta_i$ with $\alpha_i \in \sigma(A)$ and $\beta_i \in \sigma(B)$ for $i = 1, 2, \dots, m$. (The pairs $\langle\alpha_i, \beta_i\rangle$ depend on λ, of course.) It must be shown that each α_i is in $\sigma(A_0)$ and each β_i is in $\sigma(B_0)$.

Fix an m-tuple of such pairs for each λ. Observe that there must exist an infinite set of values of λ that have the same m-tuple of pairs $\langle \alpha_i, \beta_i \rangle$. Then

$$\det[(A_0 - \alpha_i) + \lambda(B_0 - \beta_i)] = 0$$

for all these values of λ and $i = 1, \ldots, m$. Since this is a polynomial equation of degree at most m in λ for each i, it is identically zero, and we deduce that its coefficients are all 0. Hence

$$\det(A_0 - \alpha_i) = 0 \text{ and } \det(B_0 - \beta_i) = 0,$$

and thus $\sigma(A_0 + \lambda B_0) \subseteq \sigma(A_0) + \lambda \sigma(B_0)$ for all λ.

Now assume that A and B have rank one. If either $A_0 = 0$ or $B_0 = 0$, or if $m \leq 1$, then the assertion of subadditivity is trivial. So assume $A_0 \neq 0$, $B_0 \neq 0$, and $m \geq 2$, and observe that $\sigma(A_0) = \sigma(A)$ and $\sigma(B_0) = \sigma(B)$. Then

$$\sigma(A_0 + B_0) \subseteq \sigma(A + B) \subseteq \sigma(A) + \sigma(B) = \sigma(A_0) + \sigma(B_0). \qquad \square$$

Corollary 4.2.3. *Sublinearity of spectrum is inherited by quotients. Subadditivity is inherited by quotients if the operators are of rank ≤ 1.*

Proof. Let $\mathcal{M}_1$ and $\mathcal{M}_2$ be invariant subspaces for a pair $\{A, B\}$ with $\mathcal{M}_1 \subset \mathcal{M}_2$. By the lemma above, we can assume that $\mathcal{M}_2$ is the whole space. We can then decompose A and B with respect to $\mathcal{M}_1$ and its orthocomplement, writing

$$A = \begin{pmatrix} A_1 & A_2 \\ 0 & A_3 \end{pmatrix} \quad \text{and} \quad B = \begin{pmatrix} B_1 & B_2 \\ 0 & B_3 \end{pmatrix}.$$

Note that subadditivity and sublinearity of spectrum on $\{A, B\}$ implies the same for their adjoints. Since A_3^* and B_3^* are, respectively, the restrictions of A^* and B^* to $\mathcal{M}_1^\perp$, it follows from Lemma 4.2.2 that $\{A_3^*, B_3^*\}$ has sublinearity or subadditivity of spectrum if $\{A, B\}$ does, and thus so does $\{A_3, B_3\}$. Therefore, the properties are inherited by quotients. $\qquad \square$

The following simple lemmas will be used more than once.

Lemma 4.2.4. *Let S be an irreducible semigroup of $n \times n$ matrices of rank ≤ 1.*

  (i) *There exist two bases $\{e_i\}$ and $\{f_j\}$ of column vectors for $\mathbb{C}^n$ such that the basis*

$$\{e_i f_j^* : i, j = 1, \ldots, n\}$$

  of $\mathcal{M}_n(\mathbb{C})$ is contained in S.

 (ii) *For each $k \leq n$, there exists a k-dimensional subspace $\mathcal{M}$ of $\mathbb{C}^n$ and a subsemigroup S_0 of S leaving $\mathcal{M}$ invariant such that $S_0|_{\mathcal{M}}$ is irreducible.*

(iii) *In particular, if $k = 2$, there exist numbers $\alpha, \beta, \gamma, \delta$ with $\alpha\delta - \beta\gamma \neq 0$ and $\beta\gamma \neq 0$ such that $S_0|_{\mathcal{M}}$ is generated by the two operators*

$$\begin{pmatrix} \alpha & 0 \\ \beta & 0 \end{pmatrix} \text{ and } \begin{pmatrix} 0 & \gamma \\ 0 & \delta \end{pmatrix}$$

with respect to an appropriate basis.

Proof. There is a nonzero matrix S_0 in $\mathcal{S}$; since S_0 has rank 1, it can be written in the form ef^* where e and f are column vectors. Irreducibility implies that the collection $\mathcal{S}e$ contains a basis $\{e_i\}$ for $\mathbb{C}^n$. Since $\mathcal{S}^*$ is also irreducible, we obtain a basis $\{f_i\}$ for $\mathbb{C}^n$ contained in $\mathcal{S}^*f$. Thus the ideal $\mathcal{S}S_0\mathcal{S}$ contains every $e_i f_j^*$. This proves (i).

To prove (ii), let $\mathcal{S}_0$ be the subsemigroup generated by

$$\{e_i f_j^* : 1 \leq i \leq k, 1 \leq j \leq n\}.$$

Then the linear span $\mathcal{M}$ of $\{e_i : i = 1, \dots, k\}$ is invariant under $\mathcal{S}_0$. To show the irreducibility of $\mathcal{S}_0|_{\mathcal{M}}$, write the matrices of $\mathcal{S}_0$ relative to the basis $\{e_i\}$ of $\mathbb{C}^n$. Observe that, for each $i \leq k$, the i-th rows of members of $\mathcal{S}_0$ contain n independent vectors, so that their initial "k-sections" contain a basis for $\mathbb{C}^k$. Thus $\mathcal{S}_0|_{\mathcal{M}}$ contains a basis for $\mathcal{M}_k(\mathbb{C})$.

If $k = 2$, apply (i) to $\mathcal{S}_0|_{\mathcal{M}}$ and assume, after a simultaneous similarity, that $\{f_1, f_2\}$ is the standard basis for $\mathbb{C}^2$. Now express $\mathcal{S}_0|_{\mathcal{M}}$ relative to the basis $\{f_1, f_2\}$ of $\mathcal{M}$ to get the four matrices

$$S_1 = \begin{pmatrix} \alpha & 0 \\ \beta & 0 \end{pmatrix}, \quad S_2 = \begin{pmatrix} 0 & \alpha \\ 0 & \beta \end{pmatrix}, \quad S_3 = \begin{pmatrix} \gamma & 0 \\ \delta & 0 \end{pmatrix}, \quad S_4 = \begin{pmatrix} 0 & \gamma \\ 0 & \delta \end{pmatrix}$$

in $\mathcal{S}_0|_{\mathcal{M}}$ with $\alpha\delta - \beta\gamma \neq 0$. We can also assume that $\beta\gamma \neq 0$ (for otherwise $\alpha\delta \neq 0$, and we can simply interchange S_1 with S_2 and S_3 with S_4). Now the semigroup $\mathcal{S}_1$ generated by S_1 and S_4 is easily seen to contain nonzero multiples of S_2 and S_3, so that $\mathcal{S}_1$ is an irreducible subsemigroup of $\mathcal{S}_0|\mathcal{M}$. This proves (iii). □

The following special result should be compared to Theorem 3.3.9.

Theorem 4.2.5. *A semigroup $\mathcal{S}$ of operators of rank ≤ 1 with subadditive spectrum is triangularizable.*

Proof. By Corollary 4.2.3 and the Triangularization Lemma (1.1.4), it suffices to show reducibility. Suppose that $\mathcal{S}$ is irreducible (and the dimension of the underlying space is at least two).

Using Lemma 4.2.4 and Corollary 4.2.3, we can assume that $\mathcal{S}$ acts on $\mathbb{C}^2$ and contains the two matrices

$$S = \begin{pmatrix} \alpha & 0 \\ \beta & 0 \end{pmatrix} \text{ and } T = \begin{pmatrix} 0 & \gamma \\ 0 & \delta \end{pmatrix}$$

with $\alpha\delta - \beta\gamma \neq 0$ and $\beta\gamma \neq 0$.

The subadditivity hypothesis applied to S and T implies that the characteristic equation

$$x^2 - (\alpha + \delta)x + \alpha\delta - \beta\gamma = 0$$

of $S + T$ has its roots in the set $\{0, \alpha, \delta, \alpha + \delta\}$. But this occurs only if either $\alpha\delta - \beta\gamma$ or $\beta\gamma$ is zero, which is a contradiction. $\qquad\square$

Example 4.2.6. *The assertion of Theorem 4.2.5 does not hold without the rank hypothesis.*

Proof. Let $\{E_{ij}\}$ be the standard basis for 2×2 matrices, and let $J = \left(\begin{smallmatrix} 1 & 0 \\ 0 & -1 \end{smallmatrix}\right)$. It is easily verified that the following 4×4 matrices form a semigroup S that is not triangularizable:

$$E_{11} \oplus I, \quad E_{22} \oplus I, \quad O \oplus I, \quad E_{12} \oplus J, \quad E_{21} \oplus J, \quad O \oplus J.$$

Subadditivity of spectra needs to be checked only for noncommuting pairs. There are exactly five such (unordered) pairs $\{A, B\}$. Not both A and B have their second direct summand equal to I, and, by the obvious symmetries of S, we can assume $A = E_{12} \oplus J$. Then either $B = E_{21} \oplus J$, in which case

$$\sigma(A + B) = \{1, -1, 2, -2\},$$

or B is one of the two matrices $E_{11} \oplus I$ and $E_{22} \oplus I$. For either of the latter cases,

$$\sigma(A + B) = \{0, 1, 2\}.$$

It is easily seen that $\sigma(A) + \sigma(B)$ contains $\sigma(A + B)$ in each case. $\qquad\square$

The following lemma is a special case of a theorem of O.J. Schmidt [1] on finite (abstract) groups.

Lemma 4.2.7. *Every minimal nonabelian finite group $\mathcal{G}$ (i.e., group such that every proper subgroup is abelian) is solvable. In particular, such a group contains a normal subgroup of prime index.*

Proof. We use induction. Thus assume that the first assertion is false, and let $\mathcal{G}$ be a counterexample of minimal order.

We first show that $\mathcal{G}$ is simple. Otherwise there is a normal subgroup $\mathcal{H}$ other than $\mathcal{G}$ and $\{I\}$, that is abelian by the minimality hypothesis. Observe that every (proper) maximal subgroup of the quotient group $\mathcal{G}/\mathcal{H}$ is of the form $\mathcal{M}/\mathcal{H}$, where $\mathcal{M}$ is a maximal subgroup of $\mathcal{G}$. Since every such $\mathcal{M}$ is abelian, so is $\mathcal{M}/\mathcal{H}$. Thus $\mathcal{G}/\mathcal{H}$ is solvable by the minimality hypothesis. This implies that $\mathcal{G}$ is solvable, which is a contradiction. Hence $\mathcal{G}$ is simple.

We next show that if $\mathcal{M}_1$ and $\mathcal{M}_2$ are distinct maximal subgroups of $\mathcal{G}$, then $\mathcal{M}_1 \cap \mathcal{M}_2 = \{I\}$. Otherwise, let $\mathcal{R} = \mathcal{M}_1 \cap \mathcal{M}_2 \neq \{I\}$ and consider its normalizer

$$\mathcal{N} = \{G \in \mathcal{G} : G^{-1}\mathcal{R}G = \mathcal{R}\}$$

in $\mathcal{G}$. Observe that $\mathcal{N}$ contains $\mathcal{M}_1$ and $\mathcal{M}_2$. Also, since $\mathcal{R}$ is not trivial, $\mathcal{N}$ must be proper by (1) above, and is thus abelian. Since $\mathcal{N}$ contains the $\mathcal{M}_i$, it follows that $\mathcal{M}_1 \cup \mathcal{M}_2$ is abelian, which is a contradiction. Thus $\mathcal{M}_1 \cap \mathcal{M}_2 = \{I\}$ for any pair $\{\mathcal{M}_1, \mathcal{M}_2\}$.

Since $\mathcal{G}$ is not solvable, it is not a p-group, so that every Sylow p-subgroup of it is contained in a maximal (and abelian) subgroup. Let the order $|\mathcal{G}|$ of $\mathcal{G}$ be $p_1^{r_1} p_2^{r_2} \cdots p_t^{r_t}$ with distinct primes $\{p_i\}$; t is at least 2. Since the order of a Sylow p_i-subgroup is $p_i^{r_i}$, such a maximal subgroup cannot contain a Sylow subgroup for all p_i. We conclude that there are at least two non-conjugate maximal subgroups.

Let $\{\mathcal{M}_1, \mathcal{M}_2, \ldots, \mathcal{M}_k\}$ be a maximal set of mutually nonconjugate maximal subgroups of $\mathcal{G}$. Fix any $\mathcal{M}_i$. For any x and y, $x^{-1}\mathcal{M}_i x$ and $y^{-1}\mathcal{M}_i y$ are also maximal subgroups. If $x^{-1}\mathcal{M}_i x = y^{-1}\mathcal{M}_i y$, then $yx^{-1}\mathcal{M}_i xy^{-1} = \mathcal{M}_i$. Let $z = xy^{-1}$, so that $z^{-1}\mathcal{M}_i z = \mathcal{M}_i$. Then the subgroup generated by z and $\mathcal{M}_i$ cannot be all of $\mathcal{G}$, for otherwise $\mathcal{M}_i$ would be a normal subgroup. Hence that generated subgroup is $\mathcal{M}_i$, by maximality, so $z \in \mathcal{M}_i$. Thus x and y are in the same coset relative to $\mathcal{M}_i$. It follows that there are $|\mathcal{G}|/|\mathcal{M}_i|$ maximal subgroups conjugate to $\mathcal{M}_i$.

The distinct maximal subgroups conjugate to $\mathcal{M}_i$ intersect in $\{I\}$, so the number of elements in $\mathcal{G}$ satisfies

$$|\mathcal{G}| = 1 + \sum_{i=1}^{k}(|\mathcal{M}_i| - 1)\frac{|\mathcal{G}|}{|\mathcal{M}_i|} = 1 + k|\mathcal{G}| - \sum_{i=1}^{k}\frac{|\mathcal{G}|}{|\mathcal{M}_i|} .$$

Since $|\mathcal{M}_i| \geq 2$ for each i, this yields $|\mathcal{G}| \geq 1 + \frac{k}{2}|\mathcal{G}|$. This implies that $k < 2$, which contradicts the fact that there are at least two $\{\mathcal{M}_i\}$.

We have proven the first assertion of the lemma, from which it follows that the commutator subgroup $\mathcal{G}_0$ of $\mathcal{G}$ is proper. Since the commutator subgroup is normal, we need only show that it has prime index. Note that $\mathcal{G}$ is generated by $\mathcal{G}_0$ and any x not in $\mathcal{G}_0$, by minimality. Pick x in $\mathcal{G} \setminus \mathcal{G}_0$ and let x^k be the smallest power of x that belongs to $\mathcal{G}_0$. By the minimality hypothesis, k is prime. $\square$

Lemma 4.2.8. *Let p be prime. If A and B are $p \times p$ matrices, where B is nonscalar and diagonal and*

$$A = \begin{pmatrix} 0 & 0 & \cdots & 0 & 1 \\ 1 & 0 & \cdots & 0 & 0 \\ 0 & 1 & \cdots & 0 & 0 \\ \vdots & & & & \\ 0 & 0 & \cdots & 1 & 0 \end{pmatrix},$$

then the pair $\{A, B\}$ is irreducible.

Proof. By passing to the algebra $\mathcal{A}$ generated by the collection, we can assume that B is a nontrivial projection, i.e., its diagonal consists of ones and zeros. It suffices to show that $\mathcal{A}$ contains a diagonal projection E of rank 1. For it is then easy to check that A shares no nontrivial invariant subspace with the collection $\{A^{-i}EA^i : 1 \leq i \leq p\}$ of rank-one projections in $\mathcal{A}$.

Note that $A^p = I$. Let E be a diagonal projection in $\mathcal{A}$ of minimal positive rank r. Then, for every i and j, the rank of the product $(A^{-i}EA^i)(A^{-j}EA^j)$ is either 0 or r. Suppose that $r > 1$. Then there must be distinct integers i and j within $p - 1$ of each other such that $A^{-i}EA^i = A^{-j}EA^j$. Thus $A^{-k}EA^k = E$ for some k satisfying $1 \leq k \leq p - 1$. Then $A^{-sk}EA^{sk} = E$ for all integers s, and, since p is prime and $A^p = I$, we get $A^{-1}EA = E$. But this implies that E is scalar, i.e., $E = I$ or $E = 0$, which is a contradiction. $\qquad\square$

Lemma 4.2.9. *Let $\mathcal{G}$ be a minimal nonabelian finite group of operators on $\mathbb{C}^n$. Then there exist primes p and q, not necessarily distinct, and a p-dimensional subspace $\mathcal{M}$ of $\mathbb{C}^n$ invariant under $\mathcal{G}$ such that $\mathcal{G}|_{\mathcal{M}}$ is, after a similarity, generated by two operators of the form*

$$A = \alpha \begin{pmatrix} 0 & 0 & \cdots & 0 & 1 \\ 1 & 0 & \cdots & 0 & 0 \\ 0 & 1 & \cdots & 0 & 0 \\ \vdots & & & & \\ 0 & 0 & \cdots & 1 & 0 \end{pmatrix} \quad and \quad B = \beta \begin{pmatrix} \theta_1 & 0 & \cdots & 0 \\ 0 & \theta_2 & & \vdots \\ \vdots & & \ddots & 0 \\ 0 & \cdots & 0 & \theta_p \end{pmatrix},$$

where B is not scalar and $\theta_i^q = 1$ for all i. Furthermore, $\alpha^{p^r} = \beta^{q^s} = 1$ for some nonnegative integers r and s.

Proof. By Theorem 3.1.5, we can assume that $\mathcal{G}$ is a unitary group. Let $\mathcal{H}$ be the normal subgroup of index p given by Lemma 4.2.7. Then $\mathcal{H}$ is a commutative group of unitaries and can therefore be assumed to be diagonal. Choose G in $\mathcal{G} \setminus \mathcal{H}$, so that $G^p \in \mathcal{H}$.

Decompose $\mathbb{C}^n$ into a sum $\mathcal{M}_1 \oplus \cdots \oplus \mathcal{M}_r$, where each $\mathcal{M}_i$ is a maximal subspace of $\mathbb{C}^n$ invariant under $\mathcal{H}$ such that $\mathcal{H}|_{\mathcal{M}_i}$ consists of scalar multiples of the identity.

Note that $r \geq 2$, because $\mathcal{H}$ does not consist entirely of scalars. Since $G\mathcal{H} = \mathcal{H}G$, $G\mathcal{M}_i$ is invariant under $\mathcal{H}$ and $\mathcal{H}|_{G\mathcal{M}_i}$ consists of scalars for each i. Thus there is a permutation τ of $\{1, \ldots, r\}$ such that $G\mathcal{M}_i = \mathcal{M}_{\tau(i)}$ for all i. Since $\mathcal{G}$ is not commutative, $\tau(i) \neq i$ for some i. Pick a vector x in such an $\mathcal{M}_i$, and let $\mathcal{M}$ be the linear span of $\{x, Gx, \ldots, G^{p-1}x\}$. Note that $\mathcal{M}$ is invariant under G. It is also invariant under $\mathcal{H}$, because each $G^j x$ is a common eigenvector for the members of $\mathcal{H}$. Let $A = G|_{\mathcal{M}}$ and $\mathcal{H}_0 = \mathcal{H}|_{\mathcal{M}}$.

Observe that x and Gx come from different maximal subspaces $\mathcal{M}_i$ and $\mathcal{M}_{\tau(i)}$, so that at least one member H of $\mathcal{H}$ is not scalar on $\mathcal{M}$. Now, the vectors $G^j x$, $0 \leq j \leq p-1$, form a basis for $\mathcal{M}$, because the corresponding subspaces $G^j \mathcal{M}_i$ are distinct. (Otherwise, for some positive $k < p$, we must have $G^k \mathcal{M}_i = \mathcal{M}_i$; since $G^p \mathcal{M}_i = \mathcal{M}_i$ and p is prime, this yields $G\mathcal{M}_i = \mathcal{M}_i$, which is a contradiction.)

Relative to the basis $\{G^j x\}$ of $\mathcal{M}$, the subgroup $\mathcal{H}_0$ is diagonal, and A has the form

$$\begin{pmatrix} 0 & 0 & \ldots & 0 & \lambda \\ 1 & 0 & \ldots & 0 & 0 \\ 0 & 1 & \ldots & 0 & 0 \\ \vdots & \vdots & & \vdots & \vdots \\ 0 & 0 & \ldots & 1 & 0 \end{pmatrix}.$$

Since $\mathcal{H}_0$ has nonscalar members, we can pick a nonscalar member B in $\mathcal{H}_0$ and assume, replacing it by one of its powers if necessary, that B^q is scalar for some prime q. If α is a p-th root of λ, then A is diagonally similar to αP, where P is a cyclic permutation, and hence A has the form described in the statement of the lemma. This similarity does not alter B, which is also of the form described in the statement of the lemma. The group $\mathcal{G}_0$ generated by A and B is irreducible by Lemma 4.2.8. Now the minimality of $\mathcal{G}$ implies that $\mathcal{G}_0 = \mathcal{G}|_{\mathcal{M}}$.

Denote the order of the scalar α by mp^r, where m is not divisible by p. Choose an integer t such that $mt = 1$ (modulo p). Then A^{mt} has the same form as A but with α replaced by $\alpha_1 = \alpha^{mt}$, and $\alpha_1^{p^r} = 1$. By minimality, $\alpha_1 = \alpha$ and $m = 1$. A similar argument can be applied to the order of β. $\square$

Theorem 4.2.10. *A finite group of matrices with sublinear spectrum is abelian (and thus diagonalizable).*

Proof. Suppose that a nonabelian finite group with sublinear spectrum exists, and pick a minimal such group. By Lemma 4.2.9, the group has an irreducible restriction $\mathcal{G}$ generated by matrices A and B as in that lemma.

Spectrum is sublinear on $\mathcal{G}$, by Lemma 4.2.2. Clearly, it is also sublinear on $\mathbb{C}\mathcal{G}$. Hence we can assume with no loss of generality that the α and β of Lemma 4.2.9 are both one. If the p and q of that lemma both equal 2, then $A = \begin{pmatrix} 0 & 1 \\ 1 & 1 \end{pmatrix}$ and $B = \begin{pmatrix} 1 & 0 \\ 0 & -1 \end{pmatrix}$ with no loss of generality; thus it is easy to verify that $\sigma(A + \lambda B) \subseteq \sigma(A) + \lambda\sigma(B)$ implies $\lambda = 0$, which is a contradiction.

We now assume that at least one of p and q is different from 2. Multiplying B by a scalar, we can also assume that $\det A = \det AB$, so that AB is similar to A (via a suitable diagonal matrix). We now use the sublinearity property on A and AB, both with spectrum $\{z : z^p = 1\}$, to deduce that, for all $\lambda \in \mathbb{C}$, the eigenvalues of

$$
T_\lambda = AB + \lambda A = \begin{pmatrix}
0 & 0 & \cdots & 0 & \lambda + \theta_p \\
\lambda + \theta_1 & 0 & \cdots & 0 & 0 \\
0 & \lambda + \theta_2 & \cdots & 0 & 0 \\
\vdots & & & & \\
0 & 0 & \cdots & \lambda + \theta_{p-1} & 0
\end{pmatrix}
$$

are all contained in $\{\lambda\phi + \psi : \phi^p = \psi^p = 1\}$.

Since the number of ordered pairs $\langle \phi, \psi \rangle$ with $\phi^p = \psi^p = 1$ is finite, there exists a fixed such pair for which $\lambda\phi + \psi$ is in $\sigma(T_\lambda)$ for infinitely many values of λ. It is easily verified that each eigenvalue of T_λ is a p-th root of the product $\prod_{j=1}^{p}(\lambda + \theta_j)$. Thus the equation

$$
\prod_{j=1}^{p}(\lambda + \theta_j) = (\lambda\phi + \psi)^p = \left(\lambda + \frac{\psi}{\phi}\right)^p
$$

holds for infinitely many values and hence all values of λ. But this implies that all the θ_j are equal, which is a contradiction. $\qquad\square$

Theorem 4.2.11. *Every semigroup of matrices with sublinear spectrum is triangularizable.*

Proof. In view of Corollary 4.2.3 and the Triangularization Lemma (1.1.4), we need only show the reducibility of such a semigroup S. Observe that $\mathbb{C}S$ and $\overline{S}$ both have sublinear spectrum and so does $\Phi(S)$ for every ring automorphism of $\mathcal{M}_n(\mathbb{C})$ induced by a field isomorphism of $\mathbb{C}$. Thus the property $\mathcal{P}$ of sublinearity for spectra satisfies the hypothesis of the Finiteness Lemma (4.1.1). Reducibility will follow if we verify the conditions of Lemma 4.1.2.

Clearly, any ideal (even any subset) of S satisfies $\mathcal{P}$. If E is an idempotent in S, and A and B any members, then spectrum is sublinear on EAE and EBE, and hence on $EAE|_{EV}$ and $EBE|_{EV}$, by Lemma 4.2.2. Thus condition (i) of Lemma 4.1.2 is satisfied. To complete the proof, we observe that conditions (ii) and (iii) are satisfied, by Theorem 4.2.10 and Theorem 4.2.5, respectively.   □

Corollary 4.2.12. *A self-adjoint semigroup of matrices with sublinear spectrum is diagonalizable (and thus commutative).*

Proof. Triangularizability of $S \cup S^*$ is obviously equivalent to diagonalizability of S for any collection S.   □

Corollary 4.2.13. *A unitary group with sublinear spectrum is abelian.*

Proof. This is a special case of the preceding corollary.   □

The following result shows that local triangularizability in semigroups implies global triangularizability.

Corollary 4.2.14. *If every pair in a semigroup S is triangularizable, then so is S itself.*

Proof. The hypothesis implies the sublinearity of spectrum for each pair. The assertion then follows directly from Theorem 4.2.11.   □

Corollary 4.2.15. *The following conditions are mutually equivalent for a semigroup S of matrices:*

   (i) *S is triangularizable;*
   (ii) *for all integers m, scalars $\lambda_1, \ldots, \lambda_m$, and members $S_1, \ldots, S_m$ of S,*

$$\sigma(\lambda_1 S_1 + \cdots + \lambda_m S_m) \subseteq \lambda_1 \sigma(S_1) + \cdots + \lambda_m \sigma(S_m) \ ;$$

  (iii) *S has sublinear spectrum;*
  (iv) *$\sigma(A + mB) \subseteq \sigma(A) + m\sigma(B)$ for all integers m and all pairs A and B in S;*
  (v) *for every pair A and B in S, there are infinitely many values of λ for which*

$$\sigma(A + \lambda B) \subseteq \sigma(A) + \lambda \sigma(B).$$

Proof. Since every assertion trivially implies the next, and (iii) implies (i) (by Theorem 4.2.11), it suffices to prove that (v) implies (iii). The proof of this is a counting argument similar to the one given for Lemma 4.2.2: Let A and B be any $n \times n$ matrices satisfying (v) and let $\{\alpha_i\}$ and $\{\beta_i\}$ be their respective eigenvalues (counting multiplicities). Then, for an infinite set of values of λ, the eigenvalues of $A + \lambda B$ coincide with $\{\alpha_i + \lambda \beta_i\}$ for a fixed n-tuple of pairs $\langle \alpha_i, \beta_i \rangle$. Hence the polynomial equation $\det(A + \lambda B - \alpha_i - \lambda\beta_i) = 0$ holds for all values of λ. Thus this n-tuple works for every linear combination of A and B.   □

The last part of the proof above shows that even the weak version (v) of sublinearity for a pair A and B implies property L. In fact, (v) can be further weakened: It is apparent from the proof that "infinitely many" can be replaced with "sufficiently many." What is "sufficient" depends on n, of course; it would suffice if (v) held for more than $n^{2n}/(n-1)!$ values of λ.

4.3 Further Multiplicative Conditions on Spectra

We have seen in Chapter 3 that submultiplicativity of spectrum on a semigroup is insufficient for triangularizability (and even for reducibility; see Example 3.4.13). A much stronger condition called property G will be discussed below. This property, as well as property L mentioned in the preceding section, are weakenings of the following more general "property P," which is necessary for triangularizability.

Definition 4.3.1. A pair of $n \times n$ matrices A and B is said to have *property P* if their eigenvalues, counting multiplicities, can be put in ordered sequences $\{\alpha_1, \ldots, \alpha_n\}$ and $\{\beta_1, \ldots, \beta_n\}$, respectively, such that for every (noncommutative) polynomial f in two variables, the eigenvalues of $f(A, B)$ are precisely the numbers $f(\alpha_i, \beta_i)$, $i = 1, \ldots, n$ (counting multiplicities). A collection is said to have *property P* if every pair in it does. The definition of *property G* is obtained by replacing "polynomial" with "monomial" or "word"; i.e., A and B have *property G* if for all nonnegative integers m, r_k, s_k with $1 \leq k \leq m$, the eigenvalues of the word

$$A^{r_1} B^{s_1} A^{r_2} B^{s_2} \cdots A^{r_m} B^{s_m}$$

are precisely the numbers $\alpha_i^{\Sigma r_k} \beta_i^{\Sigma s_k}$, $i = 1, \ldots, n$. If only linear polynomials are used in the definition of property P, we obtain *property L*.

Theorem 4.3.2. *A semigroup with property G is triangularizable.*

Proof. In view of Corollary 4.2.14, it suffices to show that if A and B are arbitrary members of the semigroup, then the subsemigroup S generated by A and B is triangularizable. Let R, S, and T be any members of S, i.e., words in A and B. By hypothesis, the two words RST and SRT have exactly the same eigenvalues with the same multiplicities. Thus $\text{tr}(RST) = \text{tr}(SRT)$, and S is triangularizable by Theorem 2.2.1   □

In contrast to the situation with property L, which could be substantially weakened (to sublinearity of spectra) and still be sufficient for triangularizability, property G is weak. The following example shows that a very slight weakening of property G may make it insufficient for triangularizability. We shall see, however, that weaker versions of this property do imply reducibility.

Example 4.3.3. *There exists a semigroup that is not triangularizable but satisfies property G except for multiplicity.*

Proof. Consider the semigroup $\mathcal{S}$ of $(n+1) \times (n+1)$ matrices given in Example 2.1.11, and let $n \geq 2$. Observe that $\sigma(S) = \{0, 1\}$ for every S in $\mathcal{S}$, where the multiplicity of 1 is either one or two, depending on S. Thus if A and B are any members, we can arrange their eigenvalues into n-tuples of the form (with $\{\alpha, \beta\} \subseteq \{0, 1\}$)

$$\{\alpha_j\} = \{1, \alpha, 0, \ldots, 0\} \text{ and } \{\beta_j\} = \{1, 0, \beta, 0, \ldots, 0\}$$

counting multiplicities. Now, the corresponding n-tuple for any word in A and B in which both letters occur is just $\{1, 0, 0, \ldots, 0\}$. It follows that the equation

$$\sigma(A^{r_1} B^{s_1} \cdots A^{r_m} B^{s_m}) = \{\alpha_j^{\Sigma r_i} \beta_j^{\Sigma s_i} : j = 1, \ldots, n\}$$

holds for all monomials. The deviation from property G is in the fact that multiplicities are not respected in this equation. $\qquad\square$

The following reducibility result uses the much weakened form of property G obtained by restricting it to short words and doing away with the multiplicity requirement. The hypothesis below should be compared to the condition $\sigma(ABA^{-1}B^{-1}) = \{1\}$ for all A and B, which implies triangularizability in the group case (by Corollary 2.2.5).

Theorem 4.3.4. *Let $\mathcal{S}$ be a semigroup. If*

$$\sigma(ABA^{k-1}B^{k-1}) \subseteq \sigma(A^k)\sigma(B^k)$$

for every pair A and B in $\mathcal{S}$ and every positive integer k, then $\mathcal{S}$ is reducible.

Proof. It is easily seen that the hypothesis is satisfied by $\mathbb{C}\mathcal{S}$, $\overline{\mathcal{S}}$, and $\Phi(\mathcal{S})$ for every induced ring automorphism Φ. Thus the Finiteness Lemma (4.1.1) is applicable. We need only show that the additional hypotheses of Lemma 4.1.2 are satisfied.

(i) Any ideal of $\mathcal{S}$ clearly satisfies the hypothesis. Also, if E is a minimal idempotent in $\mathcal{S}$, $A_1 = EAE|_{EV}$, and $B_1 = EBE|_{EV}$, then the hypothesis on EAE and EBE yields

$$\sigma(A_1 B_1 A_1^{k-1} B_1^{k-1}) \cup \{0\} \subseteq \sigma(A_1^k)\sigma(B_1^k) \cup \{0\}.$$

But all words in A_1 and B_1 are invertible by Lemma 3.1.6(i), so we deduce that

$$\sigma(A_1 B_1 A_1^{k-1} B_1^{k-1}) \subseteq \sigma(A_1^k)\sigma(B_1^k).$$

(ii) If S is a finite group, then by taking k to be the order of the group we get $\sigma(ABA^{-1}B^{-1}) = \{1\}$, and thus S is triangularizable by Corollary 2.2.5 (and in fact abelian, by finiteness).

(iii) If the maximal rank in S is one, we use $k = 1$ (i.e., the submultiplicativity of spectrum) to get reducibility, by Theorem 3.3.9.

□

The hypothesis of the theorem above is a strengthening of submultiplicativity, while that of the following result is a substantial weakening of permutability. Note that even permutability does not imply triangularizability for the general semigroup, as shown by the semigroup S of Example 2.1.11, every element of which has spectrum $\{0, 1\}$.

Theorem 4.3.5. *If, in a semigroup S, the inclusion*

$$\sigma(ABA^{k-1}B^{k-1}) \subseteq \sigma(A^k B^k)$$

holds for all pairs A, B in S and all positive integers k, then S is reducible.

Proof. The proof is very similar to the preceding proof; the only difference is in the verification of condition (iii) of Lemma 4.1.2. To do this, we show that the current hypothesis implies submultiplicativity if the maximal rank in S is 1.

Let A and B be any members of S. If either one is nilpotent, then (since its rank is at most 1) its square is zero. Using $k = 2$ in the hypothesis we get

$$\sigma((AB)^2) \subseteq \sigma(A^2 B^2) = \{0\},$$

so that AB is nilpotent. If $A^2 \neq 0$ and $B^2 \neq 0$, then (multiplying by appropriate scalars) we can assume that A and B are both idempotents of rank 1. Then

$$\sigma((AB)^2) \subseteq \sigma(A^2 B^2) = \sigma(AB).$$

Since AB has rank ≤ 1, the inclusion $\sigma((AB)^2) \subseteq \sigma(AB)$ implies that $\sigma(AB) \subseteq \{0, 1\}$. But $\{0, 1\} = \sigma(A)\sigma(B)$ (assuming that the underlying space has dimension at least two). Thus $\sigma(AB) \subseteq \sigma(A)\sigma(B)$ in all cases.

□

The question arises of whether any further reduction in the lengths of words is possible in the weak variations of property G. That is, do any relations of the form

$$\sigma(A^i B^j A^k) \subseteq \sigma(A^{i+k}) \sigma(B^j)$$

or

$$\sigma(A^i B^j A^k) \subseteq \sigma(A^{i+k} B^j),$$

assumed for all nonnegative integers i, j, k and all pairs A and B in a semigroup, guarantee reducibility? The answer is easily seen to be no. The left-hand side is simply $\sigma(A^{i+k} B^j)$ (by Lemma 1.4.2). Thus the first relation is just submultiplicativity, already proven insufficient for reducibility. The second inclusion holds identically for all operators A and B.

Observe that the hypothesis of Theorem 4.3.4 weakens property G not only by shortening the words used, but also by discounting multiplicities and prearrangements of eigenvalues. The following result shows that even if these other requirements of property G are satisfied, Theorem 4.3.4 is still best possible. Note that assuming property G for all words of the form $A^i B^j A^k$ is equivalent to assuming it for all shorter words of the form $A^i B^j$.

Theorem 4.3.6. *For every odd prime p there exists an irreducible finite group S of matrices of size $n = p^2$ with the following property: If A and B are any members of S, then there are respective orderings of the eigenvalues of A and B into n-tuples $\{\alpha_1, \ldots, \alpha_n\}$ and $\{\beta_1, \ldots, \beta_n\}$, counting multiplicities, such that*

$$\{\alpha_1^i \beta_1^j, \ldots, \alpha_n^i \beta_n^j\}$$

is the n-tuple of eigenvalues of $A^i B^j$ for all nonnegative integers i and j.

Proof. Let $\omega \neq 1$ be a primitive p-th root of unity and let

$$\Omega = \{1, \omega, \ldots, \omega^{p-1}\}.$$

Let $\mathcal{H}$ be the group generated by the two $p \times p$ matrices

$$U = \begin{pmatrix} 1 & & & & \\ & \omega & & & \\ & & \ddots & & \\ & & & & \omega^{p-1} \end{pmatrix} \text{ and } V = \begin{pmatrix} 0 & 0 & \cdots & 0 & 1 \\ 1 & 0 & \cdots & 0 & 0 \\ 0 & 1 & \cdots & 0 & 0 \\ \vdots & & & & \\ 0 & 0 & \cdots & 1 & 0 \end{pmatrix}.$$

It is easily seen that

$$\mathcal{H} = \{\omega^i U^j V^k : 0 \leq i, j, k \leq p - 1\},$$

and that $\mathcal{H}$ is irreducible. It also has submultiplicative spectrum, a fact proven in Example 3.3.7 for a group of which $\mathcal{H}$ is a subgroup. Thus the group $\mathcal{G} = \mathcal{H} \otimes \mathcal{H}$ is also irreducible with submultiplicative spectrum (see the proof of Theorem 3.3.8). We claim that $\mathcal{G}$ has the desired property.

If H is any nonscalar member of $\mathcal{H}$, then $\sigma(A) = \Omega$ (cf. Example 3.3.7). This shows that if H_1 and H_2 in $\mathcal{H}$ are not both scalars (equivalently, $H_1 \otimes H_2$ is not scalar), then $\sigma(H_1 \otimes H_2) = \Omega$, and the multiplicity of every eigenvalue is p.

Let A and B be any members of $\mathcal{G}$. Consider $A^i B^j$ with i and j integers. Since $A^p = B^p = I$, we can assume that $0 \leq i, j \leq p-1$. The claim needs to be established only for the case in which A and B do not commute. Then $A^i B^j \neq B^j A^i$ unless $i = 0$ or $j = 0$, because p is prime. Thus $A^i B^j$ is nonscalar whenever i and j are not both zero. Arrange $\sigma(A)$ and $\sigma(B)$ into n-tuples

$$\{\alpha_k\}_{k=1}^n = \{\omega, \omega, \dots, \omega; \omega^2, \omega^2, \dots, \omega^2; \dots; \omega^p, \omega^p, \dots, \omega^p\}$$

and

$$\{\beta_k\}_{k=1}^n = \{\omega, \omega^2, \dots, \omega^p; \omega, \omega^2, \dots, \omega^p; \dots; \omega, \omega^2, \dots, \omega^p\}.$$

Now, if $j \neq 0$, it is easily seen that $\{\alpha_k^i \beta_k^j\}$ is a listing of the set Ω with multiplicity p, which is $\sigma(A^i B^j)$. The case $j = 0$ can be trivially verified.

$\square$

4.4 Polynomial Conditions on Spectra

In the preceding sections we have shown that property P is too strong as a sufficient condition for triangularizability, since various weakenings of it suffice. In this section we study the restriction of property P to a single noncommutative polynomial f. A further weakening of this would be the hypothesis that, for matrices A and B,

$$\sigma(f(A, B)) \subseteq \{f(\alpha, \beta) : \alpha \in \sigma(A), \beta \in \sigma(B)\}.$$

Subadditivity and submultiplicativity of spectra are special cases, where f is taken to be $x + y$ and xy, respectively. A different, "hybrid," example is $f(x, y) = xy - yx$. In this case, the condition just amounts to the hypothesis that $AB - BA$ is nilpotent. We shall see that if this is assumed for every pair in a semigroup $\mathcal{S}$, then $\mathcal{S}$ is triangularizable. In fact, our methods will establish more general results.

Let $f(x, y)$ be a noncommutative polynomial that vanishes whenever x and y commute. If $\mathcal{S}$ is any collection of operators, then nilpotence of $f(A, B)$ for every pair $\{A, B\}$ in $\mathcal{S}$ is a necessary condition for triangularizability of $\mathcal{S}$ (by the Spectral Mapping Theorem (1.1.8)). A natural question is whether the nilpotence of f on a semigroup is sufficient for triangularizability. In this section we consider a special case, where the polynomial is linear in one of the variables.

For a given polynomial $g(x) = \sum_{j=0}^{m} a_j x^j$, we denote by f_g the noncommutative polynomial

$$f_g(x,y) = \sum_{j=0}^{m} a_j x^j y x^{m-j}.$$

The polynomial $xy - yx$, for example, is $f_g(x,y)$ for $g(x) = x - 1$.

Definition 4.4.1. Let g and f_g be defined as above. We say f_g *is nilpotent on a collection $\mathcal{S}$ of operators* if $f_g(S,T)$ is nilpotent for every pair $\{S,T\}$ in $\mathcal{S}$.

Before stating reducibility results, we note that there are many polynomials f_g that are nilpotent (even zero) on irreducible groups. For example, in the group $\mathcal{G}$ of Example 3.3.7 every member A has order p or 1, so that $yx^p - x^p y$ vanishes on $\mathcal{G}$. This polynomial is f_g with $g(x) = 1 - x^p$. Our results below show that divisibility of g by a polynomial $1 - x^p$ is essentially the only obstruction to triangularizability of semigroups of invertible operators.

We need the following simple lemma; it is a direct consequence of the fact that the cyclotomic polynomial $(x^p - 1)/(x - 1)$ is irreducible over the rationals for every prime number p.

Lemma 4.4.2. *Let p be a prime number and let $\{e_1, \ldots, e_p\}$ be a basis for $\mathbb{C}^p$. Let T be the permutation operator defined by*

$$Te_i = e_{i+1} \text{ for } i < p \text{ and } Te_p = e_1.$$

If $\mathcal{E}$ is any proper, nonempty subset of $\{1, 2, \ldots, p\}$, then $e = \sum_{j \in \mathcal{E}} e_j$ is a cyclic vector for T (i.e., the span of $\{T^n e : n \in \mathbb{N}\}$ is $\mathbb{C}^p$).

Proof. Assume with no loss of generality that 1 is not in $\mathcal{E}$. Order $\mathcal{E}$ as $r_1 < r_2 < \cdots < r_s$, where $2 \le r_1$. Let $\psi(x) = x^{r_1 - 1} + x^{r_2 - 1} + \cdots + x^{r_s - 1}$, so that $\psi(T)e_1 = e$. Then, for any polynomial ϕ, $\phi(T)e = \phi(T)\psi(T)e_1$. It suffices to show that $\phi(T)e = 0$ implies that $\phi(x)$ is divisible by the minimal polynomial $x^p - 1$ of T. Thus assume $\phi(T)\psi(T)e_1 = \phi(T)e = 0$. Since e_1 is cyclic for T, the polynomial $\phi(x)\psi(x)$ is divisible by $x^p - 1$. The proof will be completed if we show that the greatest common divisor $\delta(x)$ of $\psi(x)$ and $x^p - 1$ is constant. Observe that the coefficients of $\delta(x)$ are rational and that $\psi(1) \ne 0$. Thus $\delta(x)$ is a divisor of the cyclotomic polynomial $1 + x + \cdots + x^{p-1}$, which is irreducible over $\mathbb{Q}$, so $\delta(x)$ is constant. $\square$

Theorem 4.4.3. *Let $g(x)$ be a polynomial that is not divisible by $x^p - 1$ for any prime p. If $\mathcal{G}$ is a finite group of matrices on which f_g is nilpotent, then $\mathcal{G}$ is abelian.*

Proof. Suppose that $\mathcal{G}$ is a minimal counterexample. Then Lemma 4.2.9 is applicable. Let $\mathcal{M}, A$, and B be as in that lemma. Since f_g is clearly nilpotent on $\mathcal{G}|_{\mathcal{M}}$, and since $f_g(x, y)$ is homogeneous in x and in y, we can assume that $\mathcal{M}$ is the whole space and that $\alpha = \beta = 1$.

Let $g(x) = \sum_{j=0}^{m} a_j x^j$, so $f_g(x, y) = \sum_{j=0}^{m} a_j x^j y x^{m-j}$. By hypothesis, $f_g(A, B^k A^{-m})$ is nilpotent for every k. This means that the diagonal matrix

$$D_k = a_0 B^k + a_1 A B^k A^{-1} + \cdots + a_m A^m B^k A^{-m}$$

is nilpotent for all k. Since 0 is the only nilpotent diagonal matrix, this implies that

$$a_0 h(B) + a_1 A h(B) A^{-1} + \cdots + a_m A^m h(B) A^{-m} = 0$$

for any polynomial h. Since B is not scalar, it follows that for some h, $h(B)$ is a nontrivial diagonal idempotent E. Let e be the image under E of the (column) vector u whose components are all 1. Since $A^{-j}u = u$ for all j,

$$\begin{aligned}
g(A)e &= (a_0 + a_1 A + \cdots + a_m A^m) E u \\
&= (a_0 E + a_1 A E A^{-1} + \cdots + a_m A^m E A^{-m}) u \\
&= 0.
\end{aligned}$$

By Lemma 4.4.2, the vector e is cyclic for A. Hence $g(x)$ is divisible by the minimal polynomial of A, namely $x^p - 1$. This contradicts the hypothesis. $\square$

Before extending this proposition to more general semigroups, we study the case of semigroups of rank-one operators. It turns out that, in this case, an affirmative result can be achieved only when g is linear.

It is shown below (Theorem 4.4.12) that the rank condition is not needed in the next theorem.

Theorem 4.4.4. *Let S be a semigroup of operators of rank at most one such that $AB - BA$ is nilpotent for all A and B in S. Then S is triangularizable.*

Proof. By the Triangularization Lemma (1.1.4), we need only show that S is reducible. Supposing S irreducible, we can use Lemma 4.2.4 to obtain a subsemigroup S_0 of S and an invariant subspace $\mathcal{M}$ of S_0 such that $S_0|_{\mathcal{M}}$ contains

$$S = \begin{pmatrix} \alpha & 0 \\ \beta & 0 \end{pmatrix} \text{ and } T = \begin{pmatrix} 0 & \gamma \\ 0 & \delta \end{pmatrix}$$

with $\beta\gamma(\alpha\delta - \beta\gamma) \neq 0$. Since every pair in $S_0|_\mathcal{M}$ satisfies the hypothesis, $ST - TS$ is nilpotent. Hence

$$0 = \det(ST - TS) = \beta\gamma(\alpha\delta - \beta\gamma),$$

which is a contradiction.   □

Consider a polynomial $g(x) = \Sigma a_j x^j$. Observe that if $\Sigma a_j \neq 0$, then the nilpotence of f_g on a semigroup S implies, in particular, that $(\Sigma a_j) A^{m+1} = f_g(A, A)$ is nilpotent, and thus that S consists of nilpotent operators. This makes S triangularizable, by Levitzki's Theorem (2.1.7). In the opposite case, where the coefficients sum to 0, there are counterexamples.

Example 4.4.5. *There exists an irreducible semigroup on which f_g is nilpotent for every polynomial g of degree at least 2 the sum of whose coefficients is 0.*

Proof. Let S_0 be the semigroup consisting of zero and the standard basis for $\mathcal{M}_n(\mathbb{C})$. We need only verify that

$$f_g(A, B) = a_0 BA^m + a_1 ABA^{m-1} + \cdots + a_m A^m B$$

is nilpotent when $A^2 = B^2 = 0$, because if either A or B is idempotent, then $\{A, B\}$ is simultaneously upper or lower triangular. Now if $m > 2$, the equation $A^2 = 0$ implies $f_g(A, B) = 0$; if $m = 2$, then

$$(f_g(A, B))^2 = (a_1 ABA)^2 = a_1^2 ABA^2 BA = 0.$$

This example can be extended to contain operators of every rank; to make such as an example, adjoin all diagonal idempotents to S_0. To see that $f_g(A, B)$ is nilpotent for every pair in the larger semigroup, note that either they both belong to S_0 or they are simultaneously triangular.   □

Example 4.4.6. *Let $g(x) = \displaystyle\sum_{j=0}^{m} a_j x^j$ with $m \geq 2$ and $g(0) = g(1) = 0$. Then f_g is nilpotent on the entire semigroup of operators of rank at most 1.*

Proof. We must show $f_g(A, B)$ is nilpotent for any rank-one operators A and B. Since f_g is homogeneous, we can assume that either $A^2 = A$ or $A^2 = 0$.

If $A^2 = A$, then $f_g(A, B) = a_m AB(1 - A)$. If $A^2 = 0$, then $f_g(A, B)$ equals either zero (for $m \geq 3$) or $a_1 ABA$ (for $m = 2$). In both cases the square of $f_g(A, B)$ is 0.   □

The following proposition is needed for our affirmative results. It says that if $g(0) \neq 0$, then there is essentially only one irreducible semigroup of operators of rank at most one on which f_g is nilpotent, namely, the semigroup S_0 of Example 4.4.5.

Theorem 4.4.7. *Let g be a polynomial of degree at least two with $g(0) \neq 0$. Let S be an irreducible semigroup of $n \times n$ matrices of rank at most one on which f_g is nilpotent. Then $\mathbb{C}S$ is simultaneously similar to*

$$\mathbb{C}\{e_i e_j^* : 1 \leq i, j \leq n\},$$

where $\{e_i\}$ is the standard basis for column vectors.

Proof. Assume that $S = \overline{\mathbb{C}S}$. We claim that if E and F are distinct idempotents in S, then $EF = FE = 0$. Let E and F be such a pair and note that either their ranges or their kernels must be distinct. If $\bar{g}$ is the polynomial whose coefficients are the conjugates of those of g, then $f_{\bar{g}}$ is nilpotent on the set, S^*, of adjoints. Thus by passing to S^*, if necessary, we can assume that E and F have distinct ranges. If e and f are nonzero vectors in the ranges of E and F respectively, and if W is the span of e and f, then W is invariant under E and F, and the matrices of $E|_W$ and $F|_W$ are of the form

$$A = \begin{pmatrix} 1 & \alpha \\ 0 & 0 \end{pmatrix} \text{ and } B = \begin{pmatrix} 0 & 0 \\ \beta & 1 \end{pmatrix}.$$

It can be easily checked that $EF = FE = 0$ if and only if $AB = BA = 0$. Thus we need only show that $\alpha = \beta = 0$.

Let $g(x) = a_0 + a_1 x + \cdots + a_m x^m$, where $a_0 = g(0) \neq 0$ (and $a_m \neq 0$). Now, since $a_0 + \cdots + a_m = 0$ and $f_g(A, B)$ is nilpotent, we have

$$\det f_g(A, B) = \det \sum_{j=0}^{m} a_j A^j B A^{m-j}$$

$$= a_0 a_m \alpha \beta (\alpha\beta - 1) = 0,$$

so that $\alpha\beta(\alpha\beta - 1) = 0$. The irreducibility of S implies, via Theorem 1.2.2, that ES has the dimension of the space underlying S. Thus $ES|_W$ has a member of the form

$$C = \begin{pmatrix} \gamma & \delta \\ 0 & 0 \end{pmatrix}$$

that is independent of A, that is, $\alpha\gamma - \delta \neq 0$. Now, the two equations

$$\det f_g(B, C) = a_0 a_m \beta\delta(\gamma - \delta\beta) = 0,$$

$$\det f_g(A, BC) = a_0 a_m \alpha\beta^2 \gamma(\alpha\gamma - \delta) = 0$$

together with $\alpha\beta(\alpha\beta - 1) = 0$ show that if $\alpha\beta \neq 0$, then $\gamma = \delta = 0$, which is a contradiction. Thus we can assume, by switching E and F if necessary, that $\alpha = 0$. Then $\delta \neq 0$ and the first equation above yields $\beta(\gamma - \delta\beta) = 0$. Suppose that $\beta \neq 0$. Then $\gamma = \beta\delta$ and

$$\det f_g(C, BA) = a_0 a_m (\beta\delta)^{m+1} = 0,$$

which is a contradiction. Hence $\beta = 0$, and the claim has been proven.

To complete the proof of the theorem, we use Lemma 4.2.4 to obtain bases $\{e_i\}$ and $\{f_j\}$ with $e_j f_j^* \in \mathcal{S}$ for every i and j. Since $\mathcal{S} = \mathbb{C}\mathcal{S}$, we can assume (after a similarity) that $\{e_i\}$ coincides with the standard basis. For each i there is some j_i such that $f_{j_i}^* e_i \neq 0$. Then $T_i = e_i f_{j_i}^*$ has nonzero trace and is thus a multiple of an idempotent. The preceding claim implies that there is a unique j_i for each i, for otherwise there would be idempotents with the same range but distinct kernels, contradicting the fact that products of distinct idempotents are 0. A similar argument, with kernels and ranges interchanged, shows that the mapping $i \mapsto j_i$ is injective. Thus we can assume $j_i = i$, so that, after scaling if necessary, $f_i = e_i$ for every i.   □

Corollary 4.4.8. *Assume f_g is nilpotent on an irreducible semigroup $\mathcal{S}$ containing a rank-one operator, where g has degree at least two and $g(0) \neq 0$. Then $\mathcal{S}$ has a matrix representation in which every member has at most one nonzero entry in each row and in each column.*

Proof. There is no harm in assuming $\mathcal{S} = \overline{\mathbb{C}\mathcal{S}}$. Then the ideal $\mathcal{J} = \{\mathcal{S} \in \mathcal{S} : \text{rank } S \leq 1\}$ is irreducible (by Lemma 2.1.10) and can be assumed, by Theorem 4.4.7, to coincide with $\mathbb{C}\{e_i e_j^*\}$. Now, if $S \in \mathcal{S}$, then $S e_i e_i^*$ belongs to $\mathcal{J}$ for each i and thus equals $\lambda e_j e_i^*$ for some $\lambda \in \mathbb{C}$ and some j. Hence in the i-th column of S every coordinate is zero except possibly the j-th, which is λ. The same assertion on rows can be proven by considering $e_i e_i^* S$.   □

Theorem 4.4.9. *Let g be a polynomial with $g(0) \neq 0$ that is not divisible by $x^p - 1$ for any prime p. If $\mathcal{S}$ is a semigroup of invertible operators on which f_g is nilpotent, then $\mathcal{S}$ is triangularizable.*

Proof. If g is constant, then the hypothesis implies that every member of $\mathcal{S}$ is nilpotent, contradicting the invertibility hypothesis. Assume g is a nonconstant polynomial. By the Triangularization Lemma (1.1.4), we need only show that $\mathcal{S}$ is reducible.

It is easily checked that f_g is nilpotent on the semigroup $\Phi^{-1}(\overline{\mathbb{C}\Phi(\mathcal{S})})$ for every ring automorphism Φ induced by a field automorphism of $\mathbb{C}$. Thus the property $\mathcal{P}$, that f_g is nilpotent on $\mathcal{S}$, satisfies the hypothesis of the Finiteness Lemma (4.1.1).

If g is of degree one, then Lemma 4.1.2 is applicable by Theorems 4.4.3 and 4.4.4. Thus we can assume $g(x) = \sum_{j=0}^{m} a_j x^j$ is of degree at least two.

Let $\hat{S}$ be a maximal semigroup containing S with property $\mathcal{P}$. If S were irreducible, then $\hat{S}$ would be irreducible and, by Lemma 3.1.6, there would be a minimal idempotent E in $\hat{S}$. If the rank of this idempotent is greater than 1, the Finiteness Lemma (4.1.1) implies that the restriction of $E\hat{S}E$ to the range of E is essentially a finite group, so Theorem 4.4.3 shows that the restriction is reducible. Hence $\hat{S}$, and thus also S, is reducible by Lemma 2.1.12.

The remaining case is where there is an idempotent of rank 1 in S. In this case, we can apply Corollary 4.4.8 to get a matrix representation in which each member has at most one nonzero entry in each row and in each column. If S is irreducible, it must contain an S that is not diagonal. By hypothesis, S is invertible. By permuting the basis, if necessary, it can be assumed that S has a cyclic direct summand, say A. Multiplying A by an appropriate root of the reciprocal of its determinant and applying an appropriate diagonal similarity, we can further assume that A is a cyclic permutation matrix. Some power of A is I. Replacing A by an appropriate power of it, we can assume that $A^p = I$ for some prime p. Then A is a cyclic permutation matrix satisfying $A^p = I$, so, by permuting the basis, we can assume that A is a $p \times p$ matrix of the form

$$A = \begin{pmatrix} 0 & 0 & \cdots & 0 & 1 \\ 1 & 0 & \cdots & 0 & 0 \\ 0 & 1 & \cdots & 0 & 0 \\ \vdots & \vdots & & \vdots & \vdots \\ 0 & 0 & \cdots & 1 & 0 \end{pmatrix}.$$

The operators in S of rank at most 1 form an irreducible semigroup (Lemma 2.1.10). If such an operator has a nonzero entry on the diagonal, then it has a diagonal matrix. Thus the idempotent B with $p \times p$ matrix

$$B = \operatorname{diag}(1, 0, 0, \ldots, 0)$$

is the restriction of a member of S. Also, since A is the restriction of S, it follows from the equation $A^p = I$ that A^{-m} is the restriction of S^{mp-m}, which is also an element of S. Then

$$f_g(A, BA^{-m}) = \sum_{j=0}^{m} a_j A^j (BA^{-m}) A^{m-j}$$

$$= \sum_{j=0}^{m} a_j A^j B A^{-j}$$

$$= \operatorname{diag}(b_0, b_1, \ldots, b_{p-1}),$$

where $b_i = a_i + a_{i+p} + a_{i+2p} + \cdots$ for each i. The nilpotence of $f_g(A, BA^{-m})$ implies that every $b_i = 0$. This means that $g(x)$ is divisible by $x^p - 1$, which contradicts the hypothesis. $\qquad\square$

The proof above shows that the hypothesis that g is not divisible by $x^p - 1$ is needed only for primes $p \leq n$, where n is the dimension of the underlying space.

In the case of groups we do not need assumptions on the coefficients of g.

Corollary 4.4.10. *Let* $\{a_0, \ldots, a_k\}$ *be any scalars such that* $a_0 + a_1 x + \cdots + a_k x^k$ *is not divisible by* $x^p - 1$ *for any prime p. If* $\mathcal{G}$ *is a group of operators such that* $\sum_{j=0}^{k} a_j A^j B A^{k-j}$ *is nilpotent for all A and B in* $\mathcal{G}$, *then* $\mathcal{G}$ *is triangularizable. In particular, if such a* $\mathcal{G}$ *consists of unitary operators, then it is commutative.*

Proof. Let a_r and a_t be the first and last nonzero members of the sequence $\{a_0, \ldots, a_k\}$ respectively. Let $m = t - r$ and define g by

$$g(x) = a_r + a_{r+1}x + \cdots + a_t x^m.$$

Then g is not divisible by any $x^p - 1$, $g(0) \neq 0$, and

$$\sum_{j=0}^{k} a_j A^j B A^{m-j} = A^r f_g(A, B) A^t$$

$$= f_g(A, A^r B A^t).$$

Replacing B with $A^{-r} B A^{-t}$, we deduce from the hypothesis that f_g is nilpotent on $\mathcal{G}$, and the assertion follows from Theorem 4.4.9. $\qquad\square$

Corollary 4.4.11. *Let g be a polynomial not divisible by $x^p - 1$ for any prime p. Let $\mathcal{S}$ be a semigroup on which f_g is nilpotent. Then either of the following conditions implies reducibility:*

(i) *$\mathcal{S}$ does not contain a rank-one operator and $g(0) \neq 0$.*
(ii) *$\overline{\mathbb{C}\mathcal{S}}$ does not contain a rank-one operator.*

Proof. (i) Let A be a nonzero member of $\mathcal{S}$ with minimal rank. We observe, by passing to powers of A, that A is of the form $A_0 \oplus 0$, where A_0 is invertible. It is easily checked that $\mathcal{S}_0 = A\mathcal{S}A|_{A\mathcal{V}}$ is irreducible if $\mathcal{S}$ is. But the semigroup $\mathcal{S}_0$ consists of invertible operators, and is triangularizable by Theorem 4.4.9. Thus $\mathcal{S}$ is reducible by Lemma 2.1.12.

(ii) In this case, assuming irreducibility, pick an idempotent E of minimal rank in $\overline{\mathbb{C}\mathcal{S}}$. Then the restriction to the range of E of $E\overline{\mathbb{C}\mathcal{S}}$ is a group (by Lemma 3.1.6) and thus is triangularizable by Corollary 4.4.10. But this contradicts Lemma 2.1.12. $\qquad\square$

We now extend Theorem 4.4.4 to general semigroups.

Theorem 4.4.12. *Let S be any semigroup of operators such that $AB - BA$ is nilpotent for every pair $\{A, B\}$ in S. Then S is triangularizable.*

Proof. By the Triangularization Lemma (1.1.4), we need only show reducibility. If S does not contain any operator of rank one, then it is reducible by Corollary 4.4.11. Otherwise, the ideal of S consisting of operators of rank at most one is nonzero and reducible by Theorem 4.4.4. This implies reducibility of S (by Lemma 2.1.10). $\qquad\square$

We conclude this section with a simple corollary that does not require the nondivisibility condition on g.

Corollary 4.4.13. *Let g be any polynomial of degree m with $g(0) \neq 0$. Assume f_g is nilpotent on a semigroup S containing a rank-one operator such that*

$$\{S^{m!} : S \in S\}$$

is not a diagonalizable set. Then S is reducible.

Proof. Assume S_0 is irreducible and $S = \overline{\mathbb{C}S}$. Consider the matrix representation for S given by Corollary 4.4.8. Every member S of S is, after a permutation of the basis, a direct sum of cyclic matrices of various sizes. We show that each size is at most m, so that $S^{m!}$ is diagonal.

Suppose, if possible, that S has a $t \times t$ direct summand A with $t > m$. We can assume as before that

$$A = \begin{pmatrix} 0 & 0 & \cdots & 0 & \alpha \\ 1 & 0 & \cdots & 0 & 0 \\ 0 & 1 & \cdots & 0 & 0 \\ \vdots & \vdots & & \vdots & \vdots \\ 0 & 0 & \cdots & 1 & 0 \end{pmatrix},$$

after a permutation of basis and a diagonal similarity, where α is zero or 1. Recall that $e_i e_j^* \in S$ for all basis vectors e_i and e_j. Letting $B = e_1 e_{m+1}^*$, we deduce that

$$f_g(A, B)e_1 = \left(\sum_{j=0}^{m} a_j A^j e_1 e_{m+1}^* A^{m-j}\right) e_1$$

$$= \sum_{j=0}^{m} a_j e_{j+1} e_{m+1}^* e_{m-j+1} = a_0 e_1,$$

and thus $a_0 \in \sigma(f_g(A, B))$. Since $a_0 \neq 0$, this contradicts the nilpotence of $f_g(A, B)$. $\qquad\square$

4.5 Notes and Remarks

The Finiteness Lemma was developed in Radjabalipour-Radjavi [1]. The slightly stronger form presented here is from Radjavi [6]. Both of these papers used versions of Lemma 4.1.2. Lemmas 4.1.3, 4.1.5, and Theorem 4.1.6 are rearrangements of the elegant treatment in Guralnick [1] which, as he indicates, has antecedents in Wales-Zassenhaus [1]. Kaplansky's result given in Corollary 4.1.7 is from Kaplansky [1]; the proof presented here is slightly different from the original.

Property L has a long history. It was proposed by M. Kac and studied by Wielandt [1] and Motzkin-Taussky ([1], [3]). Motzkin-Taussky [3] showed that a finite group of matrices over a field of characteristic zero in which every pair satisfies property L is abelian. Wales-Zassenhaus [1] removed the finiteness condition, and proved the same result over an arbitrary field, so long as the eigenvalues of members of the group (when considered as matrices over the algebraic closure of the field) are all in the field. Zassenhaus [1] extended this to semigroups with certain restrictions on the field; Guralnick [1] proved it for all fields with more than two elements.

Subadditivity and sublinearity conditions were introduced in Radjavi [6]; all the results of Section 4.2 are from this paper with two exceptions. Lemma 4.2.7 is a special case of a result by Schmidt [1] which states that every minimal non-nilpotent finite group is solvable. The proof presented here was described to us by T. Laffey. Lemma 4.2.4 appears here for the first time, although variants of it are in Radjavi [6]. Some of our proofs are simpler than the original ones.

All the results in Section 4.3 are from Radjavi [6]. Theorem 4.4.12, and its special case, Theorem 4.4.4, are from Guralnick [1]. (See also Radjavi-Rosenthal-Shulman [1].) The remaining results of Section 4.4 are those of Radjavi [6] with some simplifications.

Guralnick [1] considers various restrictions of property P on semigroups as well as on rings and Lie algebras of matrices.

Lemma 4.1.1 was originally formulated for studying so-called "rationality problems." Let S be an irreducible semigroup in $\mathcal{M}_n(\mathbb{F})$ with $\mathbb{F}$ an algebraically closed field. Assume that a subfield $\mathbb{K}$ of $\mathbb{F}$ contains $\sigma(S)$ for all S in S. Is S simultaneously similar to a semigroup in $\mathcal{M}_n(\mathbb{K})$? The answer seems to be unknown in general, even for $\mathbb{F} = \mathbb{C}$; the well-known Brauer Theorem implies that the answer is affirmative if the semigroup is a finite group. Partial solutions are given in Ponizovskii [1], Omladič-Radjabalipour-Radjavi [1] and Radjabalipour-Radjavi [1]. The answer is affirmative, for example, if $\mathbb{F} = \mathbb{C}$ and $\mathbb{K}$ is a subfield of $\mathbb{R}$.

In Section 4.4 we have considered semigroups in $\mathcal{M}_n(\mathbb{C})$ on which f^n is identically zero, where f is a homogeneous, noncommutative polynomial of a certain type. If subrings of $\mathcal{M}_n(\mathbb{C})$ are considered, as opposed to subsemigroups, then, clearly, fewer polynomials can be identically zero. Rings on which a fixed noncommutative polynomial (in two or more variables) is zero, the so-called polynomial-identity rings, have been studied extensively. The entire ring $\mathcal{M}_n(\mathbb{F})$ is a polynomial-identity ring; in fact, it satisfies the polynomial equation $f_{n+1}(y^n x, y^{n-1} x, \ldots, x) = 0$, where f_k is the "standard" polynomial of degree k in k noncommuting variables x_i:

$$f_k(x_1, \ldots, x_k) = \sum_\tau (\text{sign } \tau) x_{\tau(1)} x_{\tau(2)} \cdots x_{\tau(n)},$$

with τ ranging over all permutations of $\{1, 2, \ldots, k\}$. The Amitsur-Levitzki Theorem states that $f_{2n}(x_1, \ldots, x_{2n})$ is identically zero on $\mathcal{M}_n(\mathbb{K})$, where $\mathbb{K}$ is any commutative ring. For these and other results see Procesi [1].

CHAPTER 5
Semigroups of Nonnegative Matrices

In this chapter we consider a stronger form of reducibility for collections of matrices with nonnegative entries, which we simply call *nonnegative matrices*. The results will also be applied to questions of ordinary reducibility. A substantial part of the chapter is devoted to extensions to semigroups of the Perron-Frobenius Theorem on the existence of positive eigenvectors for nonnegative matrices and symmetries of their spectra (Corollary 5.2.13 below).

5.1 Decomposability

Definition 5.1.1. A *permutation matrix* is a square matrix having one 1 in each row and column and all of its other entries zero. A collection $\mathcal{S}$ of matrices is said to be *decomposable* if there exists a permutation matrix P such that $P^{-1}\mathcal{S}P$ has a block upper-triangular form. If $P^{-1}\mathcal{S}P$ is actually triangular for some permutation matrix P, then $\mathcal{S}$ is said to be *completely decomposable*.

Equivalently, if there is a nontrivial invariant subspace spanned by a subset of the standard basis vectors, then $\mathcal{S}$ is decomposable; if there is a triangularizing chain of such subspaces, then $\mathcal{S}$ is completely decomposable.

Theorem 5.1.2. *A semigroup of nonnegative nilpotent matrices is completely decomposable.*

Proof. If a semigroup $\mathcal{S}$ satisfying the hypotheses is in block triangular form, then every diagonal block of $\mathcal{S}$ also consists of nonnegative nilpotent matrices. Thus we need only show that $\mathcal{S}$ is decomposable. This will be accomplished by verifying that, after a permutation, the first column of every S in $\mathcal{S}$ is zero.

By Levitzki's Theorem (2.1.7), $\mathcal{S}$ is triangularizable. This implies that the product of any n members of $\mathcal{S}$ is zero, where the matrices are $n \times n$. Let k be the smallest integer such that every product of k members is zero. There is nothing to prove if $k = 1$. Let $k > 1$ and pick $S_1, \ldots, S_{k-1}$ in $\mathcal{S}$ with $A = S_1 S_2 \cdots S_{k-1} \neq 0$. It follows that $SA = 0$ for all S in $\mathcal{S}$. Choose a nonzero column x of A; then $Sx = 0$ for every S. After a permutation of the basis we can assume that the first component of x is not 0 and hence is positive. Now the non-negativity of the entries in $\mathcal{S}$ and x together with $Sx = 0$ forces the first column of S to be zero for all S in $\mathcal{S}$. $\qquad\square$

Lemma 5.1.3. *Let ϕ_i denote the linear functional on $n \times n$ matrices defined by $\phi_i(M) = M_{ii}$, the (i,i) entry of M.*

(i) If there is an i such that ϕ_i is submultiplicative on a semigroup $\mathcal{S}$ of nonnegative matrices, then $\mathcal{S}$ is decomposable. Furthermore, after a suitable permutation of the basis, every S in such a semigroup $\mathcal{S}$ has the block form

$$\begin{bmatrix} R & X & Y \\ 0 & s & Z \\ 0 & 0 & T \end{bmatrix},$$

where s represents a 1×1 block and equals S_{ii}. (Either R or T may be absent.)

(ii) A semigroup $\mathcal{S}$ of nonnegative matrices is completely decomposable if and only if every ϕ_i is submultiplicative on it.

Proof. (i) We first show that $\mathcal{S}$ is decomposable. Assume, for notational simplicity, that $i = 1$. The non-negativity of entries implies

$$\phi_1(AB) \geq \phi_1(A)\phi_1(B)$$

for all A and B in $\mathcal{S}$. The hypothesis on ϕ_1 then implies equality. In other words,

$$A_{12}B_{21} + \cdots + A_{1n}B_{n1} = 0,$$

which, by non-negativity, implies $A_{1j}B_{j1} = 0$ for $j = 2, 3, \ldots, n$. If $A_{12} = \cdots = A_{1n} = 0$ for all A in $\mathcal{S}$, then the standard basis vectors indexed 2 and higher span an invariant subspace for $\mathcal{S}$, and we are done. Otherwise, pick A in $\mathcal{S}$ with a nonzero off-diagonal entry in its first row. We can assume, after a permutation, that $A_{1n} \neq 0$. This implies that $B_{n1} = 0$ for all B in $\mathcal{S}$. A further permutation allows us to assume that $\{k+1, \ldots, n\}$ is the maximal subset J of $\{2, \ldots, n\}$ such that $S_{j1} = 0$ for all j in J and all S in $\mathcal{S}$.

We now show that the standard basis vectors with index in $\{1, \ldots, k\}$ span an invariant subspace, or, equivalently, that $S_{jm} = 0$ for all S whenever $j \geq k+1$ and $2 \leq m \leq k$. Fix any such m. By maximality of J, there is a member T of $\mathcal{S}$ with $T_{m1} \neq 0$. Then for $j \geq k+1$,

$$0 = (ST)_{j1} = \sum_{i=1}^{n} S_{ji}T_{i1} \geq S_{jm}T_{m1},$$

and hence $S_{jm} = 0$.

Once $\mathcal{S}$ is decomposed, we can further decompose whichever compression of $\mathcal{S}$ contains the entries in position (i, i). Thus we can continue until the stated block form is reached.

(ii) If the semigroup is completely decomposable, then each ϕ_i is clearly multiplicative. The converse follows immediately from part (i) by induction. (Note, incidentally, that it suffices to assume that $n - 1$ of the n functionals are submultiplicative.) □

Corollary 5.1.4. *Let A be a nonnegative matrix such that every positive power of A has at least one diagonal entry equal to 0. Then A is decomposable and has 0 as an eigenvalue.*

Proof. By Lemma 5.1.3 (applied to the semigroup generated by A), it suffices to show that $\phi_i(A^m) = 0$ for some i and all positive m. Suppose that for each i there is an m_i with $\phi_i(A^{m_i}) \neq 0$. Since, for every positive r, we have

$$\phi_i(A^{rm_i}) \geq (\phi_i(A^{m_i}))^r$$

by non-negativity, this would imply, with $m = m_1 \cdots m_n$, that

$$\phi_i(A^m) \geq (\phi_i(A^{m_i}))^{m/m_i} > 0$$

for all i, contradicting the hypothesis. $\qquad\qquad\qquad\qquad\square$

In the following lemma we specify several simple conditions equivalent to decomposability.

Lemma 5.1.5. *For a semigroup S of nonnegative matrices, the following are mutually equivalent:*

  (i) *S is decomposable;*
  (ii) *$ASB = \{0\}$ for some nonzero nonnegative matrices A and B (not necessarily in S and not necessarily square);*
  (iii) *for some fixed i and j, the (i, j) entry of every member of S is zero;*
  (iv) *every sum of members of S has a zero entry;*
  (v) *some nonzero ideal of S is decomposable.*

Proof. We first establish the equivalence of (i) to (iv). Assume (i). After a permutation, S has simultaneous 2×2 block upper-triangular form in which the $(2, 1)$ block is zero. Then the block matrices

$$A = \begin{pmatrix} 0 & 0 \\ 0 & I \end{pmatrix} \text{ and } B = \begin{pmatrix} I & 0 \\ 0 & 0 \end{pmatrix}$$

yield $ASB = 0$, implying (ii).

If (ii) is assumed, then some (i, j) entry of A and some (r, s) entry of B are nonzero. It follows from the non-negativity of the matrices that the (j, r) entry of every S in S is zero, yielding (iii).

Note that (iv) is an obvious consequence of (iii). We next show that (iv) implies (i). Pick a sum $T = S_1 + \cdots + S_k$ that has a minimal number of zero entries. Note that if, for any i and j, the (i, j) entry T_{ij} of T is zero, then $S_{ij} = 0$ for all S in S, for if $S_{ij} \neq 0$, then $(T + S)_{ij} \neq 0$, by non-negativity. Thus we need only show that T is a decomposable matrix. By hypothesis, T has a zero entry, which can be assumed, without loss of generality, to be in its first column. Observe that $T_{ij} = 0$ implies $(T^2)_{ij} = 0$ by the choice of T, because $T + T^2$ is also a sum of members of $\mathcal{S}$. Now if T_{11} is zero, we are done by Lemma 5.1.3. Thus let $T_{11} \neq 0$. A permutation allows us to assume that

$$T_{i1} \neq 0 \text{ for } i \leq k \text{ and } T_{i1} = 0 \text{ for } i > k.$$

The equations $\Sigma_j T_{ij} T_{j1} = (T^2)_{i1} = 0$ for $i > k$ show, by non-negativity of the entries, that $T_{ij} = 0$ whenever $i > k$ and $j \leq k$. This yields the desired decomposition of T.

We have demonstrated the equivalence of (i) through (iv). Since (i) obviously implies (v), we must show the converse to complete the proof. Let $\mathcal{J}$ be a nonzero ideal of $\mathcal{S}$ that is decomposable. Apply (ii) above to $\mathcal{J}$ to get $A\mathcal{J}B = \{0\}$. If $\mathcal{J}B = \{0\}$, then $\mathcal{J}\mathcal{S}B = \{0\}$, so that we can pick a nonzero J in $\mathcal{J}$ and conclude the decomposability of $\mathcal{S}$ from the equation $J\mathcal{S}B = \{0\}$ together with the equivalence of (i) and (ii). Thus assume $\mathcal{J}B \neq \{0\}$. If JB is a nonzero member of $\mathcal{J}B$, then the equation $A\mathcal{S}(JB) = 0$ yields the desired result, since (i) and (ii) are equivalent.   □

Theorem 5.1.6. *Let $\mathcal{S}$ be a semigroup of invertible nonnegative matrices. For each $S \in \mathcal{S}$, let $\delta(S)$ denote the product of all the diagonal entries of S. Then the following conditions are mutually equivalent:*

(i) *$\mathcal{S}$ is completely decomposable;*

(ii) *each member of $\mathcal{S}$ is completely decomposable;*

(iii) *the diagonal of each S in $\mathcal{S}$ consists precisely of its eigenvalues repeated according to their multiplicities;*

(iv) *$\det(S) = \delta(S)$ for every S in $\mathcal{S}$;*

(v) *δ is multiplicative on $\mathcal{S}$;*

(vi) *δ is submultiplicative on $\mathcal{S}$;*

(vii) *each member of $\mathcal{S}$ becomes nilpotent if its diagonal entries are replaced by zeros.*

Proof. Since the set of diagonal entries of a matrix is unaltered under permutations of bases, each of (i) to (v) implies its successor.

We show that (vi) implies (i). Let ϕ_i be the diagonal functionals given in Lemma 5.1.3, and recall that

$$\phi_i(AB) \geq \phi_i(A)\phi_i(B)$$

for any nonnegative matrices A and B. Thus we also have $\delta(AB) \geq \delta(A)\delta(B)$ in general, so that (v) and (vi) are equivalent. Now let A be any member of $\mathcal{S}$. We claim that $\delta(A) \neq 0$. Otherwise, $\delta(A^m) = (\delta(A))^m = 0$ for every positive m, by (v), and it would follow that

$$\prod_{i=1}^{n} \phi_i(A^m) = 0$$

for every m, and thus A has 0 as an eigenvalue by Corollary 5.1.4, contradicting the invertibility hypothesis. Hence $\delta(A) > 0$. For each i, then, if $\phi_i(AB) > \phi_i(A)\phi_i(B)$, then $\delta(AB) > \delta(A)\delta(B)$, so multiplicativity of δ forces each ϕ_i to be multiplicative. Thus Lemma 5.1.3(ii) implies the complete decomposability of the semigroup generated by A, so (vi) implies (i).

To show that (iv) implies (i), just observe that, for A and B in $\mathcal{S}$,

$$\det(AB) = \delta(AB) = \prod_{i=1}^{n} \phi_i(AB) \geq \prod_{i=1}^{n} \phi_i(A) \cdot \prod_{i=1}^{n} \phi_i(B)$$
$$= \delta(A)\delta(B) = \det(A) \cdot \det(B).$$

Thus since $\det(AB) = \det A \cdot \det B$, we have $\Pi_i \phi_i(AB) \neq 0$ and $\Pi_i \phi_i(AB) = \Pi_i[\phi_i(A)\phi_i(B)]$, which, together with $\phi_i(AB) \geq \phi_i(A)\phi_i(B)$, implies the multiplicativity of each ϕ_i. Hence (i) holds by Lemma 5.1.3.

To complete the proof, it suffices to show that (vii) is equivalent to (ii). Let A be in $\mathcal{S}$ and let A_0 be the matrix obtained from A when its diagonal entries are replaced by zeros. Clearly, A is completely decomposable if and only if A_0 is. If A_0 is nilpotent, then (ii) follows from (vii) by Theorem 5.1.2. If (ii) is assumed, then A_0 is completely decomposable with zero diagonal entries, which makes it nilpotent. $\square$

The invertibility hypothesis in the theorem above is essential, as the indecomposable semigroup $\mathcal{S}_0$ of Example 2.1.11 clearly demonstrates. However, there is an affirmative result for an individual singular matrix.

Theorem 5.1.7. *Let A be a nonnegative matrix. The following conditions are mutually equivalent:*

(i) *A is completely decomposable;*
(ii) *the diagonal of A consists precisely of its eigenvalues repeated according to multiplicity;*
(iii) *A becomes nilpotent upon the replacement of its diagonal entries by zeros.*

Proof. That (i) implies (ii) is clear. The equivalence of (i) to (iii) was shown in the proof of the preceding theorem and did not use the nonsingularity hypothesis. Thus we need only show that (ii) implies (i).

Assume (ii). We shall show that the functionals ϕ_i given in Lemma 5.1.3 are all submultiplicative on the semigroup $\mathcal{S}$ generated by A. Now, the hypothesis implies that, for each m,

$$\operatorname{tr}(A^m) = \sum_i (\phi_i(A))^m.$$

Since $(\phi_i(A))^m \leq \phi_i(A^m)$ for each i, by non-negativity, and since $\Sigma_i \phi_i(A^m) = \text{tr}(A^m)$, we conclude that $\phi_i(A^m) = \phi_i((A))^m$ for each i and each m. Thus every ϕ_i is multiplicative on S, and S is completely decomposable by Lemma 5.1.3.   □

It should be noted that the conditions (iv), (v), and (vi) of Theorem 5.1.6 are not sufficient for complete decomposability in the absence of invertibility, even for a singly generated semigroup. For example, let $S = \{E\}$ be the semigroup generated by the idempotent

$$E = \frac{1}{2} \begin{bmatrix} 1 & 1 & 0 \\ 1 & 1 & 0 \\ 0 & 0 & 0 \end{bmatrix}.$$

Theorem 5.1.8. *Let A be any nonnegative matrix and let δ be as in the statement of Theorem 5.1.6. Then A is decomposable if any one of the following conditions holds on the semigroup S generated by A:*

(i) $\det S = \delta(S)$ *for every S in S;*
(ii) δ *is multiplicative on S;*
(iii) δ *is submultiplicative on S.*

Proof. Since (i) clearly implies (ii) and since (ii) and (iii) are equivalent (as we have seen in the proof of Theorem 5.1.6), we need only show that (ii) implies decomposability.

Assume (ii). If $\delta(A) = 0$, then $\delta(A^m) = 0$ for all m, and the decomposability of A follows from Corollary 5.1.4. Thus assume $\delta(A) \neq 0$. Then $\delta(A^m)$ is positive for every positive m. As in the proof of Theorem 5.1.6, this makes each of the factors ϕ_i multiplicative, resulting in complete decomposability of S by Lemma 5.1.3.   □

We shall need the following lemma, which is of some interest in itself. It gives the best possible decomposition for an idempotent matrix with nonnegative entries. Note that if such a matrix is of rank one, then it can be indecomposable: Just pick $a_i > 0$, $b_i > 0$, $i = 1, \ldots, n$, with $\Sigma a_i b_i = 1$, and let E be the matrix $(a_i b_j)$.

Lemma 5.1.9. *Let E be a nonnegative idempotent matrix of rank r.*

(i) *If E has no zero rows or columns, then there exists a permutation matrix P such that $P^{-1} E P$ has the block-diagonal form*

$$E_1 \oplus \cdots \oplus E_r,$$

where each E_i is an idempotent of rank one whose entries are all positive.

(ii) *In general, there exists a permutation matrix P such that $P^{-1}EP$ has the block-triangular form*

$$\begin{bmatrix} 0 & XF & XFY \\ 0 & F & FY \\ 0 & 0 & 0 \end{bmatrix},$$

where $F = E_1 \oplus \cdots \oplus E_r$ as in (i) above and where X and Y are matrices with nonnegative entries.

Proof. (i) If the rank of E is one, then $E = AB^t$ where A and B are nonnegative columns with $B^t A = 1$. Since E has no zero rows or columns, A and B have all their coordinates positive, and thus the entries of E are all positive. We shall prove the lemma by induction on the rank r.

We first show that if $r \geq 2$, then the range of E contains a nonzero (column) vector z with nonnegative entries and at least one zero entry. Pick any two linearly independent columns x and y of E. Since $Ex = x$ and $Ey = y$, we are done if either x or y has a zero entry. Otherwise, let

$$x = \begin{bmatrix} x_1 \\ \vdots \\ x_n \end{bmatrix} \text{ and } y = \begin{bmatrix} y_1 \\ \vdots \\ y_n \end{bmatrix}$$

and let $y_j/x_j = \max\{y_i/x_i : i = 1, \ldots, n\}$. Then the vector $z = y_j x - x_j y$ is nonzero, has nonnegative entries, and its j-th entry is zero. Since $z = Ez$, z is in the range of E.

We next show that, if $r \geq 2$, then E is decomposable. Let z be a nonzero nonnegative vector in the range of E with a minimal number of nonzero entries. After a permutation of the basis, we can assume that the entries $\{z_i\}$ of z satisfy

$$z_1 \geq \cdots \geq z_s > z_{s+1} = \cdots = z_n = 0.$$

Then the equation $Ez = z$, together with the non-negativity of entries in E and z, implies that the (i, j) entry of E is zero whenever $i \geq s + 1$ and $j \leq s$. Thus the span of the first s basis vectors is invariant under E; i.e., E is decomposable.

After a permutation, we can assume

$$E = \begin{bmatrix} F & R \\ 0 & G \end{bmatrix},$$

where F and G are idempotents of rank less than r. Now, the hypothesis implies that F has no zero columns and G has no zero rows. It follows that if A is a nonnegative matrix with $FA = 0$, then $A = 0$. Similarly, $BG = 0$ implies $B = 0$ if B has nonnegative entries. Using this we can show that $R = 0$ as follows: Since E is an idempotent, we have $FR + RG = R$, which implies, upon multiplication by F on the left and by G on the right, that $FRG = 0$. Hence $R = 0$.

Since $R = 0$, F cannot have zero rows and G cannot have zero columns. Thus both F and G satisfy the hypothesis of (i), and the proof of (i) is complete by induction.

(ii) We can assume, after applying a permutation, that

$$
E = \begin{bmatrix} 0 & X & Z \\ 0 & F & Y \\ 0 & 0 & 0 \end{bmatrix},
$$

where all the zero columns appear as the beginning columns 1 to m, and all the zero rows numbered higher than m appear at the end. Since $E^2 = E$, we have

$$
F^2 = F, \quad X = XF, \quad Y = FY, \text{ and } Z = XFY.
$$

These equations imply that the idempotent F cannot have zero rows or columns and is thus of the form given in (i) above. The above equations imply that X, Y, and Z are of the form given in the statement of the lemma. $\square$

Corollary 5.1.10. *Let E be a nonnegative idempotent matrix of rank r. Then there exist r columns $x_1, \dots, x_r$ of E whose nonnegative linear combinations include all columns of E. In the special case where E has no zero columns or rows, every column of E is a positive multiple of some x_i with $1 \leq i \leq r$.*

Proof. First assume that E has no zero rows or columns, so that $E = E_1 \oplus \cdots \oplus E_r$ as in Lemma 5.1.9(i). Since all columns of E_i are positive multiples of its first column, the desired x_i can be chosen to correspond to these first columns.

In the general case, consider the block triangular form of E given in part (ii) of Lemma 5.1.9. It is easily seen that every column of E is a nonnegative linear combination of the columns of the block

$$
\begin{bmatrix} XF \\ \\ F \\ \\ \\ 0 \end{bmatrix} = \begin{bmatrix} X_1E_1 & X_2E_2 & \cdots & X_rE_r \\ E_1 & 0 & \cdots & 0 \\ 0 & E_2 & \cdots & 0 \\ & & & \vdots \\ \vdots & \vdots & & E_r \\ 0 & 0 & \cdots & 0 \end{bmatrix},
$$

where $[X_1, \dots, X_r]$ is the partition of X conforming to $F = E_1 \oplus \cdots \oplus E_r$. Now x_i can be chosen to be the first column in the i-th block column on the right-hand side. $\square$

The following simple lemmas will be needed more than once.

Lemma 5.1.11. *If G is a group of invertible nonnegative matrices, then every member of G has exactly one nonzero entry in each of its rows and columns. Furthermore, if G is bounded, then there is a diagonal matrix D with positive diagonal entries such that $D^{-1}GD$ is a group of permutation matrices (so that G is indecomposable if and only if the corresponding permutation group is a transitive subgroup of the symmetric group).*

Proof. Each row and column of every element of G contains some nonzero entry, by invertibility. Now if a row, say the first row, of some S in G contained at least two positive entries $(S)_{1i}$ and $(S)_{1j}$, then the entries $(S^{-1})_{ik}$ and $(S^{-1})_{jk}$ of S^{-1} would necessarily be zero for $k \geq 2$. This would imply that rank $(S^{-1}) \leq n - 1$, which is a contradiction. Hence each row has exactly one nonzero entry, and the proof for columns is similar.

Next assume that G is bounded (so that, in particular, the spectral radius formula (Theorem 6.1.10) implies $\rho(G) = \rho(G^{-1}) = 1$ for all G in G). It follows from the preceding paragraph that some power of every G in G is diagonal. The only diagonal matrix in G is I, for the diagonal entries of any diagonal matrix in G are all positive and have their powers and the powers of their reciprocals bounded. Thus some power of every $G \in G$ is I. After a permutation, we can assume that members of G have the simultaneous block diagonal form

$$G = G_1 \oplus \cdots \oplus G_k,$$

with $G_i \in G_i$, where each G_i is an indecomposable group. It suffices to prove the lemma for the case of an indecomposable group, because, if D_i is a diagonal matrix such that $D_i^{-1}G_iD_i$ is a permutation group for each i, then $D = D_1 \oplus \cdots \oplus D_k$ has the desired property for the group G. Thus assume G that is indecomposable.

By Lemma 5.1.5(iii), for each i there is a G_i in G whose first column has positive i-th coordinate p_i; since $I \in G$, we can take $G_1 = I$ and $p_1 = 1$. Let

$$D = \text{diag}(p_1, \dots, p_n).$$

Then the i-th coordinate of the first column of $D^{-1}G_iD$ is 1 for every i. We must show that every nonzero entry in every member G of $D^{-1}GD$ is 1. Assume, with no loss of generality, that $D = I$, and let c be the positive (i, j) entry in G. Then it is easily verified that the $(1, 1)$ entry in $G_i^{-1}GG_j$ is c. Thus c is a diagonal entry, which forces it to be 1, since some power of $G_i^{-1}GG_j$ is I. □

Lemma 5.1.12. *Let A be a nonzero nonnegative matrix and assume that A has at least one zero column or row. Let S be the semigroup of nonnegative matrices that commute with A. Then S is decomposable.*

Proof. By using transposes, if necessary, we can assume that A has a zero column. After a permutation we can also assume that the first m columns of A are precisely the zero columns of A, so that

$$A = \begin{bmatrix} 0 & B \\ 0 & C \end{bmatrix}$$

where the upper left zero block is $m \times m$. We shall show that in the corresponding block decomposition of every S in $\mathcal{S}$, the lower left block S_{21} is zero. The relation $AS = SA$ implies that

$$BS_{21} = 0 \text{ and } CS_{21} = 0.$$

Suppose $S_{21} \neq 0$. Pick a nonzero entry, say $(S_{21})_{ij}$, of S_{21}. By non-negativity of B, C, and S_{21}, we deduce that the i-th columns of both B and C are zero. This contradicts the fact that A has no zero column except the ones exhibited above.

$\square$

We have seen that a nonnegative idempotent of rank one can be indecomposable (i.e., all of its entries can be positive). Thus there are (even singleton) indecomposable bands of nonnegative matrices. However, if every matrix in the band has rank at least two, then it is decomposable. (The singleton case follows from Lemma 5.1.9.)

Theorem 5.1.13. *Let $\mathcal{S}$ be a band of nonnegative matrices, and denote the minimal rank of members of $\mathcal{S}$ by r. If $r > 1$, then $\mathcal{S}$ is decomposable. In fact, there exists a permutation P such that $P^{-1}\mathcal{S}P$ has an $r \times r$ block-upper-triangular form.*

Proof. Fix a member Q of rank r. By Lemma 5.1.9, we can assume

$$Q = \begin{pmatrix} Q_1 & X \\ 0 & Q_2 \end{pmatrix}$$

(after a permutation), where Q_1 and Q_2 are idempotents of positive rank. Now, for arbitrary S in $\mathcal{S}$, the matrix QSQ is an idempotent of rank r having the same range and kernel as the idempotent Q, which implies $QSQ = Q$ for all S in $\mathcal{S}$. For

$$S = \begin{pmatrix} S_{11} & S_{12} \\ S_{21} & S_{22} \end{pmatrix},$$

this equation yields $Q_2 S_{21} Q_1 = 0$, which implies

$$\begin{pmatrix} 0 & 0 \\ 0 & Q_2 \end{pmatrix} \begin{pmatrix} S_{11} & S_{12} \\ S_{21} & S_{22} \end{pmatrix} \begin{pmatrix} Q_1 & 0 \\ 0 & 0 \end{pmatrix} = 0$$

for all S in $\mathcal{S}$. Thus $\mathcal{S}$ is decomposable by Lemma 5.1.5 (ii).

We have shown that $\mathcal{S}$ is decomposable; it remains to be shown that it has an $r \times r$ block upper-triangular form after a suitable permutation. After some permutation, if necessary, every S in $\mathcal{S}$ has the form

$$\begin{pmatrix} S_1 & * \\ 0 & S_2 \end{pmatrix}.$$

Note that the set of all matrices S_1 when S runs through $\mathcal{S}$ is a band $\mathcal{S}_1$. Similarly, the set of matrices S_2 forms a band $\mathcal{S}_2$. The proof by induction will be completed if we show that $r = r_1 + r_2$, where

$$r_j = \min\{\text{rank } T : T \in \mathcal{S}_j\},$$

$j = 1, 2$. Clearly, $r \geq r_1 + r_2$. To verify the reverse inequality, pick

$$S = \begin{pmatrix} S_1 & * \\ 0 & S_2 \end{pmatrix} \quad \text{and} \quad T = \begin{pmatrix} T_1 & * \\ 0 & T_2 \end{pmatrix}$$

in $\mathcal{S}$ with rank $(S_1) = r_1$ and rank $(T_2) = r_2$. Note that $S_2 \neq 0$ (and $T_1 \neq 0$), since otherwise the product of S and Q (or of T and Q) would have positive rank less than r. Thus

$$\begin{aligned} r \leq \text{rank}(ST) &= \text{tr}(ST) = \text{tr}(S_1 T_1) + \text{tr}(S_2 T_2) \\ &= \text{rank}(S_1 T_1) + \text{rank}(S_2 T_2) \\ &= \text{rank}(S_1) + \text{rank}(T_2) = r_1 + r_2. \end{aligned}$$

$\square$

We conclude this section with a simple result on submultiplicativity of spectral radius.

Theorem 5.1.14. *If the spectral radius is submultiplicative but not multiplicative on a semigroup $\mathcal{S}$ of nonnegative matrices, then $\mathcal{S}$ is decomposable.*

Proof. Suppose that $\mathcal{S}$ is indecomposable. We can assume, by continuity of spectral radius, that $\mathcal{S} = \overline{\mathbb{R}^+ \mathcal{S}}$. The set $\mathcal{N}$ of nilpotents in $\mathcal{S}$ is an ideal, by the submultiplicativity of ρ. Thus $\mathcal{N} = \{0\}$ by Lemma 5.1.5(v) and Theorem 5.1.2. By Theorem 3.4.3, since ρ is not multiplicative, $\mathcal{S}$ must contain nonzero divisors of 0. Let R and T be nonzero operators in $\mathcal{S}$ with $RT = 0$. Then $T\mathcal{S}R$ consists of nilpotents, so that $T\mathcal{S}R = \{0\}$. At least one of the two matrices R and T is zero by Lemma 5.1.5 (ii), which is a contradiction. $\square$

5.2 Indecomposable Semigroups

The main goal of this section is to extend the Perron-Frobenius Theorem to the general setting of semigroups. The assertions of this classical result, although usually stated for a single nonnegative matrix A, are really about a singly generated (norm) closed semigroup that is also closed under multiplication by positive scalars; i.e., the closure of

$$\{cA^n : c \in \mathbb{R}^+, n \in \mathbb{N}\}.$$

The principal result will be stated for a much larger class of semigroups, those whose minimal idempotents have a common range. By specializing to progressively smaller classes of semigroups (e.g., to those with a minimal idempotent in their center, and then to commutative semigroups), we obtain more information on the forms of such semigroups and the spectra of their elements.

Lemma 5.2.1. *Let S be a semigroup of nonnegative matrices and let E be a nonnegative idempotent of rank r (not necessarily in S). Then*

   (i) *relative to some basis $x_1, \ldots , x_r$ of $E\mathcal{V}$, every operator in the collection $ESE|_{E\mathcal{V}}$ has a nonnegative matrix, and*
   (ii) *this representation of $ESE|_{E\mathcal{V}}$ is indecomposable if S is indecomposable.*

Proof. Choose the basis $x_1, \ldots , x_r$ to be the r columns of E (obtained by Corollary 5.1.10) such that all the other columns of E are nonnegative linear combinations of $\{x_i\}$. Fix any $S \in S$. For each i, the vector ESx_i is a nonnegative linear combination of the columns of E and thus is such a combination of the x_i. This proves (i).

If S is indecomposable, then, by Lemma 5.1.5(iv), there exists a positive-entried matrix $T = S_1 + \cdots + S_m$ with all S_j in S. Fix i and note that Tx_i has all its coordinates positive. Thus $ETEx_i = E(Tx_i)$ is a positive linear combination of the columns of E and hence of the columns $x_1, \ldots , x_r$. Since this is true for each i, we conclude that the matrix of $ETE|_{E\mathcal{V}}$ in the basis $\{x_i\}$ has all positive entries. Thus we can apply Lemma 5.1.5 to the semigroup on $E\mathcal{V}$ generated by the matrices $ES_jE|_{E\mathcal{V}}$, $j = 1, \ldots , m$, to complete the proof of (ii). $\qquad\square$

Recall (Lemma 3.1.8) that a nonzero idempotent E in a semigroup S is *minimal* if $EF = FE = F$ for any idempotent F in S implies that either $F = E$ or $F = 0$.

The following lemma is an analogue of Lemma 3.1.6 in the present context.

Lemma 5.2.2. *Let $S = \overline{\mathbb{R}^+S}$ be an indecomposable semigroup of nonnegative matrices and let r be the minimal rank of nonzero members of S. Then the following hold:*

(i)   an idempotent E in S is of rank r if and only if it is minimal;

(ii) for each A of rank r in S there is a minimal idempotent F in S with $FA = A$;

(iii) for each i there exists a minimal idempotent in S whose i-th row is nonzero, and the same assertion is true for columns;

(iv) if E is a minimal idempotent in S, then $ESE \setminus \{0\}$ is a group with identity E. The set

$$\{ESE : S \in S, \ \rho(ESE) = 1\}$$

is a subgroup whose restriction G to the range of E is simultaneously similar, via a diagonal matrix, to a transitive group of permutation matrices.

Proof. (i): Every idempotent of rank r is obviously minimal. To show the converse, let E be a minimal idempotent in S of rank s, so that $s \geq r$. Let S_r be the ideal of S consisting of members of rank $\leq r$; S_r is indecomposable, by Lemma 5.1.5(v). Apply Lemma 5.2.1 to S_r and E to get an $s \times s$ representation of $S_0 = ES_rE|_{E\mathcal{V}}$, which is nonnegative and indecomposable. By Theorem 5.1.2 and Lemma 5.1.5 there is a non-nilpotent A in S_0. Then there is such an A with $\rho(A) = 1$. We will be done if we show that some subsequence of $\{A^n\}$ tends to an idempotent F, for then $F \in S_0$ and, since $FE = EF = F$, we conclude that $s = r$ and $E = F$.

The proof of the existence of F has nothing to do with non-negativity of matrices and resembles the argument given in the proof of Lemma 3.1.6. Use the fact that rank $(S) = r$ for all S in the closure of

$$\{cA^m : c \in \mathbb{R}^+, \ m \in \mathbb{N}\}$$

together with $\rho(A) = 1$ to express A as a direct sum of operators $B \oplus C$, where $\sigma(B)$ is on the unit circle and $\rho(C) < 1$. Then, up to a similarity, $B = U + N$, where U is unitary, N is nilpotent, and $UN = NU$. Applying the argument of Lemma 3.1.6 to powers of $B \oplus C$, we deduce that $N = 0$. Since a subsequence of $\{U^m\}$ approaches the identity operator on the range of U and $\lim_{m \to \infty} C^m = 0$, we obtain an idempotent F in S_0.

(ii): By Lemma 5.1.5, $AS \neq \{0\}$. Pick S_0 in S with $AS_0 \neq 0$ and note that, by that lemma again, $SAS_0 \neq 0$, so that the ideal SAS is nonzero and thus indecomposable (Lemma 5.1.5(v)). It follows from Theorem 5.1.2 that, for some S_1 and S_2 in S,

$$\sigma(AS_1S_2) = \sigma(S_2AS_1) \neq \{0\}.$$

Hence we can replace A with the non-nilpotent AS_1S_2 with no loss of generality. We can also assume that $\rho(A) = 1$. The rest of the proof is as in (i) above: Just observe that the idempotent F constructed from the powers of A in the preceding paragraph satisfies $FA = A$.

(iii): The indecomposable ideal $\mathcal{S}_r$ defined in (i) above has a member A whose i-th row is nonzero, by Lemma 5.1.5. Let F be the idempotent obtained in (ii) with $FA = A$ and observe that the i-th row of F is nonzero. The corresponding assertion for columns can be verified by passing to the indecomposable semigroup $\mathcal{S}^*$ and observing that E is an idempotent if and only if E^* is.

(iv): For every S in $\mathcal{S}$ with $\rho(ESE) = 1$, it follows from the minimality of r that $ESE|_{E\mathcal{V}}$ is invertible and, in fact, similar to a unitary operator as shown by the argument applied to the matrix A in the proof of (i). The multiplicativity of determinant on $\mathcal{G}$, together with the fact that $|\det G| = 1$ for all $G \in \mathcal{G}$, then shows that spectral radius is also multiplicative on $\mathcal{G}$. This implies that $\mathcal{G}$ is a semigroup. To show that it is a group, note that $\rho(ESE) = 1$ implies, as in the proof of (i), that there is a sequence $\{m_i\}$ of integers such that

$$E = \lim_{i \to \infty} (ESE)^{m_i} = \lim_{i \to \infty} (ESE)(ESE)^{m_i-1}.$$

Hence $\lim_{i \to \infty} (ESE)^{m_i-1}$ exists and defines a matrix ETE in $\mathcal{S}$ such that $(ETE)(ESE) = E$. Thus $ESE\setminus\{0\}$ is a group (with identity E), implying that $\mathcal{G}$ is a group.

Next let $x_1, \ldots, x_r$ be r independent columns of E, obtained by Corollary 5.1.10, such that every column of E is a nonnegative linear combination of $\{x_i\}$. Represent $\mathcal{G}$ relative to this basis and observe that, by Lemma 5.2.1, this representation is indecomposable and nonnegative. Lemma 3.1.6(ii) shows that $\mathcal{S}$ is bounded, since every closed subgroup of $\mathbb{R}^+\mathcal{U}$ is bounded. Lemma 5.1.11 then finishes the proof.

$\square$

We next discuss two lemmas, the second of which gives many equivalent forms of the hypothesis of the main result of this section (Theorem 5.2.6).

Recall that a *right ideal* of a semigroup $\mathcal{S}$ is a subset $\mathcal{J}$ such that $JS \in \mathcal{J}$ for all $J \in \mathcal{J}$ and $S \in \mathcal{S}$. A *minimal right ideal* is a nonzero right ideal that contains no other nonzero right ideal. There are obvious corresponding definitions and statements for left ideals.

Lemma 5.2.3. *Let $\mathcal{S} = \overline{\mathbb{R}^+\mathcal{S}}$ be an indecomposable semigroup of nonnegative matrices. The following hold:*

  (i) *Every nonzero right ideal of $\mathcal{S}$ contains a minimal right ideal;*

  (ii) *Every minimal right ideal is of the form $E\mathcal{S}$ for some minimal idempotent E in $\mathcal{S}$.*

Proof. Let $\mathcal{J}$ be a right ideal and let J be a nonzero member of $\mathcal{J}$. Let r be the minimal nonzero rank in $\mathcal{S}$, so that

$$\mathcal{S}_0 = \{S \in \mathcal{S} : \text{rank } S \leq r\}$$

is indecomposable by Lemma 5.1.5, which also implies that $JS_0 \neq \{0\}$. Thus we can assume that $J \in S_0$. By Lemma 5.2.2(ii), there is a minimal idempotent E with $J = EJ$. By part (iv) of that lemma, there is an element T in ESE with $(EJ)T = E$. Thus $E \in J$ and $ES \subseteq J$.

To complete the proof we need only show that if E is a minimal idempotent, then the right ideal ES is minimal. To see this, let J_0 be any nonzero right ideal contained in ES. By the preceding paragraph, J_0 contains FS for some minimal idempotent F. Thus $F = EA$ for some A in S, implying that E and F have the same range. This yields $FE = E$ (an equation that holds for any two idempotents with the same range). Hence

$$ES = FES \subseteq FS \subseteq J_0 .$$

$\square$

Lemma 5.2.4. *Let $S = \overline{\mathbb{R}^+ S}$ be an indecomposable semigroup of nonnegative matrices and let $\mathcal{E}$ denote the set of all its minimal idempotents. The following conditions are mutually equivalent:*

  (i) *$EF = F$ for every E and F in $\mathcal{E}$ (i.e., all minimal idempotents have the same range);*
 (ii) *all nonzero minimal-rank members of S have the same range;*
(iii) *$SE = ESE$ for some E in $\mathcal{E}$;*
 (iv) *$SE = ESE$ for all E in $\mathcal{E}$;*
  (v) *$SE \subseteq ES$ for some E in $\mathcal{E}$;*
 (vi) *$SE \subseteq ES$ for all E in $\mathcal{E}$;*
(vii) *S leaves the range of some E in $\mathcal{E}$ invariant;*
(viii) *S leaves the range of every E in $\mathcal{E}$ invariant;*
 (ix) *some minimal right ideal is an ideal;*
  (x) *every minimal right ideal is an ideal;*
 (xi) *S has a unique minimal right ideal.*

Proof. The equality $EF = F$ for every pair in $\mathcal{E}$ is easily seen to be equivalent to the assertion that $FS = ES$, which proves the equivalence of (i) and (xi) by Lemma 5.2.3. Also, (i) and (ii) are equivalent by Lemma 5.2.2 (ii).

If (ii) holds and $E \in \mathcal{E}$, then every nonzero member of SE has the same range as E, and thus $SE = ESE$. This proves that (ii) implies (iv). It is obvious that (iv) implies (vi) and that (vi) implies (viii). To show that (viii) implies (x), consider any minimal right ideal ES and observe that the hypothesis of invariance gives $SE = ESE$ for all S, which yields

$$SES = ESES \subseteq ES.$$

To derive (xi) from (x), let E and F be in $\mathcal{E}$. Then $ESF \neq \{0\}$ by Lemma 5.1.5(ii). Pick A in S with $EAF \neq 0$. Then $EAFS$ is a right ideal contained in the minimal ideal ES, and thus $EAFS = ES$. Now (x) implies that FS is an ideal, so that

$$ES = (EA)FS \subseteq FS,$$

and thus $ES = FS$.

It is clear that (ix) implies (v), which implies (vii). Also, (vii) implies (iii), which implies (ix). Since (x) trivially implies (ix), the cycle will be completed if we prove that (ix) implies (i). For this, assume that the minimal right ideal ES is an ideal and let F be an arbitrary member of $\mathcal{E}$. By Lemma 5.1.5(ii) there is a T in S with $FTE \neq 0$, so that FTE has the same range as F. Then

$$FTE \in SE \subseteq SES = ES,$$

showing that the ranges of E and F coincide. $\square$

Lemma 5.2.5. *Let $S = \overline{\mathbb{R}^+S}$ be an indecomposable semigroup of nonnegative matrices satisfying any one, and thus all, of the conditions in Lemma 5.2.4. Let $\mathcal{R}$ be the common range of the minimal idempotents in S. Then*

  (i) *the only nilpotent element of S is zero;*
  (ii) *spectral radius is multiplicative on S, so that*

$$\mathcal{S}_0 = \{S/\rho(S) : 0 \neq S \in \mathcal{S}\}$$

  is a subsemigroup on which ρ is identically one, and
  (iii) *there exists a nonnegative idempotent P with range $\mathcal{R}$, in the closed convex hull of S that has no zero rows or columns and satisfies*

$$S|_{\mathcal{R}} = PSP|_{\mathcal{R}}.$$

Proof. We first prove (iii). By part (iii) of Lemma 5.2.2, there exist minimal idempotents $E_1, \dots, E_n$ in S (where n is the size of matrices in S) such that the i-th column of E_i is nonzero for each i. Any row of any E_i that is zero is also zero in all minimal idempotents, because they all have the same range. Thus, by Lemma 5.2.2 (iii) again, no E_i has any zero rows.

The equality of ranges implies that $E_i E_j = E_j$ for all i and j. It follows that the average $P = (E_1 + \cdots + E_n)/n$ of the E_i is itself an idempotent with their common range $\mathcal{R}$. Thus $PE = E$ and $EP = P$ for every minimal idempotent E in S. Now $\mathcal{R}$ is invariant under S by Lemma 5.2.4(vi), so that $SP = PSP$ as well as $SE = ESE$ for all S in S and every minimal idempotent E. If $x \in \mathcal{R}$, then $Ex = Px = x$ and hence

$$S\,|_{\mathcal{R}} = ESE|_{\mathcal{R}} = PSP|_{\mathcal{R}}.$$

Since P clearly does not have any zero rows or columns, this proves (iii).

To establish (i), note that if $N \in S$ is nilpotent, then so is its restriction to $\mathcal{R}$. If $N \neq 0$, choose any minimal idempotent E in S; then $NE \neq 0$, since E has no zero rows, and thus part (iv) of Lemma 5.2.2 would imply the existence of M in $\mathcal{S}$ such that

$$MNE = EME \cdot ENE = E$$

(since, by Lemma 5.2.4(iii), the range of E is invariant under $\mathcal{S}$). Thus the restriction of N to the range of E is invertible, which is a contradiction. Thus (i) holds.

We next show that $\rho(S) = \rho(SP)$ for every S in $\mathcal{S}$, where P is as above. Suppose this is not true and, since $\rho(SP)$ is at most $\rho(S)$, assume that $\rho(SP) < 1 = \rho(S)$. Observe that the sequence $\{S^m\}$ is bounded, for otherwise some subsequence of $\{S^m/\|S^m\|\}$ would converge to a nonzero nilpotent member of $\mathcal{S}$ by continuity of spectrum, which contradicts the preceding paragraph. Thus there is a subsequence $\{S^{m_i}\}$ converging to some A in $\mathcal{S}$ with $\rho(A) = 1$. Then

$$AP = \lim_{i \to \infty} S^{m_i} P = \lim_{i \to \infty} (SP)^{m_i} = 0$$

by the assumption $\rho(SP) < 1$. Since P can be written as a direct sum of matrices with positive entries (by Lemma 5.1.9), this forces A to be zero, which is a contradiction.

Finally, to prove (ii), observe that for S, T in $\mathcal{S}$, $\rho(ST) = \rho(STP)$ by the preceding paragraph. Thus

$$\rho(ST) = \rho(STP) = \rho(PSP \cdot PTP) = \rho(PSP)\rho(PTP)$$

by part (iv) of Lemma 5.2.2, since the range of P is the same as the range of a minimal idempotent in $\overline{\mathbb{R}^+\mathcal{S}}$. Hence

$$\rho(ST) = \rho(SP)\rho(TP) = \rho(S)\rho(T).$$

$\square$

The next theorem has many consequences.

Theorem 5.2.6. *Let $\mathcal{S}$ be an indecomposable semigroup of nonnegative matrices and denote the minimal positive rank in $\overline{\mathbb{R}^+\mathcal{S}}$ by r. If $\overline{\mathbb{R}^+\mathcal{S}}$ has a unique minimal right ideal (or satisfies any of the other conditions of Lemma 5.2.4), then the following hold:*

(i) *there is a vector x with positive entries, unique up to scalar multiples, such that*

$$Sx = \rho(S)x$$

for all S in $\mathcal{S}$;

(ii) *every S in $\mathcal{S}$ has at least r eigenvalues of modulus $\rho(S)$, counting multiplicities; these are all of the form $\rho(S)\theta$ with $\theta^{r!} = 1$;*

(iii) *after a permutation of the basis, S has an $r \times r$ block partition such that the block matrix $(S_{ij})_{i,j=1}^r$ of each nonzero member S of $\mathcal{S}$*

has exactly one nonzero block in each block row and in each block column;

(iv) if, for any S in $\mathcal{S}$, the block matrix (S_{ij}) has a cyclic pattern (i.e., there exists a permutation $\{i_1, \ldots, i_r\}$ of $\{1, \ldots, r\}$ such that the nonzero blocks of S are precisely $S_{i_1,i_2}, S_{i_2,i_3}, \ldots, S_{i_r,i_1}$), then $\sigma(S)$ is invariant under the rotation about the origin by the angle $2\pi/r$;

(v) $r = 1$ if and only if some member of $\mathcal{S}$ has at least one positive column.

Proof. We can assume, with no loss of generality, that $\mathcal{S} = \overline{\mathbb{R}^+ \mathcal{S}}$ except in the proof of (v). Let $\mathcal{R}$ and P be as in Lemma 5.2.5. By Lemma 5.1.9, there are r positive idempotents P_i of rank one such that, after a permutation,

$$P = P_1 \oplus \cdots \oplus P_r.$$

Pick a positive vector v_i in the range of P_i for each i. Then the nonnegative vectors

$$x_1 = \begin{bmatrix} v_1 \\ 0 \\ \vdots \\ 0 \\ 0 \end{bmatrix}, \quad \cdots, \quad x_r = \begin{bmatrix} 0 \\ 0 \\ \vdots \\ 0 \\ v_r \end{bmatrix}$$

form a basis for $\mathcal{R}$. After a diagonal similarity, the group

$$\mathcal{G} = \{S|_{\mathcal{R}} : S \in \mathcal{S}, \rho(S) = 1\}$$

is a transitive subgroup of permutations on $\{v_1, \ldots, v_r\}$, by Lemmas 5.2.2 and 5.2.5. Thus we can scale the v_i to assume further that $\mathcal{G}$ is actually a permutation group. It is easy to see that $x = x_1 + \cdots + x_r$ is a common fixed vector for $\mathcal{G}$. Using Lemma 5.2.5 and the equation $Px = x$, we deduce that, for $S \in \mathcal{S}$ with $\rho(S) = 1$,

$$Sx = SPx = PSPx = x.$$

In other words, $Sx = \rho(S)x$ for all S in $\mathcal{S}$.

Now, if y is any vector (not even assumed nonnegative) such that $Sy = \rho(S)y$ for every $S \in \mathcal{S}$, we prove that y is a scalar multiple of x. To see this, note that $Py = y$ (since, as constructed in the proof of Lemma 5.2.5, P is the average of minimal idempotents in $\mathcal{S}$), so that $y = a_1 x_1 + \cdots + a_r x_r$ for some scalars a_i. Since $\mathcal{G}$ acts transitively on the x_i and since $Gy = y$ for every $G \in \mathcal{G}$, we deduce that $a_i = a_j$ for all i and j. This proves that y is a multiple of x, which is the uniqueness assertion of (i).

Consider the partition $(S_{ij})_{i,j=1}^r$ of matrices conforming to the decomposition $P = P_1 \oplus \cdots \oplus P_r$. Fix $S \in \mathcal{S}$ and let τ be the permutation of

$\{1, \ldots, r\}$ that corresponds to the matrix $PSP/\rho(S)$ in the group $\mathcal{G}$; i.e., for each k,

$$PSPx_k = \rho(S)x_{\tau(k)}.$$

Note that the (i,j) block of PSP is $P_iS_{ij}P_j$, so that the i-th block in the block column $PSPx_j$ is

$$P_iS_{ij}P_jv_j = \sum_j (PSP)_{ij}v_j = \rho(S)u,$$

where u denotes the i-th block in the column $x_{\tau(j)}$. Thus $u = v_{\tau(j)}$ if $i = \tau(j)$, and $u = 0$ otherwise. It follows that the only nonzero blocks $P_iS_{ij}P_j$ are those with $i = \tau(j)$. The proof of (iii) is now completed by observing that $S_{ij} = 0$ if and only if $P_iS_{ij}P_j = 0$, which is an immediate consequence of the positivity of P_i and P_j.

To prove (ii), note that the $r!$-th power of every S in $\mathcal{G}$ is block diagonal and

$$(PSP)^{r!} = P$$

by what we have proved. This implies that $\sigma(PSP)$ contains at least r numbers θ with $\theta^{r!} = 1$. Since the range of P is invariant under S, we have $\sigma(PSP) \subseteq \sigma(S)$. The conclusion for a general S follows by considering $S/\rho(S)$.

For the proof of (iv) we can assume, after a permutation, that the nonzero blocks are precisely $S_{1,2}, S_{2,3}, \ldots, S_{r,1}$. Let $\theta = e^{2\pi i/r}$ and

$$D = \theta I_1 \oplus \theta^2 I_2 \oplus \cdots \oplus \theta^r I_r,$$

where I_i is the identity matrix of the same size as S_{ii}. Then $SD = \theta DS$, and

$$\sigma(S) = \sigma(\theta DSD^{-1}) = \theta\sigma(DSD^{-1}) = \theta\sigma(S).$$

Thus a rotation by the angle $2\pi/r$ leaves $\sigma(S)$ unchanged.

Finally, it is easy to verify (v): It follows from (iii) that if any member of S has a positive column, then $r = 1$. For the converse, pick any rank-one idempotent E in $\overline{\mathbb{R}^+S}$ and a nonzero column y of E. The equation $Ex = x$ implies that y is a multiple of the positive vector x and is thus positive itself. Since E is a limit of members of $\mathbb{R}^+S$, it follows that some member S of S has a positive column.   □

A stronger hypothesis than that of the preceding theorem is that the minimal idempotent (rather than right ideal) be unique. Before using this hypothesis in our next corollary, we give a lemma listing some equivalent conditions.

Lemma 5.2.7. *Let* $S = \overline{\mathbb{R}^+ S}$ *be an indecomposable semigroup of nonnegative matrices. Denote its center by* $\mathcal{Z}$ *and its subset of minimal idempotents by* $\mathcal{E}$. *The following are mutually equivalent:*

(i) $\mathcal{E}$ *is a singleton;*

(ii) $\mathcal{E} \cap \mathcal{Z} \neq \emptyset$;

(iii) $\mathcal{E} \subseteq \mathcal{Z}$;

(iv) $SE = ES$ *for some* $E \in \mathcal{E}$;

(v) $SE = ES$ *for every* $E \in \mathcal{E}$.

Proof. If E is the unique element in $\mathcal{E}$, then $\mathcal{S}^*$ also has a unique minimal idempotent E^*. By Lemma 5.2.4, the ranges of E and E^* are invariant under $\mathcal{S}$ and $\mathcal{S}^*$, respectively, so that $SE = ESE = ES$. This proves that (i) implies (iii).

Assume (ii) and let $E \in \mathcal{E} \cap \mathcal{Z}$. Condition (v) of Lemma 5.2.4 is clearly satisfied, so $EF = F$ for every $F \in \mathcal{E}$, by 5.2.4(i). But $EF = FE$ and $FE = E$, so $E = F$. Thus (ii) implies (i). Clearly, (iii) implies (ii). Thus the equivalence of (i), (ii), and (iii) is proved.

Since it is obvious that (iii) implies (v) and (v) implies (iv), we need only show that (iv) implies (ii). But if $SE = ES$, then the range of E is invariant under $\mathcal{S}$, so that $SE = ESE$ for every $S \in \mathcal{S}$. Also, if $Ex = 0$ and $S \in \mathcal{S}$, then $ESx = S_1 Ex$ for some $S_1 \in \mathcal{S}$ implies $ESx = 0$, so the kernel of E is also invariant under $\mathcal{S}$. Hence $ES = ESE = SE$ for every $S \in \mathcal{S}$. $\qquad\square$

Corollary 5.2.8. *Let* S *be an indecomposable semigroup of nonnegative matrices such that* $\overline{\mathbb{R}^+ S}$ *has a unique minimal idempotent* E. *Then all the conclusions of Theorem 5.2.6 hold. Moreover,*

(i) *no nonzero member of* S *has a zero column or row;*

(ii) *the adjoint semigroup* $\mathcal{S}^*$ *also has a common positive eigenvector* y, *unique up to scalar multiples, such that* $S^* y = \rho(S) y$ *for all* $S \in \mathcal{S}$, *and*

(iii) *the rank of the minimal idempotent is 1 if and only if some member of* S *is positive.*

Proof. Without loss of generality, assume $\mathcal{S} = \overline{\mathbb{R}^+ S}$. Since ES is the only minimal right ideal, by Lemma 5.2.3(ii), Theorem 5.2.6 applies. To prove (i), note that E has no zero rows or columns, by Lemma 5.1.12. Thus E coincides with the idempotent P of Lemma 5.2.5, where each P_i in the decomposition $P = P_1 \oplus \cdots \oplus P_r$ is positive. Now, in the corresponding block partition S_{ij} of any given $S \in \mathcal{S}$, we have

$$P_i S_{ij} = S_{ij} P_j$$

for all i and j. An easy computation shows that the positivity of P_i and P_j implies that either $S_{ij} = 0$ or no row or column of S_{ij} is zero. Since one S_{ij} is nonzero in S, the proof of (i) is completed.

The adjoint of a matrix A is an idempotent if and if only if A is. Thus the indecomposable semigroup S^* also has a unique minimal idempotent, which proves (ii).

If $r = 1$, then $E = P = P_1$ is positive. Thus the proof of (iii) is completed by (v) of Theorem 5.2.6. $\qquad\square$

Example 5.2.9. *There are semigroups that have a unique minimal right ideal, as in the hypothesis of Theorem 5.2.6, but do not have a unique idempotent (and therefore do not satisfy the hypothesis of Corollary 5.2.8).*

Proof. For the case $r = 1$, consider the semigroup S_1 of all $n \times n$ matrices with exactly one nonzero column v, where v is the vector whose entries are all 1. Then S_1 consists of all the nonzero idempotents of $\overline{\mathbb{R}^+ S_1}$. Since all the nonzero idempotents in S have the same range, Lemma 5.2.4 implies that S has a unique minimal right ideal.

For $r > 1$, let $\mathcal{P}$ be the group of all $r \times r$ permutation matrices and take the tensor product $S = S_1 \otimes \mathcal{P}$. Then every nonzero member of S has rank r, and the (minimal) idempotents of $\overline{\mathbb{R}^+ S}$ are precisely the n matrices $S \otimes I$ with $S \in S_1$. $\qquad\square$

Corollary 5.2.10. *Let S be an indecomposable semigroup of nonnegative matrices. If S is normal (i.e., $AS = SA$ for every A in S), then all the conclusions of Corollary 5.2.8 hold.*

Proof. It must be shown that $\overline{\mathbb{R}^+ S}$ has a unique minimal idempotent. First, $\overline{\mathbb{R}^+ S}$ has minimal idempotents (by Lemma 5.2.2). Then condition (iv) of Lemma 5.2.7 holds, so there is a unique minimal idempotent by 5.2.7(i). $\qquad\square$

We need the following simple lemma on permutation groups.

Lemma 5.2.11. *Let $\mathcal{G}$ be an abelian, transitive group of $n \times n$ permutation matrices. For each $G \in \mathcal{G}$ there exists a positive integer m dividing n such that, after a permutation of the basis, G is the direct sum of m copies of the cyclic permutation*

$$
G_0 = \begin{pmatrix}
0 & 0 & \cdots & 0 & 1 \\
1 & 0 & \cdots & 0 & 0 \\
0 & 1 & \cdots & 0 & 0 \\
\vdots & \vdots & & \vdots & \vdots \\
0 & 0 & \cdots & 1 & 0
\end{pmatrix}
$$

of size n/m.

Proof. If G is cyclic, there is nothing to prove. Otherwise, permute the basis if necessary to obtain a decomposition $G = G_1 \oplus \cdots \oplus G_m$, where each G_i is cyclic and, except possibly for size, looks like G_0 above. It suffices to show that G_1 and G_2 have the same size. Assume that G_2 has larger size than G_1, and consider the block partition $(S_{ij})_{i,j=1}^m$ for each $S \in \mathcal{G}$ conforming to the partition of G.

By transitivity, $\mathcal{G}$ has a member S with $S_{12} \neq 0$. Since S_{12} is not square, it has nontrivial kernel. Hence if $\mathcal{B}$ denotes the basis underlying G_2 (and permuted by G_2), then the subset

$$\mathcal{B}_1 = \{u \in \mathcal{B} : S_{12}u = 0\}$$

is nonempty and proper. But it follows from $GS = SG$ that $G_1 S_{12}u = S_{12}G_2u$ for every $u \in \mathcal{B}_1$, and thus $G_2\mathcal{B}_1 \subseteq \mathcal{B}_1$. This contradicts the cyclicity of G_2. □

Corollary 5.2.12. *Let S be a commutative indecomposable semigroup of nonnegative matrices. Then $\overline{\mathbb{R}^+ S}$ has a unique minimal idempotent E, whose rank we denote by r, and all the conclusions of Theorem 5.2.6 and Corollary 5.2.8 hold. Furthermore, the following assertions are true:*

(i) *$S^r E = \rho(S)^r E$ for every S in S, so that the spectrum of $S/\rho(S)$ contains the r-th roots of unity whenever $S \neq 0$;*

(ii) *for each nonzero S in S, if $k = k(S)$ denotes the least positive integer k such that $S^k E = \rho(S)^k E$, then k divides r;*

(iii) *the multiplicity of each k-th root of unity in the spectrum of $S/\rho(S)$ is at least r/k;*

(iv) *$\sigma(S)$ is invariant under rotation about the origin by the angle $2\pi/k(S)$, for each S in S.*

Proof. By Lemma 5.2.7, $\overline{\mathbb{R}^+ S}$ has a unique minimal idempotent, so we can apply Theorem 5.2.6 and Corollary 5.2.8. To prove the additional claims, note that the group $\mathcal{G}$ that is defined in the proof of Theorem 5.2.6 is commutative. It follows from Lemma 5.2.11 that to each member G of $\mathcal{G}$ there corresponds a smallest integer k, which is just n/m of the lemma, such that $(PSP)^k = P$, where P is the unique minimal idempotent of $\overline{\mathbb{R}^+ S}$. This proves (i), (ii), and (iii).

The form given for G in Lemma 5.2.11 also implies that the block partition $((PSP)_{ij})$ of the corresponding PSP can be rearranged (by a permutation of basis blocks) to take the form

$$S_1 \oplus \cdots \oplus S_m,$$

where each S_i is a $k \times k$ block matrix with cyclic pattern. As in the proof of Theorem 5.2.6, we conclude that $\sigma(S_i)$ is invariant under rotation by $2\pi/k$. Hence, so is $\sigma(S)$. □

In the statement of the Perron-Frobenius Theorem, which follows, we assume $\rho(A) = 1$. This is merely a normalization, since indecomposability implies $\rho(A) \neq 0$ (by Theorem 5.1.2).

Corollary 5.2.13. (The Perron-Frobenius Theorem) *Let A be an indecomposable nonnegative matrix with $\rho(A) = 1$. Denote by r the minimal rank of nonzero members of $\overline{\mathbb{R}^+S}$, where S is the semigroup generated by A. Then the following hold:*

(i) *The sequence $\{A^{rj}\}_{j=1}^{\infty}$ converges to an idempotent E of rank r;*

(ii) *if $r > 1$, there is a permutation matrix P such that $P^{-1}AP$ has the block form*

$$\begin{bmatrix} 0 & 0 & \cdots & 0 & A_r \\ A_1 & 0 & \cdots & 0 & 0 \\ 0 & A_2 & \cdots & 0 & 0 \\ \vdots & \vdots & & \vdots & \vdots \\ 0 & 0 & \cdots & A_{r-1} & 0 \end{bmatrix}$$

(with square diagonal blocks);

(iii) *there is a positive column vector x, unique up to a scalar multiple, such that $Ax = x$;*

(iv) *the set $\{\lambda \in \sigma(A) : |\lambda| = 1\}$ consists precisely of all the r-th roots of unity; each member of the set is a simple eigenvalue;*

(v) *$\sigma(A)$ is invariant under the rotation about the origin by the angle $2\pi/r$;*

(vi) *1 is dominant in $\sigma(A)$ if and only if some power of A has all its entries positive. This occurs precisely when $\{A^j\}$ is convergent.*

Proof. By Corollary 5.2.12, there is a unique minimal idempotent E of rank r in $\overline{\mathbb{R}^+S}$. Applying part (ii) of that Corollary to A, and using the indecomposability of A, we deduce that $k(A) = r$. This proves (ii) and (v) by Corollary 5.2.12. The assertion (iii) is just (i) in Theorem 5.2.6.

Now, $E = \lim_{m \to \infty} c_m A^{n_m}$ for some sequences $\{c_m\} \subseteq \mathbb{R}^+$ and $\{n_m\} \subseteq \mathbb{N}$. But by continuity of spectrum we can assume $c_m = 1$ for all m. Thus the rank of E is precisely the number of modulus-one members of $\sigma(A)$, counting multiplicities. This, together with (iii) of Corollary 5.2.12, proves (iv).

Since $\sigma(A(I - E))$ lies inside the unit circle, we have $\lim_{j \to \infty} A^j(I - E) = 0$. Also, $A^r E = E$. Thus

$$\lim_{j \to \infty} A^{rj} = \lim_{j \to \infty} A^{rj}E + \lim_{j \to \infty} A^{rj}(I - E) = E,$$

proving (i).

Finally, note that the dominance of 1 in $\sigma(A)$ is equivalent, by (iv) above, to $r = 1$. This occurs if and only if $\lim_{j \to \infty} A^j = E$. Also, $r = 1$ is equivalent to positivity of all entries of E, which occurs precisely when a sufficiently large power of A is positive. This proves (vi). □

It is easy to construct examples to show that the hypothesis of Corollary 5.2.8 is strictly weaker than that of Corollary 5.2.10: The semigroup of all doubly stochastic matrices (i.e., nonnegative matrices whose rows and columns each sum to 1) is probably the simplest one. The full group of permutation matrices is an example of a normal but not commutative semigroup. This shows that Corollary 5.2.10 has a strictly weaker hypothesis than that of Corollary 5.2.12.

Corollary 5.2.14. *If S is an indecomposable semigroup of nonnegative matrices such that $\overline{\mathbb{R}^+S}$ has a unique minimal right ideal, then*

   (i) *the spectral radius is multiplicative on S, and*
  (ii) *the normalized semigroup $S_0 = \{S \in S : \rho(S) = 1\}$ can be embedded in the semigroup $\hat{S}$ of stochastic matrices; more precisely, there is an invertible, diagonal, nonnegative matrix D such that $D^{-1}S_0D \subseteq \hat{S}$.*

Proof. We use (i) of Theorem 5.2.6, from which (i) follows directly. Now, if the coordinates of the positive vector x of (i) are denoted by $a_1, \dots, a_n$, we define $D = \mathrm{diag}\,(a_1, \dots, a_n)$ and let y denote the column vector whose entries are all 1. Then $Dy = y$ for all y in $D^{-1}S_0D$. Thus $D^{-1}S_0D$ consists of stochastic matrices. □

5.3 Connections with Reducibility

Most of the indecomposable semigroups we have considered are reducible. Some are even triangularizable (e.g., commutative ones). Those that are triangularizable are fairly special.

Corollary 5.3.1. *Let S be an indecomposable semigroup of nonnegative matrices that is triangularizable. Then ρ is multiplicative on S.*

Proof. Triangularizability of S implies submultiplicativity of ρ, so Theorem 5.1.14 yields the corollary. □

There are some reducibility results that use nonnegativity of matrices.

Theorem 5.3.2. *If S is a semigroup of nonnegative matrices such that $\overline{\mathbb{R}^+S}$ contains no matrix of rank one, then S is reducible.*

Proof. We can assume that $S = \overline{\mathbb{R}^+ S}$ and that S is indecomposable. We apply Lemma 5.2.2 to get an idempotent E in S of minimal positive rank. By that lemma, $ESE|_{E\mathcal{V}}$ is simultaneously similar to multiples of permutation matrices. Since the vector $(1, 1, \ldots, 1)$ is fixed by every permutation matrix, $ESE|_{E\mathcal{V}}$ has a one-dimensional invariant subspace. Thus S is reducible, by Lemma 2.1.12. $\qquad\square$

Corollary 5.3.3. *If S is semigroup of nonnegative matrices and if, for every S in $\mathcal{S}$, $\sigma(S)$ contains at least two elements with modulus $\rho(S)$, counting multiplicities, then S is reducible.*

Proof. It follows from Lemma 3.1.2 that every member of $\overline{\mathbb{R}^+ S}$ satisfies the spectral hypothesis, so $\overline{\mathbb{R}^+ S}$ contains no matrix of rank 1 and Theorem 5.3.2 applies. $\qquad\square$

The following result generalizes the fact that the semigroup of all stochastic matrices has a nontrivial (one-dimensional) invariant subspace.

Theorem 5.3.4. *A semigroup of nonnegative matrices is reducible if spectral radius is submultiplicative on it.*

Proof. Assume that $S = \overline{\mathbb{R}^+ S}$ and that S is indecomposable, so that, by Theorem 5.1.13, ρ is multiplicative. Also, in view of Theorem 5.3.2, we can assume that S contains a member E of rank one, which can be taken to be an idempotent by Lemma 5.2.2(ii). Now consider the ideal SES. Since every member T of this ideal is of rank at most one, we have $\text{tr}(T) = \rho(T)$ by non-negativity. Hence trace is multiplicative on SES, forcing the ideal to be reducible (in fact, triangularizable) by Corollary 2.2.4. Thus S is reducible by Lemma 2.1.10. $\qquad\square$

It should be noted that the hypothesis of the theorem above does not imply decomposability, as demonstrated by the group of all permutation matrices.

Corollary 5.3.5. *Let S be a semigroup of nonnegative matrices whose spectra all lie on circles centered at 0. Then S is reducible.*

Proof. Observe that $|\det S| = \rho(S)^n$ for every S in $\mathcal{S}$, where n is the size of the matrices. Thus the multiplicativity of the determinant function implies that of ρ, and Theorem 5.3.4 applies. $\qquad\square$

Corollary 5.3.6. *A semigroup of nonnegative matrices is reducible if spectral radius is permutable on it.*

Proof. Permutability of spectral radius is equivalent to its submultiplicativity, by Theorem 3.4.8, so this follows from Theorem 5.3.4.   □

5.4 Notes and Remarks

There is a vast literature on non-negativity (also called "positivity") and decomposability (also called "reducibility" by many authors). The three books Berman-Plemmons [1], Bapat-Raghavan [1], and Minc [1] are among the standard texts on the subject; their union covers most of the known results.

Much of the material in Section 5.1 is presented here for the first time. Exceptions are as follows: Theorem 5.1.2 is a special case of known results (see Choi-Nordgren-Radjavi-Rosenthal-Zhong [1]). Lemma 5.1.9, on the form of a nonnegative idempotent, can be found in Berman-Plemmons [1]. Lemma 5.1.5 and Theorem 5.1.13 are from Marwaha [1]. Corollary 5.1.10 was used in Radjavi [4]. Lemma 5.1.11, whose special cases are implicit in all proofs of the Perron-Frobenius Theorem, is from Radjavi [7].

Section 5.2 is an expansion of the material in Radjavi [7], designed to put the classical Perron-Frobenius Theorem in perspective. This beautiful result was first proved in Perron [1] for matrices with positive entries, and then generalized in Frobenius [2] to nonnegative matrices. There are several different proofs in the literature; standard proofs are given in Minc [1] and in Schaefer [1]. A graph-theoretical approach is taken in Berman-Plemmons [1] and a game-theoretical one in Bapat-Raghavan [1].

A natural question arose from the Perron-Frobenius Theorem: What sequences $\{\lambda_j\}_{j=1}^n$ of complex numbers with $|\lambda_1| \geq \cdots \geq |\lambda_n|$ are attainable as sequences of eigenvalues (counting multiplicities) of nonnegative $n \times n$ matrices? Clearly, one can reduce this question to the case of indecomposable matrices. In fact, by the Perron-Frobenius Theorem, the question can be further reduced to the primitive case; i.e., the case where the integer r in Theorem 5.2.13 is 1. With this reduction, the only obvious information, then, is that λ_1 has to be positive and equal to the spectral radius, and that $\lambda_1 > |\lambda_2|$. The obvious information is far from sufficient. A complete and deep answer to the question was obtained in Boyle-Handelman [1], where methods of symbolic dynamics were employed.

The simple results in Corollary 5.3.1 and Theorem 5.3.2 seem to be new. Corollary 5.3.3 is a special case of a result in Radjavi [4].

CHAPTER 6
Compact Operators and Invariant Subspaces

We begin this chapter with a review of the basic properties of bounded linear operators on Banach spaces (referring to standard texts for some of the proofs). We give the definition of compactness of an operator and prove the Fredholm alternative. Hilden's simple proof of Lomonosov's Theorem that compact operators have hyperinvariant subspaces is presented.

For a nonzero eigenvalue λ of the compact operator K, we discuss F. Riesz's classic decomposition of the space into the direct sum of the span of the kernels of $(K - \lambda)^n$ and the intersection of the ranges of $(K - \lambda)^n$. We also develop the basic properties of the trace on the trace-class operators.

6.1 Operators on Banach Spaces

Definition 6.1.1. A *Banach space* is a complete normed vector space. We shall restrict our attention to Banach spaces over the complex scalars.

We will be concerned with continuous linear operators on Banach spaces; continuity is equivalent to the following.

Definition 6.1.2. A linear transformation T mapping the Banach space $\mathcal{X}$ into the Banach space $\mathcal{Y}$ is *bounded* if there is a constant k such that $\|Tf\| \leq k\|f\|$ for all $f \in \mathcal{X}$. If T is bounded, then the norm of T, denoted by $\|T\|$, is defined by $\|T\| = \sup\{\|Tf\| : f \in \mathcal{X} \text{ and } \|f\| \leq 1\}$. The set of bounded linear operators mapping $\mathcal{X}$ into itself is denoted by $\mathcal{B}(\mathcal{X})$.

Theorem 6.1.3. *If $T : \mathcal{X} \to \mathcal{Y}$ is linear, then T is continuous if and only if T is bounded.*

Proof. This is straightforward and can be found in any standard introductory functional analysis text (see, for example, Conway [1, p. 68], Rudin [2, p. 24], Naimark [1, p. 74]), Dunford-Schwartz [1, p. 54]). $\qquad\square$

Definition 6.1.4. If $T \in \mathcal{B}(\mathcal{X})$, then the spectrum of T, denoted by $\sigma(T)$, is defined by

$$\sigma(T) = \{\lambda \in \mathbb{C} : T - \lambda \text{ is not invertible}\}.$$

(We use $T - \lambda$ as an abbreviation for $T - \lambda I$, with the I the identity operator on $\mathcal{X}$.)

Thus $\sigma(T)$ contains the set of eigenvalues of T, and is equal to that set if $\mathcal{X}$ is finite-dimensional.

Theorem 6.1.5. *For $T \in B(\mathcal{X})$, $\sigma(T)$ is a nonempty compact subset of* $\mathbb{C}$.

Proof. This theorem includes the statement that finite matrices over $\mathbb{C}$ have eigenvalues, and thus that $\mathbb{C}$ is algebraically closed. Thus it is not surprising that the proof uses Liouville's Theorem (applied to the functions $z \mapsto F((T - z)^{-1})$ for F a continuous linear functional on $B(\mathcal{X})$; see any standard functional analysis text, such as Conway [1, p. 146], Dunford-Schwartz [1, p. 567], Naimark [1, p. 174], or Rudin [2, Theorem 10.13]). $\qquad\square$

Definition 6.1.6. The *point spectrum* of the operator T, denoted by $\Pi_0(T)$, is

$$\{\lambda \in \mathbb{C} : (T - \lambda)f = 0 \quad \text{for some} \quad f \neq 0\}.$$

(That is, the point spectrum is the set of eigenvalues.) The *approximate point spectrum* of T, denoted by $\Pi(T)$, is

$$\{\lambda \in \mathbb{C} : \{(T - \lambda)f_n\} \to 0 \text{ for some } \{f_n\} \text{ with } \|f_n\| = 1 \text{ for all } n\}.$$

If the dimension of $\mathcal{X}$ is finite, then clearly $\sigma(T) = \Pi_0(T)$ for each $T \in B(\mathcal{X})$. In the infinite-dimensional case, $\Pi_0(T)$ is empty for some T. As the next theorem shows, $\Pi(T)$ cannot be empty.

By the *boundary* of the compact set $\sigma(T)$ we mean the intersection of $\sigma(T)$ and the closure of its complement in $\mathbb{C}$.

Theorem 6.1.7. *For each $T \in B(\mathcal{X})$, the boundary of $\sigma(T)$ is contained in $\Pi(T)$.*

Proof. We need the preliminary fact that λ in the complement of $\Pi(T)$ implies that the range of $T - \lambda$ is closed. To see this, suppose $\lambda \notin \Pi(T)$. Then there is an $\varepsilon > 0$ such that $\|(T - \lambda)f\| \geq \varepsilon\|f\|$ for all $f \in \mathcal{X}$. If $\{(T - \lambda)g_n\} \to h$, then

$$\|(T - \lambda)(g_n - g_m)\| \geq \varepsilon\|g_n - g_m\|$$

implies that $\{g_n\}$ is a Cauchy sequence, and hence converges to some g. Then $(T - \lambda)g = h$, so the range of $T - \lambda$ is closed.

Now let λ be in the boundary of $\sigma(T)$. We show that assuming λ is not in $\Pi(T)$ leads to a contradiction. Suppose, then, that $\lambda \notin \Pi(T)$. If we could show that the range of $T - \lambda$ is dense, this would be a contradiction, for, by the above, the range of $T - \lambda$ would then be $\mathcal{X}$; thus $T - \lambda$ would be surjective and injective and hence invertible, contradicting the fact that $\lambda \in \sigma(T)$.

It remains to be shown that $T - \lambda$ has dense range if $\lambda \notin \Pi(T)$ and is in the boundary of $\sigma(T)$. Fix $g \in \mathcal{X}$. Choose $\{\lambda_n\}$ not in $\sigma(T)$ such that $\{\lambda_n\} \to \lambda$, and, for each n, let $f_n = (T - \lambda_n)^{-1}g$.

If $\{f_n\}$ is bounded, then $\{(T - \lambda)f_n\} \to g$ (since $(T - \lambda)f_n$ is close to $(T - \lambda_n)f_n = g$), and we are done. To show that $\{f_n\}$ is bounded, assume the contrary: Suppose $\{\|f_{n_i}\|\} \to \infty$. Then $\lambda \notin \Pi(T)$ implies there is an $\varepsilon > 0$ such that $\|(T - \lambda)f\| \geq \varepsilon\|f\|$ for all $f \in \mathcal{X}$. Now,

$$(T - \lambda_n)f_n = (T - \lambda)f_n + (\lambda - \lambda_n)f_n,$$

so

$$g - (\lambda - \lambda_n)f_n = (T - \lambda)f_n.$$

Thus

$$\|(T - \lambda)f_n\| \leq \|g\| + |\lambda - \lambda_n| \, \|f_n\|,$$

and

$$\varepsilon\|f_n\| \leq \|g\| + |\lambda - \lambda_n| \, \|f_n\|.$$

Therefore, $\varepsilon \leq \|g\|/\|f_{n_i}\| + |\lambda - \lambda_{n_i}|$, which is a contradiction, since $\{\|g\|/\|f_{n_i}\|\}$ and $\{|\lambda - \lambda_{n_i}|\}$ both approach 0 as n_i approaches ∞. $\square$

Corollary 6.1.8. *For each $T \in \mathcal{B}(\mathcal{X})$, $\Pi(T) \neq \emptyset$.*

Proof. It is an easy exercise in "applied point-set topology" to show that every nonempty compact subset of the plane has nonempty boundary. $\square$

Definition 6.1.9. The *spectral radius of* T, denoted by $\rho(T)$, is $\sup\{|\lambda| : \lambda \in \sigma(T)\}$.

Theorem 6.1.10. (The Spectral Radius Formula) *For every $T \in \mathcal{B}(\mathcal{X})$,*

$$\rho(T) = \lim_{n \to \infty} \|T^n\|^{\frac{1}{n}}.$$

Proof. This beautiful and useful formula, obviously related to the "root test" for convergence of power series, is proven in most introductory functional analysis books (see, for example, Conway [1, p. 197], Dunford-Schwartz [1, p. 567], or Rudin [2, Theorem 10.13]). $\square$

6.2 Compact Operators

Definition 6.2.1. A subset S of $\mathcal{X}$ is *bounded* if there is an N such that
$$\|f\| \leq N \quad \text{for all} \quad f \in S.$$

It is not very difficult to show that the unit ball, $\{f \in \mathcal{X} : \|f\| \leq 1\}$, of an infinite-dimensional Banach space is not compact (see Theorem 6.2.4 below). Thus infinite-dimensional spaces contain sets that are closed and bounded but not compact.

Note that a linear operator is bounded if and only if it sends bounded sets into bounded sets.

Definition 6.2.2. The linear transformation K in $\mathcal{B}(\mathcal{X})$ is *compact* if the closure of KS is compact whenever S is a bounded subset of $\mathcal{X}$.

Since $\mathcal{X}$ is, in particular, a metric space, the definition of compactness can be rephrased: K is compact if $\{Kf_n\}$ has a convergent subsequence whenever $\{f_n\}$ is a bounded sequence.

The following lemma will be useful in establishing the basic properties of spectra of compact operators.

Lemma 6.2.3. *If $\mathcal{M}$ is a proper closed subspace of a Banach space and $\varepsilon > 0$, then there is a vector f such that $\|f\| = 1$ and $\|f - g\| \geq 1 - \varepsilon$ for all $g \in \mathcal{M}$.*

Proof. Choose any h not in $\mathcal{M}$ and let $d = \inf\{\|h - g\| : g \in \mathcal{M}\}$. Then $d > 0$, since $\mathcal{M}$ is closed. For each positive integer n there is a $g_n \in \mathcal{M}$ with $\|h - g_n\| \leq d + \frac{1}{n}$. Let $f_n = \frac{1}{\|h-g_n\|}(h - g_n)$. For any $g \in \mathcal{M}$, then,

$$\|f_n - g\| = \left\|\frac{1}{\|h - g_n\|}(h - g_n) - g\right\|$$

$$= \frac{1}{\|h - g_n\|}\left\|h - g_n - \|h - g_n\|g\right\|.$$

Since $g_n + \|h - g_n\|g$ is in $\mathcal{M}$, $\left\|h - g_n - \|h - g_n\|g\right\|$ is at least d. Also, $\|h - g_n\| \leq d + \frac{1}{n}$ implies $\frac{1}{\|h-g_n\|} \geq \frac{1}{d+\frac{1}{n}}$. Thus $\|f_n - g\| \geq \frac{d}{d+\frac{1}{n}}$. For n sufficiently large, $\frac{d}{d+\frac{1}{n}} \geq 1 - \varepsilon$; let $f = f_n$ for such an n. $\square$

One consequence of this lemma is the following.

Theorem 6.2.4. *If $\mathcal{X}$ is an infinite-dimensional Banach space, then the unit ball of $\mathcal{X}$ is not compact.*

Proof. Let f_1 be any vector of norm 1, and let $\mathcal{M}_1$ be the set of multiples of f_1. Then $\mathcal{M}_1$ is a closed subspace, so, by the preceding lemma, there is an f_2 of norm 1 such that the distance from f_2 to $\mathcal{M}_1$ is at least $\frac{1}{2}$. Let $\mathcal{M}_2$ be the linear span of $\{f_1, f_2\}$. Then $\mathcal{M}_2$ is closed (it is not hard to show that all finite-dimensional subspaces of a Banach space are closed), so the lemma implies that there is an f_3 of norm 1 whose distance from $\mathcal{M}_2$ is at least $\frac{1}{2}$. Inductively define a sequence $\{f_n\}$ such that $\|f_n\| = 1$ for all n and $\|f_n - f_m\| \geq \frac{1}{2}$ when $m < n$. Then $\{f_n\}$ has no Cauchy subsequences, so no subsequence of $\{f_n\}$ converges. Hence the unit ball is not compact. $\square$

Corollary 6.2.5. *If K is compact and $\lambda \neq 0$, then $\{f : Kf = \lambda f\}$ is finite-dimensional.*

Proof. Suppose the eigenspace were infinite-dimensional; then its unit ball would not be compact, by the previous theorem. The image of this unit ball under K is simply its multiple by λ, so it would be closed but not compact, contradicting the compactness of K. $\square$

The nonzero spectrum of a compact operator is point spectrum, as will be shown below (6.2.8). We begin with the following.

Lemma 6.2.6. *If K is compact, then*

$$\Pi(K)\setminus\{0\} \subset \Pi_0(K).$$

Proof. Let $\lambda \in \Pi(K)\setminus\{0\}$. By definition,

$$\|(K - \lambda)f_n\| \to 0$$

for some $\{f_n\}$ with $\|f_n\| = 1$. Then $\{Kf_{n_i}\}$ converges for some subsequence $\{f_{n_i}\}$, and $\{(K - \lambda)f_{n_i}\} \to 0$ implies that $\{\lambda f_{n_i}\}$ converges. Since $\lambda \neq 0$, $\{f_{n_i}\} \to f$ for some f. Then clearly $(K - \lambda)f = 0$ and $\|f\| = 1$. $\square$

The next lemma will be useful in characterizing the spectra of compact operators.

Lemma 6.2.7. *If K is compact and $\varepsilon > 0$, then*

$$\Pi_0(K) \cap \{\lambda : |\lambda| \geq \varepsilon\}$$

is finite.

Proof. Suppose, to the contrary, that there is an infinite set $\{\lambda_n\}$ such that $|\lambda_n| \geq \varepsilon$ and $Kf_n = \lambda_n f_n$ for $\|f_n\| = 1$. For each k, let $\mathcal{M}_k$ denote the linear span of $\{f_1, \ldots, f_k\}$. By Lemma 6.2.3, for each $k > 1$ there is a $g_k \in \mathcal{M}_k$ such that $\|g_k\| = 1$ and the distance from $\mathcal{M}_{k-1}$ to g_k is at least $\frac{1}{2}$. If $k_1 > k_2$, then g_{k_1} has the form $\alpha f_{k_1} + f$, where $f \in \mathcal{M}_{k_1-1}$. Also,

$Kg_{k_2} \in \mathcal{M}_{k_1-1}$ and $Kg_{k_1} = \lambda_{k_1}\alpha f_{k_1} + Kf = \lambda_{k_1}g_{k_1} + (K - \lambda_{k_1})f$. Since $(K - \lambda_{k_1})f \in \mathcal{M}_{k_1-1}$, $\|Kg_{k_1} - Kg_{k_2}\|$ is at least $|\lambda_{k_1}|$ times the distance from g_{k_1} to $\mathcal{M}_{k_1-1}$, which is at least $\frac{\varepsilon}{2}$. Thus no subsequence of $\{Kg_k\}$ converges, contradicting the compactness of K. $\square$

Theorem 6.2.8. (**The Fredholm Alternative**) *If K is a compact operator on an infinite-dimensional space, then*

$$\sigma(K) = \{0\} \cup \Pi_0(K).$$

Proof. We have shown that, for compact operators, nonzero approximate point spectrum is point spectrum (Lemma 6.2.6). Thus Theorem 6.1.7 implies that the boundary of $\sigma(K)$ is contained in $\{0\} \cup \Pi_0(K)$. This set is at most countable, by Lemma 6.2.7, so every point of it is a boundary point (as a little "applied point set topology" of the plane shows). Also, a compact operator on an infinite-dimensional space cannot be invertible, so $0 \in \sigma(K)$. $\square$

Corollary 6.2.9. *The spectrum of a compact operator is countable, and if it is infinite, it consists of $\{0\}$ and a sequence converging to 0.*

Proof. This follows immediately from Lemma 6.2.7 and Theorem 6.2.8.

$\square$

6.3 Invariant Subspaces for Compact Operators

Definition 6.3.1. A *linear manifold* in a Banach space is a subset that is closed under addition and multiplication by complex numbers; a *subspace* is a linear manifold that is also closed in the topological sense. A subset $\mathcal{M}$ is *invariant* for an operator T if $Tf \in \mathcal{M}$ whenever $f \in \mathcal{M}$; $\mathcal{M}$ is invariant for a collection of operators if it is invariant for every operator in the collection. A subset is *hyperinvariant* for an operator T if it is invariant under every operator that commutes with T. A subspace is *trivial* if it is $\{0\}$ or the whole space.

There are bounded linear operators on Banach spaces that have no nontrivial invariant subspaces (See Enflo [1], Read [1], [2], [3]). The *invariant subspace problem* is the question whether every bounded linear operator on Hilbert space has a nontrivial invariant subspace. Although there are many affirmative results under various hypotheses (see the discussion in the Notes and Remarks section at the end of this chapter), this problem remains unsolved. As will be shown below, compact operators on Banach spaces have nontrivial invariant subspaces; this theorem is the foundation for all the results on triangularization of collections of compact operators.

Definition 6.3.2. A subset $\mathcal{J}$ of $B(\mathcal{X})$ is *transitive* if $\{Tf : T \in \mathcal{J}\}$ is dense in $\mathcal{X}$ whenever f is a nonzero vector.

Definition 6.3.3. A *subalgebra* of $B(\mathcal{X})$ is a subset of $B(\mathcal{X})$ that is closed under addition, multiplication, and multiplication by complex numbers. An *algebra of operators* is a subalgebra of $B(\mathcal{X})$ for some $\mathcal{X}$.

If $\mathcal{A}$ is an algebra of operators and f is a vector, then $\{Af : A \in \mathcal{A}\}$ is obviously a linear manifold that is invariant under $\mathcal{A}$, so its closure is an invariant subspace. Thus if $\mathcal{A}$ is an algebra of operators, $\mathcal{A}$ is transitive if and only if $\mathcal{A}$ has no nontrivial invariant subspaces.

Theorem 6.3.4(Lomonosov's Theorem). *Every nonzero compact operator has a nontrivial hyperinvariant subspace.*

Proof. Let K be a compact operator other than 0, and let $\mathcal{A}$ denote the commutant of K (i.e., $\mathcal{A} = \{A \in B(\mathcal{X}) : AK = KA\}$). Every eigenspace of K is clearly invariant under $\mathcal{A}$, so if K has a nontrivial eigenspace, we are done. By the Fredholm Alternative (Theorem 6.2.8), this is the case if there is a nonzero point in $\sigma(K)$.

Thus we can assume that $\rho(K) = 0$; by the spectral radius formula (Theorem 6.1.10), this is equivalent to $\{\|K^n\|^{\frac{1}{n}}\} \to 0$. We will show that this is not compatible with $\mathcal{A}$ being transitive.

Suppose that $\mathcal{A}$ is transitive. By replacing K with $\frac{1}{\|K\|}K$, we can assume that $\|K\| = 1$. Choose a vector f_0 such that $\|Kf_0\| > 1$ (necessarily then, $\|f_0\| > 1$) and let $\mathcal{S} = \{f : \|f - f_0\| < 1\}$ be the open unit ball centered at f_0. Since $\|K\| = 1$ and $\|Kf_0\| > 1$, the closure $\overline{K\mathcal{S}}$ does not contain 0.

Now, K compact implies that $\overline{K\mathcal{S}}$ is compact. For each $A \in \mathcal{A}$,

$$A^{-1}(\mathcal{S}) = \{f : \|Af - f_0\| < 1\}$$

is an open subset of $\mathcal{X}$. Since $\mathcal{A}$ is transitive,

$$\bigcup_{A \in \mathcal{A}} A^{-1}(\mathcal{S}) = \mathcal{X} \backslash \{0\},$$

so $\{A^{-1}(\mathcal{S}) : A \in \mathcal{A}\}$ is an open covering of the compact set $\overline{K\mathcal{S}}$. Thus there is a finite subset $\{A_1, A_2, \ldots, A_n\}$ of $\mathcal{A}$ such that

$$\overline{K\mathcal{S}} \subset \bigcup_{i=1}^{n} A_i^{-1}(\mathcal{S}).$$

In particular, $K f_0 \in A_i^{-1}(\mathcal{S})$ for some A_{i_1}; i.e., $A_{i_1} K f_0 \in \mathcal{S}$. Thus $K A_{i_1} K f_0 \in K\mathcal{S}$, so some $A_{i_2}^{-1}(\mathcal{S})$ contains $K A_{i_1} K f_0$. Then $A_{i_2} K A_{i_1} K f_0$ is in $\mathcal{S}$. After m steps, this process yields

$$(A_{i_m} K A_{i_{m-1}} K \cdots A_{i_2} K A_{i_1} K f_0) \in \mathcal{S}.$$

Now, K commutes with every A_i, so for each m there exist $\{A_{i_1}, \dots, A_{i_m}\} \subset \mathcal{A}$ such that

$$(A_{i_m} A_{i_{m-1}} \cdots A_{i_1} K^m f_0) \quad \text{is in} \quad \mathcal{S}.$$

Recall that the A_i are chosen from the finite set $\{A_1, \dots, A_n\}$. Let $c = \max\{\|A_i\| : i = 1, \dots, n\}$. Then

$$\|A_{i_m} A_{i_{m-1}} \cdots A_{i_1} K^m\| \leq c^m \|K^m\| = \|(cK)^m\|.$$

The operator cK also has spectral radius 0, so $\{\|(cK)^m\|^{\frac{1}{m}}\} \to 0$. In particular, $\{\|(cK)^m\|\} \to 0$. Thus

$$\{\|A_{i_m} A_{i_{m-1}} \cdots A_{i_1} K^m f_0\|\} \to 0,$$

which implies that 0 is in the closure of $\mathcal{S}$. This is a contradiction. $\square$

The above theorem and the next corollary are generalized in Corollary 7.4.12 below.

Corollary 6.3.5. *Every operator that commutes with a nonzero compact operator has a nontrivial invariant subspace.*

Proof. A hyperinvariant subspace is invariant under every operator in the commutant, so this follows from Theorem 6.3.4. $\square$

Corollary 6.3.6. (The Aronszajn-Smith Theorem) *Every compact operator has a nontrivial invariant subspace.*

Proof. A hyperinvariant subspace is, in particular, invariant. $\square$

Corollary 6.3.7. *A commutative family of compact operators has a nontrivial invariant subspace.*

Proof. If K is an operator other than 0 in the family, then K has a nontrivial hyperinvariant subspace by Lomonosov's Theorem (Theorem 6.3.4); such a subspace is invariant under the family. $\square$

The following notation is useful.

Notation 6.3.8. *If $\mathcal{F}$ is any collection of subspaces, then Alg $\mathcal{F}$ denotes the set of operators that leave all the subspaces in $\mathcal{F}$ invariant. (For any $\mathcal{F}$, Alg $\mathcal{F}$ is clearly a unital subalgebra of $\mathcal{B}(\mathcal{X})$.) If $\mathcal{S}$ is any set of operators,* then Lat S denotes the collection of all subspaces that are invariant under all the operators in S. (For any S, Lat S is a complete lattice under inclusion, where the infimum of any subset is the intersection and the supremum is the closed linear span.)

6.4 The Riesz Decomposition of Compact Operators

The Fredholm alternative states that nonzero points in the spectrum of a compact operator are eigenvalues (Theorem 6.2.8). In this section we extend this result: If $\lambda \in \sigma(K)$ and $\lambda \neq 0$, then the space $\mathcal{X}$ decomposes into a direct sum of invariant subspaces, one of which is finite-dimensional, with the property that the spectrum of the restriction of K to the finite-dimensional subspace is $\{\lambda\}$ and the spectrum of the restriction of K to the other subspace is $\sigma(K) \setminus \{\lambda\}$. We begin with several definitions and lemmas.

Definition 6.4.1. The subspaces $\mathcal{M}$ and $\mathcal{N}$ of $\mathcal{X}$ are *complementary* (or *are complements of each other*) if $\mathcal{M} \cap \mathcal{N} = \{0\}$ and $\mathcal{M} + \mathcal{N} = \mathcal{X}$, where $\mathcal{M} + \mathcal{N} = \{f + g : f \in \mathcal{M}, \ g \in \mathcal{N}\}$. If $\mathcal{M}$ and $\mathcal{N}$ are complementary, then $\mathcal{X}$ is said to be the *direct sum* of $\mathcal{M}$ and $\mathcal{N}$.

Note that saying that $\mathcal{M}$ and $\mathcal{N}$ are complementary subspaces of $\mathcal{X}$ is equivalent to saying that every vector in $\mathcal{X}$ has a unique representation in the form $f + g$ with $f \in \mathcal{M}$ and $g \in \mathcal{N}$.

Definition 6.4.2. An *idempotent* or a *projection* is a bounded linear operator P satisfying $P^2 = P$.

If P is an idempotent, let $\mathcal{M} = \{h \in \mathcal{X} : \ Ph = h\}$ and $\mathcal{N} = \{h \in \mathcal{X} : \ Ph = 0\}$. Then $\mathcal{M}$ and $\mathcal{N}$ are complementary subspaces of $\mathcal{X}$, since $h \in \mathcal{X}$ implies $h = Ph + (1 - P)h$.

Definition 6.4.3. If P is an idempotent, $\mathcal{M} = \{h \in \mathcal{X} : \ Ph = h\}$, and $\mathcal{N} = \{h \in \mathcal{X} : \ Ph = 0\}$, then P is said to be the *projection* on $\mathcal{M}$ along $\mathcal{N}$.

Theorem 6.4.4. *If $\mathcal{M}$ and $\mathcal{N}$ are complementary subspaces, then there is an idempotent P that is the projection on $\mathcal{M}$ along $\mathcal{N}$.*

Proof. Fix $h \in \mathcal{X}$; then $h = f + g$ with $f \in \mathcal{M}$ and $g \in \mathcal{N}$. Define $Ph = f$. The uniqueness of the direct sum decomposition implies that P is linear and that $P^2 h = P(Ph) = Ph$. Also, $\mathcal{M} = \{h : Ph = h\}$ and $\mathcal{N} = \{h : Ph = 0\}$.

It remains to be shown that P is bounded. This follows from the Closed Graph Theorem (see any introductory functional analysis text for a proof of the closed graph theorem). For if $\{h_n\} \to h$ and $\{Ph_n\} \to k$, let $h_n = f_n + g_n$ with $f_n \in \mathcal{M}$ and $g_n \in \mathcal{N}$. Then $Ph_n = f_n$, so $\{f_n\} \to k$. Since $\{h_n\}$ converges, it follows that $\{g_n\}$ converges to some g. Thus $\{h_n\} \to k + g$. Since $\mathcal{M}$ and $\mathcal{N}$ are closed, $k \in \mathcal{M}$ and $g \in \mathcal{N}$. Thus

$$Ph = P(k + g) = k,$$

so P has a closed graph and is therefore bounded. $\qquad\square$

There exist subspaces of Banach spaces that do not have complements (see Rudin [2], pp. 128-130); these are subspaces that are not the ranges of idempotents. If a subspace is the range of an idempotent, then the invariance of the subspace under an operator T is equivalent to a simple algebraic relation.

Theorem 6.4.5. *If P is the projection on $\mathcal{M}$ along $\mathcal{N}$ and $T \in \mathcal{B}(\mathcal{X})$, then*

   (i) *$\mathcal{M}$ is invariant under T if and only if $TP = PTP$, and*
  (ii) *$\mathcal{M}$ and $\mathcal{N}$ are both invariant under T if and only if $PT = TP$.*

Proof. For (i), simply note that $TP = PTP$ is equivalent to $PTf = Tf$ whenever $Pf = f$.

For (ii), note that $1 - P$ is the projection on $\mathcal{N}$ along $\mathcal{M}$. Then, by (i), $\mathcal{M}$ and $\mathcal{N}$ are both invariant under T if and only if

$$TP = PTP \quad \text{and} \quad T(1 - P) = (1 - P)T(1 - P),$$

and multiplying out the second equation shows that it is equivalent to $PT = PTP$. Thus (ii) follows. $\qquad\square$

Theorem 6.4.6. *Every finite-dimensional subspace of a Banach space has a complementary subspace.*

Proof. Let $\{e_1, e_2, \ldots, e_n\}$ be a basis for $\mathcal{M}$. Then there are linear functionals $\{\phi_1, \ldots, \phi_n\}$ on $\mathcal{M}$ such that $f \in \mathcal{M}$ implies $f = \sum_{j=1}^n \phi_j(f)e_j$.

On finite-dimensional spaces every linear functional is bounded. Use the Hahn-Banach Theorem to extend each of the $\{\phi_j\}$ to a bounded linear functional on $\mathcal{X}$. Define the linear operator P by $Pf = \sum_{j=1}^n \phi_j(f)e_j$ for each $f \in \mathcal{X}$. Then P is obviously an idempotent with range $\mathcal{M}$. $\qquad\square$

Notation 6.4.7. For an operator T, let $\ker T$ denote the kernel of T $(= \{f : Tf = 0\})$ and let $\operatorname{ran} T$ denote the range of T $(= \{g : g = Tf$ for some $f \in \mathcal{X}\})$.

Theorem 6.4.8. *If K is compact and $\lambda \neq 0$, then* ran $(K - \lambda)$ *is closed.*

Proof. Let $\mathcal{M} = \ker(K - \lambda)$. Then $\mathcal{M}$ is finite-dimensional (by Corollary 6.2.5), and therefore has a complement (by Theorem 6.4.6), say $\mathcal{N}$. Then $(K - \lambda)\mathcal{X} = (K - \lambda)\mathcal{N}$. If $K - \lambda$ were not bounded below on $\mathcal{N}$ then there would be a nonzero $f \in \mathcal{N}$ such that $(K - \lambda)f = 0$ (by the proof of Theorem 6.2.6). Thus $K - \lambda$ is bounded below on $\mathcal{N}$, and therefore $(K - \lambda)\mathcal{N}$ is closed (as in the first part of the proof of Theorem 6.1.7). Hence $(K - \lambda)\mathcal{X}$ is closed. □

Corollary 6.4.9. *If K is compact, $\lambda \neq 0$, and n is a positive integer, then* ran$(K - \lambda)^n$ *is closed.*

Proof. Since the range of $(K - \lambda)^n$ is the range of $K - \lambda$ on the invariant subspace ran $(K - \lambda)^{n-1}$, this follows from Theorem 6.4.8 and a trivial induction. □

The essence of the Riesz Decomposition Theorem for compact operators is contained in the next lemma.

Lemma 6.4.10. *If K is compact and $\lambda \neq 0$, then, for n sufficiently large,*

$$\ker(K - \lambda)^{n+1} = \ker(K - \lambda)^n$$

and

$$\text{ran}(K - \lambda)^{n+1} = \text{ran}(K - \lambda)^n.$$

Proof. To prove the first assertion, note that $\ker(K - \lambda)^{n+1} \supset \ker(K - \lambda)^n$ for all n. If there were an infinite number of n for which the kernels were distinct, we could get an infinite sequence $\{f_{n_i}\}$ of unit vectors such that $(K - \lambda)^{n_i+1}f_{n_i} = 0$ and the distance of f_{n_i} to the kernel of $(K - \lambda)^{n_i}$ is at least $\frac{1}{2}$ (using Lemma 6.2.3). Then if $n_i > n_j$,

$$\|Kf_{n_i} - Kf_{n_j}\| = \|(K - \lambda)f_{n_i} - Kf_{n_j} + \lambda f_{n_i}\|.$$

Now, $(K - \lambda)f_{n_i}$ is in the kernel of $(K - \lambda)^{n_i}$, as is Kf_{n_j}, so

$$\|Kf_{n_i} - Kf_{n_j}\| \geq \frac{|\lambda|}{2}.$$

Thus $\{Kf_{n_i}\}$ has no Cauchy subsequence, contradicting the compactness of K.

The proof for ranges is essentially the same. Clearly,

$$\text{ran}(K - \lambda)^{n+1} \subseteq \text{ran}(K - \lambda)^n$$

for all n. If there were an infinite number of n for which the inclusion was proper, there would be an infinite sequence $\{g_{n_i}\}$ of unit vectors such that

$g_{n_i} \in \text{ran} (K - \lambda)^{n_i}$ but the distance from g_{n_i} to the range of $(K - \lambda)^{n_i+1}$ is at least $\frac{1}{2}$. Then if $n_i > n_j$,

$$\|Kg_{n_i} - Kg_{n_j}\| = \|Kg_{n_i} - (K - \lambda)g_{n_j} + \lambda g_{n_j}\|.$$

Now, $\text{ran} (K - \lambda)^{n_i} \subseteq \text{ran} (K - \lambda)^{n_j}$, so Kg_{n_i} and $(K - \lambda)g_{n_j}$ are both in $\text{ran} (K - \lambda)^{n_j+1}$. Thus $\|Kg_{n_j} - Kg_{n_j}\| \geq \frac{|\lambda|}{2}$, contradicting the compactness of K.

We have established that the sequence of kernels and the sequence of ranges are both finite, and the lemma follows. (Note that $\ker (K-\lambda)^{N+1} = \ker (K - \lambda)^N$ implies $\ker (K - \lambda)^{N+j} = \ker (K - \lambda)^N$ for $j = 1, 2, 3, \ldots$, and the corresponding result also holds for ranges.) □

Theorem 6.4.11. *(The Riesz Decomposition Theorem for Compact Operators).* *If K is a compact operator and $\lambda \neq 0$, there is a positive integer N such that $\ker (K - \lambda)^{N+1} = \ker (K - \lambda)^N$ and $\text{ran} (K - \lambda)^{N+1} = \text{ran} (K - \lambda)^N$. For such an N, $\ker (K - \lambda)^N$ and $\text{ran} (K - \lambda)^N$ are complementary invariant subspaces of K.*

Proof. The existence of N follows immediately from the previous lemma. To show that the subspaces are complements, we first show that the intersection of $\ker (K - \lambda)^N$ and $\text{ran} (K - \lambda)^N$ is $\{0\}$: If $h = (K - \lambda)^N f$ and $(K - \lambda)^N h = 0$, then $(K - \lambda)^{2N} f = 0$, so $(K - \lambda)^N f = 0$ (by the equality of kernels), and $h = 0$.

We must show that every vector can be represented as a sum of a vector in $\ker(K - \lambda)^N$ and a vector in $\text{ran} (K - \lambda)^N$. Given $h \in \mathcal{X}$ there is a g such that $(K - \lambda)^N h = (K - \lambda)^{2N} g$. Then $h = (h - (K - \lambda)^N g) + (K - \lambda)^N g$. Since $(K - \lambda)^N (h - (K - \lambda)^N g) = (K - \lambda)^N h - (K - \lambda)^{2N} g = 0$, this is the decomposition of h. □

Corollary 6.4.12. *Let K be a compact operator and $\lambda \neq 0$. If*

$$\mathcal{N}_\lambda = \bigvee_{n=1}^\infty \ker (K - \lambda)^n$$

and

$$\mathcal{R}_\lambda = \bigcap_{n=1}^\infty \text{ran} (K - \lambda)^n,$$

then $\mathcal{N}_\lambda$ and $\mathcal{R}_\lambda$ are complementary invariant subspaces of K such that the spectrum of the restriction of K to $\mathcal{N}_\lambda$ is $\{\lambda\}$ and the spectrum of the restriction of K to $\mathcal{R}_\lambda$ is $\sigma(K) \setminus \{\lambda\}$.

Proof. For N as in the previous theorem,

$$\mathcal{N}_\lambda = \ker (K - \lambda)^N \quad \text{and} \quad \mathcal{R}_\lambda = \text{ran} (K - \lambda)^N.$$

The subspace $\mathcal{N}_\lambda$ is finite-dimensional, and the restriction of $K - \lambda$ to $\mathcal{N}_\lambda$ is nilpotent, so $\sigma(K|_{\mathcal{N}_\lambda}) = \{\lambda\}$. It is easily seen that the spectrum of the direct sum of two operators is the union of the spectra. Also, if $Kf = \lambda f$, then $f \in \mathcal{N}_\lambda$, so $\sigma(K|_{R_\lambda}) = \sigma(K) \setminus \{\lambda\}$. $\qquad\square$

Corollary 6.4.13. *If $K, \lambda, \mathcal{N}_\lambda$, and $\mathcal{R}_\lambda$ are as in the previous corollary, and if M is any invariant subspace of K, then M decomposes as a direct sum $M = \mathcal{N} + \mathcal{R}$, where $\mathcal{N} \subseteq \mathcal{N}_\lambda$ and $\mathcal{R} \subseteq \mathcal{R}_\lambda$.*

Proof. Recall that the notation $(K - \lambda)|_M$ represents the restriction of $(K - \lambda)$ to M. By Corollary 6.4.12, if

$$\mathcal{N} = \bigvee_{n=1}^{\infty} \ker\left((K - \lambda)|_M\right)^n$$

and

$$\mathcal{R} = \bigcap_{n=1}^{\infty} \operatorname{ran}\left((K - \lambda)|_M\right)^n,$$

then M is the direct sum of $\mathcal{N}$ and $\mathcal{R}$. Clearly, $\mathcal{N} \subseteq \mathcal{N}_\lambda$ and $\mathcal{R} \subseteq \mathcal{R}_\lambda$. $\quad\square$

Definition 6.4.14. If K is a compact operator and λ is a nonzero eigenvalue of K, and if N_λ and R_λ are as defined in Corollary 6.4.12, then the projection on N_λ along R_λ (see Definition 6.4.3) is called *the Riesz projection for K corresponding to λ.*

Corollary 6.4.15. *If λ is a nonzero eigenvalue of the compact operator K, and if P is the Riesz projection for K corresponding to λ, then every invariant subspace of K is invariant under P.*

Proof. If M is invariant under K, then $M = \mathcal{N} + \mathcal{R}$ as in Corollary 6.4.13. Thus $PM = \mathcal{N}$ is contained in M. $\qquad\square$

A stronger result is proven in Chapter 7 (Lemma 7.5.7).

6.5. Trace-Class Operators on Hilbert Space

Definition 6.5.1. A *Hilbert space* is a Banach space on which there is an inner product, denoted by $(\cdot, \cdot)$ (linear in the first variable and conjugate-linear in the second variable), such that $(f, f) = \|f\|^2$ for every vector f.

If $\mathcal{H}$ is a Hilbert space and $A \in B(\mathcal{H})$, it is easily seen that there is a unique operator, called the adjoint of A and denoted by A^*, that satisfies $(Af, g) = (f, A^* g)$ for all f and g in $\mathcal{H}$. It is readily shown that $\|A^*\| = \|A\|$ and $\|A^* A\| = \|A\|^2$ for every $A \in B(\mathcal{H})$. Operators that commute with their adjoints are well understood.

Definition 6.5.2. The operator A is *normal* if $A^* A = A A^*$, *self-adjoint* if $A = A^*$, and *unitary* if $A^* = A^{-1}$.

The main reason that we wish to consider operators on Hilbert space is to generalize the results of Section 2.2 (which depend upon the trace of a matrix) to an infinite-dimensional setting. There is no satisfactory extension of the trace to all operators, but there is a very complete extension to a set of compact operators called "the trace class." The following discussion is intended to develop the basic properties of trace-class operators as concisely as possible. We require several preliminary results.

Theorem 6.5.3. *The spectral radius of a normal operator is equal to its norm.*

Proof. Let A be normal. By Theorem 6.1.10, $\rho(A) = \lim_{n \to \infty} \|A^n\|^{\frac{1}{n}}$. Thus it suffices to show that $\|A^2\| = \|A\|^2$ for A normal (for then $\|A^{2^k}\| = \|A\|^{2^k}$ for all k, so $\rho(A) = \|A\|$). Using the fact that $\|B^* B\| = \|B\|^2$ for every B together with $A^* A = A A^*$ yields

$$\|A^2\|^2 = \|(A^2)^* A^2\| = \|(A A^*)^* (A A^*)\| = \|A A^*\|^2 = (\|A\|^2)^2,$$

so $\|A^2\| = \|A\|^2$. $\qquad\square$

We need another property of normal operators.

Theorem 6.5.4. *If A is normal and $Af = \lambda f$, then $A^* f = \bar{\lambda} f$.*

Proof. This follows from the fact that

$$\|(A - \lambda)f\| = \|(A^* - \bar{\lambda})f\|,$$

which is immediately verified by squaring and expanding the resulting inner products. $\qquad\square$

Corollary 6.5.5. *Eigenvectors corresponding to distinct eigenvalues of a normal operator are orthogonal to each other.*

Proof. If A is normal, $Af = \alpha f$, and $Ag = \beta g$ with $\alpha \neq \beta$, we can assume that $\alpha \neq 0$. Then

$$(\alpha f, \beta g) = (Af, Ag) = (A^* Af, g),$$

or

$$\alpha\bar{\beta}(f,g) \ = \ \bar{\alpha}\alpha(f,g).$$

Thus $\bar{\beta}(f,g) \ = \ \bar{\alpha}(f,g)$, so $(f,g) = 0$, since $\overline{\beta} \neq \overline{\alpha}$. □

In finite dimensions, a normal matrix is unitarily equivalent to a diagonal matrix, or, in other words, there is an orthonormal basis consisting of eigenvectors of the normal matrix. This is also true for compact normal operators on any Hilbert space.

Theorem 6.5.6. (The Spectral Theorem for Compact Operators)
If K is a compact normal operator, then there is an orthonormal set $\{e_n\}_{n=1}^{\infty}$ and a sequence $\{\lambda_n\}_{n=1}^{\infty}$ of complex numbers such that

$$Kf \ = \ \sum_{n=1}^{\infty} \lambda_n (f, e_n) e_n$$

for every vector f.

Proof. We first show that there is an orthonormal basis consisting of eigenvectors of K. Choose, by Zorn's Lemma, a maximal orthonormal set of eigenvectors of K, say $\{f_\alpha\}$. It must be shown that the span of $\{f_\alpha\}$ is $\mathcal{H}$. If not, let $\mathcal{M}$ denote its orthocomplement. Then $\mathcal{M}$ is invariant under K, for if $(f, f_\alpha) = 0$ and $Kf_\alpha = \lambda_\alpha f_\alpha$, then $(Kf, f_\alpha) = (f, K^*f_\alpha) = (f, \bar{\lambda}_\alpha f_\alpha)$ (by Theorem 6.5.4), so $(Kf, f_\alpha) = 0$. Since $\mathcal{M}$ is invariant under both K and K^* (it is easily seen that $\mathcal{M} \in \operatorname{Lat} K$ if and only if $\mathcal{M}^\perp \in \operatorname{Lat} K^*$), the restriction of K to $\mathcal{M}$ is also normal. Thus $\|K|_{\mathcal{M}}\| = \rho(K|_{\mathcal{M}})$ by Theorem 6.5.3, so either $K|_{\mathcal{M}} = 0$ or $\sigma(K|_{\mathcal{M}})$ contains a $\lambda \neq 0$. By Theorem 6.2.8, any such λ is an eigenvalue, so there cannot be such a λ (by the maximality of $\{f_\alpha\}$). Similarly, if $K|_{\mathcal{M}} = 0$ and $\mathcal{M} \neq \{0\}$, every vector in $\mathcal{M}$ would be an eigenvector. Hence $\mathcal{M} = \{0\}$, so $\{f_\alpha\}$ is an orthonormal basis.

Now, for each α, $Kf_\alpha = \lambda_\alpha f_\alpha$ for some complex number λ_α. Since each eigenspace corresponding to a nonzero eigenvalue is finite-dimensional, and since K has at most a countable number of eigenvalues (Corollary 6.2.9), at most countably many of the $\{\lambda_\alpha\}$ are nonzero. Relabeling those as $\{\lambda_n\}$ and the corresponding eigenvectors as $\{e_n\}$ gives the theorem. □

Note that only a finite number of the $\{\lambda_n\}$ in the above are nonzero if K has finite rank. In any case, $\{\lambda_n\} \to 0$ (by Corollary 6.2.9).

Note that K^*K is a compact self-adjoint operator whenever K is a compact operator.

Definition 6.5.7. If K is a compact operator and $\{\lambda_n\}$ is the set of eigenvalues of K^*K, then $\lambda_n \geq 0$ (since $\lambda_n = (K^*Ke_n, e_n) = \|Ke_n\|^2$), so we can define $s_n = \sqrt{\lambda_n}$. The $\{s_n\}$ are the *singular values* of K. We can and do assume that $s_1 \geq s_2 \geq s_3 \cdots$. (Of course, $\{s_n\} \to 0$ by Corollary 6.2.9.)

Note that the operator P defined by $Pf = \sum_{n=1}^{\infty} s_n(f, e_n)e_n$ has the property that $P^2 = K^*K$; P is usually denoted by $(K^*K)^{\frac{1}{2}}$.

Theorem 6.5.8. (The Singular Value Decomposition) *If K is a compact operator with singular values $\{s_n\}$, then there exist orthonormal sets $\{e_n\}$ and $\{f_n\}$ such that*

$$Kf = \sum_{n=1}^{\infty} s_n(f, e_n)f_n$$

for every vector f.

Proof. We begin with the spectral decomposition of the operator P discussed above:

$$Pf = \sum_{n=1}^{\infty} s_n(f, e_n)e_n,$$

where $P^2 = K^*K$.

Note that for each $f \in \mathcal{H}$,

$$\|Kf\|^2 = (Kf, Kf) = (K^*Kf, f) = (P^2 f, f) = (Pf, Pf) = \|Pf\|^2.$$

Thus if $\mathcal{R}_1$ is the range of P and $\mathcal{R}_2$ is the range of K, we can define a mapping $U : \mathcal{R}_1 \to \mathcal{R}_2$ by $UPf = Kf$, and U is isometric (preserves norms). Let $Ue_n = f_n$; then the $\{f_n\}$ are orthonormal, and $Kf = \sum_{n=1}^{\infty} s_n(f, e_n)f_n$ for every f. $\square$

Definition 6.5.9. The *Schatten p-class*, denoted by $\mathcal{C}_p$, is the set of all compact operators whose sequence of singular values is in the space ℓ^p (i.e., satisfies $\sum_{n=1}^{\infty} s_n^p < \infty$). The class $\mathcal{C}_1$ is the *trace class*.

The *trace norm* $\|K\|_1$ of the operator K with singular values $\{s_n\}$ is defined to be $\sum_{n=1}^{\infty} s_n$. It must be shown (see Theorem 6.5.15 below) that $\|\cdot\|_1$ is a norm. We first show that a trace can be defined on the trace class.

The following lemma will be useful both in defining the trace and in establishing that the trace norm is a norm.

Lemma 6.5.10. *If K is in the trace class and if $\{g_m\}$ and $\{h_m\}$ are orthonormal sets, then $\sum_{m=1}^{\infty} |(Kg_m, h_m)| \leq \|K\|_1$.*

Proof. We compute as follows:

$$
\sum_{m=1}^{\infty} |(Kg_m, h_m)| = \sum_{m=1}^{\infty} \left| \left(\sum_{n=1}^{\infty} s_n (g_m, e_n) f_n, h_m \right) \right|
$$

$$
\leq \sum_{m=1}^{\infty} \sum_{n=1}^{\infty} s_n |(g_m, e_n)| \, |(f_n, h_m)|
$$

$$
= \sum_{n=1}^{\infty} s_n \sum_{m=1}^{\infty} |(g_m, e_n)| \, |(f_n, h_m)|
$$

$$
\leq \sum_{n=1}^{\infty} s_n \left(\sum_{m=1}^{\infty} |(g_m, e_n)|^2 \right)^{\frac{1}{2}} \left(\sum_{m=1}^{\infty} |(f_n, h_m)|^2 \right)^{\frac{1}{2}}
$$

by the Cauchy-Schwarz inequality.

Note that $\sum_{m=1}^{\infty} |(g_m, e_n)|^2$ and $\sum_{m=1}^{\infty} |(f_n, h_m)|^2$ are at most 1, since $\{g_m\}$ and $\{h_m\}$ are orthonormal and $\|e_n\| = \|f_n\| = 1$. Hence

$$
\sum_{m=1}^{\infty} |(Kg_m, h_m)| \leq \sum_{n=1}^{\infty} s_n = \|K\|_1. \qquad \square
$$

We can now show that the trace can be computed with respect to any orthonormal basis.

Theorem 6.5.11. *If K is a trace-class operator with singular value decomposition given by $Kf = \sum_{n=1}^{\infty} s_n (f, e_n) f_n$, and if $\{g_m\}$ is any orthonormal basis, then*

$$
\sum_{m=1}^{\infty} (Kg_m, g_m) = \sum_{n=1}^{\infty} s_n (f_n, e_n).
$$

Proof. The previous lemma implies that $\sum_{m=1}^{\infty} (Kg_m, g_m)$ converges absolutely, so its terms can be re-arranged without changing the sum. Thus

$$
\sum_{m=1}^{\infty} (Kg_m, g_m) = \sum_{m=1}^{\infty} \sum_{n=1}^{\infty} s_n (g_m, e_n)(f_n, g_m)
$$

$$
= \sum_{n=1}^{\infty} s_n \sum_{m=1}^{\infty} (g_m, e_n)(f_n, g_m)
$$

$$
= \sum_{n=1}^{\infty} s_n \sum_{m=1}^{\infty} \overline{(e_n, g_m)}(f_n, g_m)
$$

$$
= \sum_{n=1}^{\infty} s_n (f_n, e_n),
$$

since $\{g_m\}$ is an orthonormal basis.   □

Definition 6.5.12. If K is a trace-class operator, then the *trace of K*, denoted by $\operatorname{tr}(K)$, is defined by

$$\operatorname{tr}(K) \;=\; \sum_{m=1}^{\infty} (K g_m, g_m),$$

where $\{g_m\}$ is any orthonormal basis.

Corollary 6.5.13. *For K in the trace class, $\operatorname{tr}(K)$ is well-defined and $|\operatorname{tr}(K)| \le \|K\|_1$.*

Proof. This follows immediately from Lemma 6.5.10 and Theorem 6.5.11.   □

Theorem 6.5.14. *The trace norm, defined on the trace-class operators as the sum of the singular values, is a norm.*

Proof. The only property that is not very straightforward is the triangle inequality. For this, suppose that K_1 and K_2 are in the trace class, and that $K_1 + K_2$ has singular value decomposition given by $(K_1 + K_2)f = \sum_{n=1}^{\infty} s_n(f, e_n) f_n$. Note that

$$((K_1 + K_2)e_n, f_n) = (s_n(e_n, e_n) f_n, f_n) = s_n.$$

Thus

$$\|K_1 + K_2\|_1 \;=\; \sum_{n=1}^{\infty} s_n$$

$$= \sum_{n=1}^{\infty} ((K_1 + K_2)e_n, f_n)$$

$$= \sum_{n=1}^{\infty} |(K_1 e_n, f_n) + (K_2 e_n, f_n)|$$

$$\le \sum_{n=1}^{\infty} |(K_1 e_n, f_n)| + \sum_{n=1}^{\infty} |(K_2 e_n, f_n)|.$$

Now, by Lemma 6.5.10,

$$\sum_{n=1}^{\infty} |(K_1 e_n, f_n)| \;\le\; \|K_1\|_1$$

and

$$\sum_{n=1}^{\infty} |(K_2 e_n, f_n)| \;\le\; \|K_2\|_1,$$

so

$$\|K_1 + K_2\|_1 \;\leq\; \|K_1\|_1 + \|K_2\|_1. \qquad\qquad \square$$

There is an interesting characterization of singular values that is very useful. Recall that $s_1 \geq s_2 \geq s_3 \cdots$.

Lemma 6.5.15. *If K is a compact operator and s_n is the n-th singular value of K, then*

$$s_n \;=\; \inf\{\|K - F\| : \; F \text{ is an operator of rank at most } \; n-1\}.$$

Proof. An operator of rank $n-1$ can be obtained within s_n of K from the singular decomposition: If $Kf = \sum_{j=1}^{\infty} s_j(f, e_j)f_j$, define F by $Ff = \sum_{j=1}^{n-1} s_j(f, e_j)f_j$. Then $s_n = \|K - F\|$.

To finish the proof it must be shown that $\|K - G\| \geq s_n$ whenever the rank of G is at most $n-1$. To see this, choose any vector f of norm 1 in the span of $\{e_1, e_2, \ldots, e_n\}$ such that $Gf = 0$ (such a vector exists, since the rank of G is less than n). Then

$$\|K - G\| \;\geq\; \|(K - G)f\| \;=\; \|Kf\|$$

$$= \|\sum_{j=1}^{n} s_j(f, e_j)f_j\|$$

$$= \left(\sum_{j=1}^{n} s_j^2 |(f, e_j)|^2\right)^{\frac{1}{2}}$$

$$\geq \left(\sum_{j=1}^{n} s_n^2 |(f, e_j)|^2\right)^{\frac{1}{2}} \;=\; s_n. \qquad \square$$

Theorem 6.5.16. *If K is a trace-class operator and A is bounded, then*

$$\|KA\|_1 \;\leq\; \|A\| \, \|K\|_1 \quad \text{and} \quad \|AK\|_1 \;\leq\; \|A\| \|K\|_1.$$

Proof. We actually prove a stronger statement: Let $\{s_n\}, \{t_n\}$, and $\{u_n\}$ be, respectively, the singular values of K, KA, and AK. We show that $t_n \leq \|A\|s_n$ and $u_n \leq \|A\|s_n$ for every n (then summing over n establishes the theorem). For this, fix n and choose F as in the first part of the proof of Lemma 6.5.15, so that $s_n = \|K - F\|$. Then, by Lemma 6.5.15,

$$t_n \;\leq\; \|KA - FA\| \;\leq\; \|K - F\| \, \|A\| \;=\; \|A\|s_n,$$

and

$$u_n \;\leq\; \|AK - AF\| \;\leq\; \|A\| \, \|K - F\| \;=\; \|A\|s_n. \qquad \square$$

Corollary 6.5.17. *The set of trace-class operators is a two-sided ideal of* $\mathcal{B}(\mathcal{H})$.

Proof. This is immediate from the above theorem.   □

The characteristic property of the trace as a linear functional is that the trace of AB is the same as the trace of BA; i.e., the trace of every commutator is 0. In the present context, that can be shown as follows.

Theorem 6.5.18. *If A and B are bounded operators such that AB and BA are both in the trace class, then* $\mathrm{tr}(AB) = \mathrm{tr}(BA)$.

Proof. Fix any orthonormal basis $\{g_m\}_{m=1}^{\infty}$. Then

$$\mathrm{tr}(BA) = \sum_{m=1}^{\infty}(BAg_m, g_m) = \sum_{m=1}^{\infty}(Ag_m, B^*g_m).$$

Now,

$$(Ag_m, B^*g_m) = \sum_{n=1}^{\infty}(Ag_m, g_n)\overline{(B^*g_m, g_n)},$$

so

$$\mathrm{tr}(BA) = \sum_{m=1}^{\infty}\sum_{n=1}^{\infty}(Ag_m, g_n)\overline{(B^*g_m, g_n)}$$

$$= \sum_{m=1}^{\infty}\sum_{n=1}^{\infty}(g_m, A^*g_n)\overline{(g_m, Bg_n)}$$

$$= \sum_{m=1}^{\infty}\sum_{n=1}^{\infty}(Bg_n, g_m)\overline{(A^*g_n, g_m)}$$

$$= \sum_{n=1}^{\infty}\left(\sum_{m=1}^{\infty}(Bg_n, g_m)\overline{(A^*g_n, g_m)}\right)$$

$$= \sum_{n=1}^{\infty}(Bg_n, A^*g_n) = \sum_{n=1}^{\infty}(ABg_n, g_n) = \mathrm{tr}(AB).\quad\square$$

6.6 Notes and Remarks

The name "Banach space" seems to be entirely justified by the history of the subject. Although many of the basic concepts were developed earlier in the context of spaces of functions and although some of the abstract concepts preceded him, Banach proved a number of the fundamental theorems and gave the first systematic exposition (Banach [1]); see Dunford-Schwartz [1, pp. 79 - 94] for a brief history. The concept of compactness of an operator was apparently first introduced by Hilbert [1]. Much of the basic theory is due to F. Riesz ([1], [2]). The Fredholm alternative was proven for integral operators in Fredholm [1].

The results of Section 6.3 on existence of invariant subspaces have an interesting history. In 1949, von Neumann proved (but did not publish) the theorem that compact operators on Hilbert space have nontrivial invariant subspaces. A few years later, Aronszajn-Smith [1] rediscovered this result and extended it to Banach spaces. Their proof uses a very interesting technique of approximation by finite-rank operators (see also Davidson [1, Theorem 3.1], Radjavi-Rosenthal [2, Theorem 5.3]). It took another dozen years and the use of nonstandard analysis to extend the proof to apply to polynomially compact operators (i.e., operators A such that $p(A)$ is compact for some polynomial p); see Bernstein-Robinson [1]. Several incremental improvements were subsequently made, but the extension by Lomonosov [1] to Theorem 6.3.4 (and, even more surprisingly, to Corollary 7.4.12) took the operator theory community by storm. Before Lomonosov [1] appeared in print, H.M. Hilden, of the University of Hawaii, discovered the extraordinarily simple proof of Theorem 6.3.4 presented in the text. News of this proof circulated so quickly that Hilden had no need to publish it. (See Radjavi-Rosenthal [4] for additional historical remarks.)

The beautiful decomposition of compact operators presented in Section 6.4 is due to Riesz [1]. Riesz [2] also developed an analytic functional calculus that applies to all bounded operators and that provides a decomposition whenever the spectrum of an operator is disconnected; see Dunford-Schwartz [1, pp. 566 - 577], Radjavi-Rosenthal [2, Chapter 2], Conway [1, pp. 199 - 205] or any of a number of other texts.

Hilbert space is also aptly named: Much of the basic work, including the spectral theorem for self-adjoint operators, was done by Hilbert [1]. The spectral theorem for compact normal operators (Theorem 6.5.6) has a generalization to bounded (and even just closed) normal operators. Although noncompact normal operators need not have any eigenvalues, the spectral theorem expresses normal operators as spectral integrals with respect to certain projections (the spectral projections) onto invariant subspaces; see Halmos [2], Dunford-Schwartz [2, pp. 895 - 920], Conway [1, Chap. 9], Rudin [2, Chap. 12] or Radjavi-Rosenthal [2, Chap. 1]

We have included only the most fundamental results on s-numbers, the trace, and the associated collections of compact operators. A very nice and somewhat more complete discussion is given in Davidson [1], and a thorough treatment can be found in Ringrose [5]. Other excellent expositions, which contain much additional material, can be found in Gohberg-Krein [1, Chaps. 2 and 3] and Simon [1].

CHAPTER 7
Algebras of Compact Operators

In this chapter we show that most of the results of Chapter 1 generalize from finite dimensions to algebras of compact operators on infinite-dimensional Banach spaces. The definition of triangularizability is rephrased to apply to both the finite- and infinite-dimensional cases.

Ringrose's Theorem (7.2.3) shows how the spectrum of a compact operator can be computed from any generalized triangular form of the operator, and yields a spectral mapping theorem (Theorem 7.2.6) for triangularizable families of compact operators.

Lomonosov's Lemma (7.3.1) provides infinite-dimensional analogues of McCoy's Theorem (Theorem 1.3.4) and Burnside's Theorem (Theorem 1.2.2). Ringrose's and Lomonosov's Theorems lead to satisfactory generalizations of the finite-dimensional theorems relating triangularizability to commutativity modulo the radical (Corollary 7.3.8) and to approximate commutativity (Theorems 7.6.2 and 7.6.4). A block triangularization theorem (Theorem 7.5.9) yields a number of sufficient conditions for triangularizability of algebras of compact operators.

7.1 The Definition of Triangularizability

To say that a linear transformation T has an upper triangular matrix with respect to a basis $\{e_1, e_2, \ldots, e_n\}$ of a finite-dimensional space is equivalent to saying that the subspaces $\mathcal{M}_j = \bigvee_{i=1}^{j}\{e_i\}$ are invariant under T for $j = 1, 2, \ldots, n-1$. If $\mathcal{M}_0 = \{0\}$ and $\mathcal{M}_n = \bigvee_{i=1}^{n}\{e_i\}$ (i.e., the entire space), then the collection $\{\mathcal{M}_j\}_{j=0}^{n}$ is a chain (i.e., totally ordered collection under inclusion) of subspaces that is maximal in the sense that there is no chain of subspaces that properly includes $\{\mathcal{M}_j\}_{j=0}^{n}$. (To see this, suppose that a chain contained an additional subspace, say $\mathcal{M}$. Let $\mathcal{M}_{j_0}$ be the largest $\mathcal{M}_j$ contained in $\mathcal{M}$. Then, since $\mathcal{M}$ is comparable with all the $\mathcal{M}_j$, $\mathcal{M} \subseteq \mathcal{M}_{j_0+1}$. Since $\mathcal{M}_{j_0} \subseteq \mathcal{M} \subseteq \mathcal{M}_{j_0+1}$ and the dimension of $\mathcal{M}_{j_0+1}$ is 1 more than that of $\mathcal{M}_{j_0}$, it follows that either $\mathcal{M} = \mathcal{M}_{j_0}$ or $\mathcal{M} = \mathcal{M}_{j_0+1}$.)

Conversely, every maximal chain of subspaces of a finite-dimensional space has the above form. For if $\mathcal{C}$ is such a chain, then $\{0\} \in \mathcal{C}$ (or else it could be added). By finite dimensionality, the collection of elements of $\mathcal{C}$ other than $\{0\}$ has a smallest element; call it $\mathcal{M}_1$. Then $\mathcal{M}_1$ must be one-dimensional, or else there would be subspaces between $\{0\}$ and $\mathcal{M}_1$ that could be added to $\mathcal{C}$. Similarly, there is a smallest element of $\mathcal{C}$ greater than $\mathcal{M}_1$; call it $\mathcal{M}_2$. Then $\mathcal{M}_2$ is two-dimensional. If the dimension of the space is n, this process can obviously be continued to produce elements $\{\mathcal{M}_j\}_{j=1}^{n}$ of $\mathcal{C}$ with the dimension of $\mathcal{M}_j$ equal to j. If e_1 is any nonzero vector in $\mathcal{M}_1$ and, for $j = 2, \ldots, n$, e_j is a vector in $\mathcal{M}_j$ that is not in $\mathcal{M}_{j-1}$, and if $\mathcal{M}_0 = \{0\}$, then $\mathcal{C} = \{\mathcal{M}_j\}_{j=0}^{n}$.

By the above discussion, the following definition reduces to the previous one (Definition 1.1.2) in the case where $\mathcal{X}$ is finite-dimensional.

Definition 7.1.1. A family $\mathcal{F}$ of bounded linear operators on a Banach space $\mathcal{X}$ is *triangularizable* if there is a chain $\mathcal{C}$ that is maximal as a chain of subspaces of $\mathcal{X}$ and that has the property that every subspace in $\mathcal{C}$ is invariant under all the operators in $\mathcal{F}$. Any such chain of subspaces is said to be a *triangularizing chain* for the family $\mathcal{F}$.

Note that the position of the quantifier is very important in the definition of triangularizability. A trivial Zorn's Lemma argument shows that for every family $\mathcal{F}$ of operators there is a chain of subspaces (perhaps just $\{\{0\}, \mathcal{X}\}$) that is maximal as a chain of common invariant subspaces of the operators in $\mathcal{F}$.

Some of the examples of triangularizable families of operators on infinite-dimensional spaces are similar to those on finite-dimensional spaces.

Example 7.1.2. *Let $\{e_n\}_{n=1}^{\infty}$ be an orthonormal basis for a Hilbert space $\mathcal{H}$ and let $\mathcal{M}_j = \bigvee_{i=1}^{j} \{e_i\}$ for every j. Then every subset of* $\mathrm{Alg}\left(\{\mathcal{M}_j\}_{j=1}^{\infty}\right)$ *(see 6.3.8) is triangularizable.*

Proof. If $\mathcal{M}_0 = \{0\}$, then it can easily be seen (following the lines of the discussion preceding Definition 7.1.1 of the finite-dimensional case) that $\{\mathcal{M}_j\}_{j=0}^{\infty} \cup \{\mathcal{H}\}$ is a maximal chain of subspaces of $\mathcal{H}$. □

With $\mathcal{M}_j$ as in the above example, $\mathrm{Alg}\left(\{\mathcal{M}_j\}_{j=1}^{\infty}\right)$ is the algebra of *upper triangular operators with respect to the basis* $\{e_j\}_{j=1}^{\infty}$. This algebra is quite analogous to the algebra of upper triangular matrices with respect to a given basis in the finite-dimensional case. There are, however, triangularizable families that look quite different.

Example 7.1.3. *Fix p with $1 \leq p < \infty$, and let $\mathcal{L}^p(0,1)$ denote the Banach space consisting of all (equivalence classes modulo sets of measure 0 of) complex-valued measurable functions on the closed interval $[0,1]$ such that $\int_0^1 |f(x)|^p dx$ is finite. For $f \in \mathcal{L}^p(0,1)$, $\|f\|$ is defined as $\left(\int_0^1 |f(x)|^p dx\right)^{\frac{1}{p}}$.*

Define a chain of subspaces as follows. For each $t \in [0,1]$, let $\mathcal{M}_t = \{f \in \mathcal{L}^p(0,1) : f = 0 \text{ a.e. on } [0,t]\}$. Then $\{\mathcal{M}_t\}_{t \in [0,1]}$ is a maximal chain of subspaces (so every subset of $\mathrm{Alg}\left(\{\mathcal{M}_t\}_{t \in [0,1]}\right)$ is triangularizable). (Note that the ordering on $\{\mathcal{M}_t\}$ is opposite to the usual ordering of t in $[0,1]$).

Proof. To see that $\{\mathcal{M}_t\}_{t\in[0,1]}$ is maximal, note first that $\mathcal{M}_0 = \mathcal{L}^p(0,1)$ and $\mathcal{M}_1 = \{0\}$.

Now suppose that $\mathcal{M}$ is a subspace of $\mathcal{L}^p(0,1)$ that is comparable with every $\mathcal{M}_t$. Let $t_0 = \inf\{t : \mathcal{M}_t \subseteq \mathcal{M}\}$. We claim that $\mathcal{M} = \mathcal{M}_{t_0}$. To show that $\mathcal{M}_{t_0} \subseteq \mathcal{M}$, suppose that $f \in \mathcal{M}_{t_0}$. For each positive integer n, define f_n by

$$f_n(x) = \begin{cases} f(x) & \text{for } x \geq t_0 + \frac{1}{n}, \\ 0 & \text{for } 0 \leq x \leq t_0 + \frac{1}{n}. \end{cases}$$

Then $\{f_n\} \to f$ in $\mathcal{L}^p(0,1)$, and each f_n is in $\mathcal{M}_{t_0+\frac{1}{n}}$, which is contained in $\mathcal{M}$, so $f \in \mathcal{M}$.

Let $t_1 = \sup\{t : \mathcal{M}_t \supseteq \mathcal{M}\}$. It is easily shown (using an argument similar to the above) that $\mathcal{M}_{t_1} \supseteq \mathcal{M}$. If $\mathcal{M}_{t_0} \neq \mathcal{M}_{t_1}$ there would be a t_2 such that $\mathcal{M}_{t_1} \subseteq \mathcal{M}_{t_2} \subseteq \mathcal{M}_{t_0}$. But then $\mathcal{M}_{t_2}$ would not be comparable with $\mathcal{M}$ (if $\mathcal{M}_{t_2} \supseteq \mathcal{M}$, then $t_2 \leq t_1$, and if $\mathcal{M}_{t_2} \subseteq \mathcal{M}$, then $t_2 \geq t_0$). Hence $\mathcal{M}_{t_0} = \mathcal{M}_{t_1} = \mathcal{M}$. ☐

Definition 7.1.4. A *Volterra-type integral operator* on $\mathcal{L}^p(0,1)$ is an operator K of the form

$$(Kf)(x) = \int_0^x k(x,y)f(y)dy,$$

where k (the *kernel* of the operator) is a fixed bounded measurable function on $[0,1] \times [0,1]$. It can be shown that every Volterra-type integral operator is compact (see, for example, Conway [1, p. 43] for a nice proof in the case where $p = 2$). The *Volterra operator* V on $\mathcal{L}^p(0,1)$ is the Volterra-type operator whose kernel is identically 1 (i.e., $(Vf)(x) = \int_0^x f(y)dy$). A *multiplication operator* on $\mathcal{L}^p(0,1)$ is an operator A of the form

$$(Af)(x) = \phi(x)f(x),$$

where ϕ is a bounded measurable function on $[0,1]$.

Note that all multiplication operators and all Volterra-type integral operators are in the algebra $\mathrm{Alg}\left(\{\mathcal{M}_t\}_{t\in[0,1]}\right)$ of Example 7.1.3, and hence are simultaneously triangularizable.

Definition 7.1.5. A chain of subspaces is *complete* if it is closed under arbitrary intersections and spans.

If an element of a complete chain is not the span of its predecessors, then its immediate predecessor can be defined as follows.

Definition 7.1.6. If C is a chain of subspaces and $\mathcal{M} \in C$, then $\mathcal{M}_-$ is defined by

$$\mathcal{M}_- = \bigvee \{\mathcal{N} \in C : \mathcal{N} \subseteq \mathcal{M} , \mathcal{N} \neq \mathcal{M}\} .$$

If $\mathcal{M}_- \neq \mathcal{M}$, then $\mathcal{M}_-$ is the *predecessor* of $\mathcal{M}$ in C.

Definition 7.1.7. The subspace chain C is *continuous* if $\mathcal{M}_- = \mathcal{M}$ for all $\mathcal{M} \in C$ and is *discrete* if $\mathcal{M} \in C$ and $\mathcal{M} \neq \{0\}$ implies $\mathcal{M}_- \neq \mathcal{M}$.

Note that the chain of Example 7.1.2 is discrete and that of Example 7.1.3 is continuous. There are, of course, chains that are neither discrete nor continuous. One way of constructing such chains is by patching together examples like 7.1.2 and 7.1.3 using direct sums.

Example 7.1.8. *Let the Hilbert space $\mathcal{H}$ have an orthonormal basis $\{e_n\}_{n=1}^{\infty}$ and let the Hilbert space $\mathcal{K}$ be the direct sum of $\mathcal{H}$ and $\mathcal{L}^2(0,1)$. The collection*

$$\{\mathcal{M}_n \oplus \{0\} : n = 1, 2, 3, \dots\} \cup \{\mathcal{H} \oplus \mathcal{M}_t : t \in [0,1]\}$$

is a maximal subspace chain, where the $\{\mathcal{M}_n\}$ are defined as in Example 7.1.2 and the $\{\mathcal{M}_t\}$ are defined as in Example 7.1.3.

Proof. The maximality of the chain of subspaces is an immediate consequence of the maximality of the chains of Examples 7.1.2 and 7.1.3.   □

The following characterization of maximal chains of subspaces is useful.

Theorem 7.1.9. *A chain of subspaces of a Banach space is maximal as a subspace chain if and only if it satisfies the following three conditions:*

(i) *it contains $\{0\}$ and $\mathcal{X}$,*
(ii) *it is complete in the sense that it is closed under arbitrary intersections and spans, and*
(iii) *if $\mathcal{M}$ is in the chain and $\mathcal{M}_- \neq \mathcal{M}$, then the quotient space $\mathcal{M}/\mathcal{M}_-$ is one-dimensional.*

Proof. That a maximal subspace chain satisfies the three conditions is obvious: Intersections and spans of members of a chain are comparable with every subspace in the chain, and if the dimension of $\mathcal{M}/\mathcal{M}_-$ is greater than 1, there are subspaces $\mathcal{N}$ between $\mathcal{M}_-$ and $\mathcal{M}$ that are not in the chain although they are comparable with every member of the chain.

For the converse, let C be a chain satisfying the conditions and let $\mathcal{M}$ be comparable with every subspace in C. The proof that $\mathcal{M}$ is in C is similar to the proofs of the special cases in Examples 7.1.2 and 7.1.3 above.

Define $\mathcal{M}_0 = \bigvee \{\mathcal{N} \in \mathcal{C} : \mathcal{N} \subseteq \mathcal{M}\}$ and $\mathcal{M}_1 = \cap \{\mathcal{N} \in \mathcal{C} : \mathcal{N} \supseteq \mathcal{M}\}$.
Then $\mathcal{M}_0$ and $\mathcal{M}_1$ are in $\mathcal{C}$, so if $\mathcal{M} \notin \mathcal{C}$ we would have $\mathcal{M}_0 \subseteq \mathcal{M} \subseteq \mathcal{M}_1$
with both inclusions proper. Thus the dimension of $\mathcal{M}_1/\mathcal{M}_0$ would be
greater than 1. This is impossible, for if $\mathcal{N} \in \mathcal{C}$ and $\mathcal{N}$ is a proper subset of
$\mathcal{M}_1$, then $\mathcal{N}$ would have to be contained in $\mathcal{M}_0$. Then the immediate predecessor of $\mathcal{M}_1$ in $\mathcal{C}$, $\mathcal{M}_{1-}$, would be contained in $\mathcal{M}_0$, and the dimension
of $\mathcal{M}_1/\mathcal{M}_{1-}$ would be greater than 1, contradicting (iii). $\qquad \square$

We require an infinite-dimensional triangularization lemma analogous to
the finite-dimensional one.

Definition 7.1.10. A property of families of operators on a Banach space
is said to be *inherited by quotients* if for each family $\mathcal{F}$ having the property,
and for every distinct pair $\mathcal{M}$, $\mathcal{N}$ in Lat $\mathcal{F}$ with $\mathcal{N} \subseteq \mathcal{M}$, the family $\hat{\mathcal{F}}$ also
has the property, where $\hat{\mathcal{F}}$ is the set of all quotient operators $\hat{A}$ on $\mathcal{M}/\mathcal{N}$
for $A \in \mathcal{F}$.

Lemma 7.1.11. (The Triangularization Lemma) *If $\mathcal{P}$ is a property of
families of operators that is inherited by quotients, and if every family satisfying $\mathcal{P}$ has a nontrivial invariant subspace, then every family satisfying
$\mathcal{P}$ is triangularizable.*

Proof. Let $\mathcal{F}$ be a family satisfying $\mathcal{P}$, and let $\mathcal{C}$ be a chain of invariant
subspaces of $\mathcal{F}$ that is maximal as a chain in Lat $\mathcal{F}$ (such a chain exists by
Zorn's Lemma). We must show that $\mathcal{C}$ is maximal as a subspace chain.

We use Theorem 7.1.9. It is clear that $\mathcal{C}$ satisfies conditions (i) and (ii)
of 7.1.9.

Suppose that $\mathcal{C}$ did not satisfy (iii); then let $\mathcal{M} \in \mathcal{C}$ with the dimension
of $\mathcal{M}/\mathcal{M}_-$ greater than 1. Since $\mathcal{P}$ is inherited by quotients and implies
the existence of invariant subspaces, the family $\hat{\mathcal{F}}$ has a nontrivial invariant
subspace $\hat{\mathcal{L}}$ in $\mathcal{M}/\mathcal{M}_-$. Let

$$\mathcal{L} = \left\{ x \in \mathcal{X} : x + \mathcal{M}_- \in \hat{\mathcal{L}} \right\}.$$

Then $\mathcal{L}$ is an invariant subspace of $\mathcal{F}$ that is properly between $\mathcal{M}_-$ and
$\mathcal{M}$. This $\mathcal{L}$ could be added to $\mathcal{C}$, contradicting the fact that $\mathcal{C}$ was maximal
as a chain in Lat $\mathcal{F}$. $\qquad \square$

7.2 Spectra from Triangular Forms

It follows easily from the above that every compact operator on a Banach
space is triangularizable; more generally, this holds for commutative families of compact operators.

Theorem 7.2.1. *Every commutative set of compact operators is triangularizable.*

Proof. Every quotient operator obtained from a compact operator is compact (see, for example, Ringrose [5, p. 51]), so the property of being a commutative family of compact operators is inherited by quotients. We have seen (Corollary 6.3.7) that commutative families of compact operators have nontrivial invariant subspaces. Hence the Triangularization Lemma (Lemma 7.1.11) applies. $\square$

We will see that for families of compact operators, as in the finite-dimensional case, triangularizability has spectral implications similar to those for commutativity. A crucial foundation of these results is Ringrose's Theorem (Theorem 7.2.3 below) describing the spectrum in terms of any triangularizing chain. We begin with a definition.

Definition 7.2.2. Let $\mathcal{C}$ be any triangularizing chain for a compact operator K. For each $\mathcal{M} \in \mathcal{C}$ the *diagonal coefficient of K corresponding to $\mathcal{M}$*, denoted by $\lambda_{\mathcal{M}})[\lambda_{\mathcal{M}}]$, is defined as follows: If $\mathcal{M}_- = \mathcal{M}$, then $\lambda_{\mathcal{M}} = 0$; if $\mathcal{M}_- \neq \mathcal{M}$, then $\lambda_{\mathcal{M}}$ is the (necessarily unique) complex number such that $(K - \lambda_{\mathcal{M}} I)\mathcal{M} \subseteq \mathcal{M}_-$. (Note that $\lambda_{\mathcal{M}}$ is the number in the spectrum of the operator $\hat{K}$ on the one-dimensional space $\mathcal{M}/\mathcal{M}_-$.)

Theorem 7.2.3. (Ringrose's Theorem) *If K is a compact operator on an infinite-dimensional Banach space and $\mathcal{C}$ is any triangularizing chain for K, then*

$$\sigma(K) = \{0\} \cup \{\lambda_{\mathcal{M}} : \mathcal{M} \in \mathcal{C}\}.$$

Proof. Since K is a compact operator on an infinite-dimensional space, $0 \in \sigma(K)$. Also, if $\lambda_{\mathcal{M}} \neq 0$, then $(K - \lambda_{\mathcal{M}})\mathcal{M} \subseteq \mathcal{M}_-$. This implies that $\lambda_{\mathcal{M}}$ is in the spectrum of the restriction of K to $\mathcal{M}$, and hence is an eigenvalue of this restriction (Theorem 6.2.8) and therefore also of K. Thus $\{0\} \cup \{\lambda_{\mathcal{M}} : \mathcal{M} \in \mathcal{C}\}$ is contained in $\sigma(K)$.

For the other inclusion, suppose that λ is a nonzero number in $\sigma(K)$. An $\mathcal{M} \in \mathcal{C}$ such that $\lambda = \lambda_{\mathcal{M}}$ can be explicitly defined as follows: Let

$$\mathcal{E} = \{f \in \mathcal{X} : Kf = \lambda f \quad \text{and} \quad \|f\| = 1\},$$

and define

$$\mathcal{M} = \cap\{\mathcal{N} \in \mathcal{C} : \mathcal{N} \cap \mathcal{E} \neq \emptyset\}.$$

Each $\mathcal{N} \cap \mathcal{E}$ is a closed and bounded subset of a finite-dimensional space, since the eigenspace is finite-dimensional, and is hence compact. Since $\mathcal{C}$ is a chain, the intersection of a finite number of $\mathcal{N} \cap \mathcal{E}$ satisfying $\mathcal{N} \cap \mathcal{E} \neq \emptyset$ is just the smallest one, and hence is nonempty. Hence $\mathcal{M} \cap \mathcal{E}$ is nonempty, by the finite-intersection property.

We must show that $\mathcal{M}_- \neq \mathcal{M}$. If $\mathcal{L}$ is a proper subspace of $\mathcal{M}$ and $\mathcal{L}$ is invariant under K, then $\mathcal{L} \cap \mathcal{E} = \emptyset$ implies $\bigvee_{n=1}^{\infty} \mathrm{null} \left((K - \lambda)|_{\mathcal{L}} \right)^n = \{0\}$. Hence, by Corollary 6.4.13, $\mathcal{L} \subseteq \bigcap_{n=1}^{\infty} \mathrm{ran} \left((K - \lambda) \right)^n$. Thus $\mathcal{M}_- \subseteq \bigcap_{n=1}^{\infty} \mathrm{ran}(K - \lambda)^n$. Since $\mathcal{M}$ contains an eigenvector of K corresponding to λ, $\mathcal{M}$ is not contained in $\bigcap_{n=1}^{\infty} \mathrm{ran}(K - \lambda)^n$. Thus $\mathcal{M}_- \neq \mathcal{M}$.

Now choose $f \in \mathcal{M}$ such that $\|f\| = 1$ and $Kf = \lambda f$. If $\hat{K}$ is the quotient operator on $\mathcal{M}/\mathcal{M}_-$, then

$$\hat{K}[f + \mathcal{M}_-] = [Kf + \mathcal{M}_-]$$
$$= [\lambda f + \mathcal{M}_-]$$
$$= \lambda [f + \mathcal{M}_-] .$$

But $\hat{K}[f + \mathcal{M}_-] = \lambda_{\mathcal{M}}[f + \mathcal{M}_-]$, by definition. Thus $\lambda = \lambda_{\mathcal{M}}$. $\quad\square$

In addition to playing a crucial role in most triangularization problems for compact operators, Ringrose's Theorem provides immediate proofs that certain compact operators are quasinilpotent.

Corollary 7.2.4. *A compact operator is quasinilpotent if it has a continuous chain of invariant subspaces.*

Proof. The diagonal coefficients computed with respect to a continuous chain are all 0, so the corollary follows directly from the theorem. $\quad\square$

Example 7.2.5. *Every Volterra-type integral operator is quasinilpotent.*

Proof. As indicated in the discussion after Definition 7.1.4, every Volterra-type integral operator has a continuous chain of invariant subspaces, so Corollary 7.2.4 applies. $\quad\square$

The consequence of Ringrose's Theorem that relates to simultaneous triangularization is the following.

Theorem 7.2.6. (Spectral Mapping Theorem for Families of Compact Operators) *If $\{K_1, \ldots, K_n\}$ is a triangularizable collection of compact operators and p is any polynomial in n (possibly noncommuting) variables, then*

$$\sigma \left(p(K_1, K_2, \ldots, K_n) \right) \subseteq p \left(\sigma(K_1), \sigma(K_2), \ldots, \sigma(K_n) \right)$$

(where the right-hand side is defined as

$$\{p(\lambda_1, \lambda_2 \ldots, \lambda_n) : \lambda_j \in \sigma(K_j) \text{ for all } j\}) \; .$$

Proof. It p has a constant term, it can be subtracted from both sides of the inclusion; thus we can assume that the constant term is 0. Then $p(K_1, K_2, \ldots, K_n)$ is a compact operator. Fix a triangularizing chain for the collection. If $\lambda_{\mathcal{M}}$ is a nonzero diagonal coefficient of $p(K_1, K_2, \ldots, K_n)$, then $p(\hat{K}_1, \hat{K}_2, \ldots, \hat{K}_n)[f] = \lambda_{\mathcal{M}}[f]$ for all $[f]$ in $\mathcal{M}/\mathcal{M}_-$. For each j there is a λ_j such that $\hat{K}_j[f] = \lambda_j[f]$ for $[f]$ in $\mathcal{M}/\mathcal{M}_-$. Clearly, $p(\lambda_1, \lambda_2, \ldots, \lambda_n) = \lambda_{\mathcal{M}}$. Also, for each j, λ_j is a diagonal coefficient of K_j, so $\lambda_j \in \sigma(K_j)$. Thus $p(\lambda_1, \lambda_2, \ldots, \lambda_n) = \lambda_{\mathcal{M}}$ implies that

$$\lambda_{\mathcal{M}} \in p\left(\sigma(K_1), \sigma(K_2), \ldots, \sigma(K_n)\right).$$

$\square$

We will see (Theorem 7.3.6) that the converse of this spectral mapping theorem holds: If

$$\sigma\left(p(K_1, K_2, \ldots, K_n)\right) \subseteq p\left(\sigma(K_1), \sigma(K_2), \ldots, \sigma(K_n)\right)$$

for all (noncommutative) polynomials p, then $\{K_1, K_2, \ldots, K_n\}$ is triangularizable.

It is sometimes useful to have a form of Ringrose's Theorem (7.2.3) that applies to "block triangularizations" (i.e., triangularizations where the dimensions of the $\mathcal{M}/\mathcal{M}_-$ may be greater than 1).

Theorem 7.2.7. *Let K be a compact operator on an infinite-dimensional space and let $\mathcal{C}$ be a complete chain of invariant subspaces of K. For each $\mathcal{M} \in \mathcal{C}$ for which $\mathcal{M}_- \neq \mathcal{M}$, define $K_{\mathcal{M}}$ to be the quotient operator on $\mathcal{M}/\mathcal{M}_-$ induced by K. Then*

$$\sigma(K) = \{0\} \cup \{\sigma(K_{\mathcal{M}}) : \mathcal{M} \in \mathcal{C} \text{ and } \mathcal{M}_- \neq \mathcal{M}\} \; .$$

Proof. If $\lambda \in \sigma(K_{\mathcal{M}})$ and $\lambda \neq 0$, then $(K - \lambda)\mathcal{M}$ is not equal to $\mathcal{M}$, so λ is an eigenvalue of $K\big|_{\mathcal{M}}$ and hence of K.

The reverse inclusion can be obtained from Ringrose's Theorem as follows. Suppose that the dimension of $\mathcal{M}/\mathcal{M}_-$ is greater than 1. Then $K_{\mathcal{M}}$ is triangularizable (by Theorem 7.2.1). We can "lift" the subspaces in a triangularizing chain from $\mathcal{M}/\mathcal{M}_-$ to $\mathcal{X}$ (by making $\mathcal{M}_- + \mathcal{L}$ correspond to the subspace $\mathcal{L}$ of $\mathcal{M}/\mathcal{M}_-$) to fill in the gap between $\mathcal{M}_-$ and $\mathcal{M}$. If $\mathcal{C}'$ is the chain of subspaces of $\mathcal{X}$ that is the union of all of the lifted chains and $\mathcal{C}$, then $\mathcal{C}'$ is a triangularizing chain for K. Suppose, then, that $\lambda \in \sigma(K)$ and $\lambda \neq 0$. By the ordinary version of Ringrose's Theorem (7.2.3), $\lambda = \lambda_{\mathcal{N}}$ for some $\mathcal{N} \in \mathcal{C}'$. If $\mathcal{N}$ is in the original chain $\mathcal{C}$, then $\lambda_{\mathcal{N}} \in \sigma(K_{\mathcal{N}})$ (by Theorem 7.2.3) and the proof is finished. Otherwise, let

$\mathcal{M} = \cap\{\mathcal{L} \in \mathcal{C} : \mathcal{L} \supset \mathcal{N}\}$ and let $\mathcal{M}_- = \bigvee\{\mathcal{K} \in \mathcal{C} : \mathcal{K}$ is properly contained in $\mathcal{M}\}$. Clearly, $\mathcal{N} \subseteq \mathcal{M}$. Also, $\mathcal{M}_- \subseteq \mathcal{N}_-$, since a subspace in $\mathcal{C}$ that is properly contained in $\mathcal{M}$ is also properly contained in $\mathcal{N}$. Thus $\lambda_\mathcal{N}$ is a diagonal coefficient of $K_\mathcal{M}$, and hence is in $\sigma(K_\mathcal{M})$. □

There is a sharpening of Ringrose's Theorem that is sometimes required: The nonzero diagonal coefficients have the same multiplicity as the corresponding eigenvalues, in the sense of the following definitions.

Definition 7.2.8. If K is a compact operator and λ is a nonzero eigenvalue of K, then the *algebraic multiplicity of* λ is the dimension of $\bigvee_{k=1}^{\infty} \mathrm{null}(K - \lambda)^k$ (which is the same as the rank of the corresponding Riesz projection; see Definition 6.4.14). If $\mathcal{C}$ is a triangularizing chain for K, then the *diagonal multiplicity of* λ *with respect to* $\mathcal{C}$ is the number of subspaces $\mathcal{M}$ in $\mathcal{C}$ such that $\lambda_\mathcal{M} = \lambda$.

Theorem 7.2.9. *The diagonal multiplicity of each nonzero eigenvalue with respect to any triangularizing chain of a compact operator is equal to its algebraic multiplicity.*

Proof. Let K be compact, λ a nonzero eigenvalue, and P the corresponding Riesz projection. Suppose that $\mathcal{C}$ is any triangularizing chain for K. Then, since $\mathrm{Lat}\,K \subset \mathrm{Lat}\,P$ (Corollary 6.4.15), $\mathcal{C}$ also triangularizes the finite-rank projection P. With respect to the decomposition $\mathcal{X} = \mathcal{N}_\lambda + \mathcal{R}_\lambda$ (see Corollary 6.4.12), $P|_{\mathcal{N}_\lambda}$ is the identity and $K|_{\mathcal{N}_\lambda}$ is unitarily equivalent to an upper-triangular matrix with all entries on the main diagonal equal to λ. Also, $P|_{\mathcal{R}_\lambda} = 0$, and $\lambda \notin \sigma\left(K|_{\mathcal{R}_\lambda}\right)$. For each $\mathcal{M} \in \mathcal{C}$, let $k_\mathcal{M}$ and $p_\mathcal{M}$ denote, respectively, the corresponding diagonal coefficients of K and P.

We show first that $k_\mathcal{M} = \lambda$ if and only if $p_\mathcal{M} = 1$. To see this, note that, for any λ,

$$\sigma(K + tP) = (\sigma(K) \setminus \{\lambda\}) \cup \{\lambda + t\}$$

for each real number t. Clearly, the diagonal coefficient of $K + tP$ corresponding to $\mathcal{M}$ is $k_\mathcal{M} + tp_\mathcal{M}$. Suppose that $k_\mathcal{M} = \lambda$. Then $\lambda + tp_\mathcal{M}$ is a diagonal coefficient and hence an eigenvalue of $K + tP$ (by Ringrose's Theorem 7.2.3). If $p_\mathcal{M} \neq 1$ and $t \neq 0$, then $\lambda + tp_\mathcal{M} \neq \lambda + t$, so $\lambda + tp_\mathcal{M}$ would have to be in $\sigma(K)$. Since $\sigma(K)$ is countable, we can choose a t such that $\lambda + tp_\mathcal{M}$ is not in $\sigma(K)$. This shows that $p_\mathcal{M}$ must be 1. Similarly, if $p_\mathcal{M} = 1$ and $k_\mathcal{M} \neq \lambda$, then $k_\mathcal{M} + t$ is in $\sigma(K)$. But t could be chosen such that $(k_\mathcal{M} + t) \notin \sigma(K)$, so $k_\mathcal{M}$ must be λ.

Thus the diagonal multiplicity of λ is equal to the diagonal multiplicity of 1 in the representation of the Riesz projection P. All that remains to be shown is that the latter is equal to the rank of P. For each $\mathcal{M}$ such that $p_{\mathcal{M}} = 1$, let $f_{\mathcal{M}}$ be a unit vector in $\mathcal{M}$ that is not in $\mathcal{M}_-$, and let $\phi_{\mathcal{M}}$ be a bounded linear functional on $\mathcal{X}$ that is 0 on $\mathcal{M}_-$ and takes the value 1 on $f_{\mathcal{M}}$. Define the operator $Q_{\mathcal{M}}$ by $Q_{\mathcal{M}}(f) = \phi_{\mathcal{M}}(f)f_{\mathcal{M}}$ for all $f \in \mathcal{X}$. Then $Q_{\mathcal{M}}$ is a rank-one operator that leaves every subspace in the given chain invariant. Moreover, the diagonal coefficient of $Q_{\mathcal{M}}$ corresponding to $\mathcal{M}$ is 1, and all other diagonal coefficients of $Q_{\mathcal{M}}$ are 0. Note that the trace of $Q_{\mathcal{M}}$ is 1, since it is a rank-one operator with 1 as an eigenvalue. Now let Q be the sum of all such $Q_{\mathcal{M}}$ over the $\mathcal{M}$ such that $p_{\mathcal{M}} = 1$. Then Q is also triangularized by $\mathcal{C}$, and $P - Q$ has all diagonal coefficients 0. Thus $P - Q$ has spectrum $\{0\}$, and, since it has finite rank, $P - Q$ is nilpotent. Hence the trace of $P - Q$ is 0, so the trace of P is equal to that of Q. But the trace of P is its rank (since P is a finite-rank projection), and the trace of Q is the number of $p_{\mathcal{M}}$ that are equal to 1, so the result follows.   □

For some applications, we need to know that spectrum is continuous on compact operators (as in the finite-dimensional case treated in Lemma 3.1.2).

Theorem 7.2.10. *If A is a bounded operator on a Banach space and λ is an isolated point in $\sigma(A)$, and if $\{\|A - A_n\|\} \to 0$, then for every $\varepsilon > 0$ there is an N such that $\sigma(A_n) \cap \{z : |z - \lambda| < \varepsilon\}$ is nonempty for $n \geq N$.*

Proof. We merely outline the proof; for a detailed exposition see the original paper of Newburgh [1] or the book of Aupetit [3, p. 51]. Let Γ be a circle of radius less than ε about λ that does not intersect $\sigma(A)$ and whose interior intersects $\sigma(A)$ in $\{\lambda\}$. A contour integral $P = \dfrac{1}{2\pi i}\displaystyle\int_{\Gamma}(z - A)^{-1}dz$ can be defined as a limit of Riemann sums, as in complex function theory (see, for example, Dunford-Schwartz [1, pp. 566–577], Conway [1, pp. 199–205], or Radjavi-Rosenthal [2, Chap.2]). It can be shown (see the references above) that P is an idempotent that commutes with A such that the spectrum of the restriction of A to the range of P is $\{\lambda\}$. (In the case where A is compact, P is the Riesz projection of Definition 6.4.14.) It is not hard to see that, if P_n is defined by $P_n = \dfrac{1}{2\pi i}\displaystyle\int_{\Gamma}(z - A_n)^{-1}dz$, then P_n is well-defined and close to P if n is large. Thus, in particular, there is an N such that $P_n \neq 0$ for $n \geq N$, from which it can be established that $\sigma(A_n)$ intersects the interior of Γ.   □

Corollary 7.2.11. *A norm limit of a sequence of compact quasinilpotent operators is quasinilpotent.*

Proof. If $\{\|K_n - K\|\} \to 0$ and K is not quasinilpotent, then, since the nonzero spectrum of a compact operator is isolated (Corollary 6.2.9), it would follow from the previous theorem that K_n is not quasinilpotent for n sufficiently large. □

Corollary 7.2.12. *If $\{K_n\}$ is a sequence of compact operators such that $\{\|K_n - K\|\} \to 0$, then $\{\rho(K_n)\} \to \rho(K)$.*

Proof. This follows from Theorem 7.2.10 together with the easily-established fact that spectrum is semicontinuous on all bounded operators in the sense that $\sigma(K_n)$ is eventually contained in any open set containing $\sigma(K)$ (see Halmos [3, Problem 103]). □

Theorem 7.2.13. *If $\{K_n\}$ is a sequence of compact operators such that $\{\|K_n - K\|\} \to 0$, then*

$$\sigma(K) = \left\{ \lambda : \lambda = \lim_{n \to \infty} \{\lambda_n\} \text{ where } \lambda_n \in \sigma(K_n) \text{ for all } n \right\}.$$

Proof. This follows by an argument as suggested for the previous corollary; for a detailed proof see Dunford-Schwartz [2, p. 1091]. □

There is a powerful theorem concerning trace-class operators that is required for an extension of the theorem on permutability of trace to infinite dimensions (Theorem 8.6.9). The definitions and basic results on traces are in Section 6.5.

Theorem 7.2.14(Lidskii's Theorem). *If K is a trace-class operator, then the trace of K is the sum of the eigenvalues of K counting multiplicity.*

Proof. This theorem is surprisingly hard to prove — it was not discovered until 1959; see Lidskii [1]. An improvement of Lidskii's proof is given in Gohberg-Krein [1, p. 101]. A beautiful proof using Ringrose's Theorem was discovered by Erdos [2] (see Davidson [1, p. 36] for an elegant treatment). Another nice proof using triangularization is due to Power [1]. □

Corollary 7.2.15. *The trace of a trace-class operator is the sum of its diagonal coefficients relative to any triangularizing chain.*

Proof. This is an immediate consequence of Lidskii's Theorem (7.2.14) and the version of Ringrose's Theorem that includes multiplicity (Theorem 7.2.9). □

7.3 Lomonosov's Lemma and McCoy's Theorem

Lomonosov's original proof of existence of hyperinvariant subspaces for compact operators (Theorem 6.3.4) was based on the following lemma; this lemma also has many applications to triangularization.

Lemma 7.3.1. (Lomonosov's Lemma) *If $\mathcal{J}$ is a transitive convex subset of $\mathcal{B}(\mathcal{X})$, and if K is any compact operator on $\mathcal{X}$ (not necessarily in $\mathcal{J}$) other than 0, then there exists an $A \in \mathcal{J}$ such that 1 is an eigenvalue of AK.*

Proof. We will construct nonnegative functions $\beta_1, \beta_2, \ldots, \beta_n$ that sum to 1 and operators $A_1, A_2, \ldots, A_n$ in $\mathcal{J}$ such that the function Φ defined by

$$\Phi(f) = \left(\sum_{j=1}^{n} \beta_j(Kf)A_j \right) Kf$$

has a fixed point f different from 0; then A can be taken to be $\sum_{j=1}^{n} \beta_j(Kf)A_j$.

The first part of the proof is the same as the first part of the proof of Lomonosov's Theorem (Theorem 6.3.4); i.e., assuming $\|K\| = 1$ and $\|Kf_0\| > 1$, let $\mathcal{S} = \{f : \|f - f_0\| < 1\}$. As in 6.3.4, there exists a subset $\{A_1, A_2, \ldots, A_n\}$ of $\mathcal{J}$ such that

$$\overline{K\mathcal{S}} \subseteq \bigcup_{i=1}^{n} A_i^{-1}(\mathcal{S}) .$$

We now construct the functions $\{\beta_j\}$. For $f \in \overline{K\mathcal{S}}$ and $j = 1, 2, \ldots, n$, define

$$\alpha_j(f) = \max \left\{ 0, 1 - \|A_j f - f_0\| \right\} .$$

Note that, for each $f \in \overline{K\mathcal{S}}$, there is at least one j such that $\|A_j f - f_0\| < 1$, so that $\alpha_j(f) > 0$. Thus we can define functions β_j from $\overline{K\mathcal{S}}$ into the nonnegative real numbers by

$$\beta_j(f) = \frac{\alpha_j(f)}{\sum\limits_{j=1}^{n} \alpha_j(f)}$$

for $j = 1, 2, \ldots, n$.

Note that each β_j is a continuous function mapping $\overline{K\mathcal{S}}$ into $[0, 1]$, and $\sum_{j=1}^{n} \beta_j(f) = 1$ for all $f \in \overline{K\mathcal{S}}$.

We now define the function Φ taking $\bar{\mathcal{S}}$ into $\mathcal{X}$ by

$$\Phi(f) = \sum_{j=1}^{n} \beta_j(Kf)A_j Kf .$$

(Note that Φ is defined on $\bar{\mathcal{S}}$, since $K\bar{\mathcal{S}} \subseteq \overline{K\mathcal{S}}$.) Clearly, Φ is continuous.

We must show that $\Phi(\bar{S}) \subseteq \bar{S}$. For this, fix any $f \in S$. Then

$$\|\Phi(f) - f_0\| = \left\| \sum_{j=1}^{n} \beta_j(Kf) A_j Kf - f_0 \right\|$$

$$= \left\| \sum_{j=1}^{n} \beta_j(Kf) A_j Kf - \sum_{j=1}^{n} \beta_j(Kf) f_0 \right\|$$

$$= \left\| \sum_{j=1}^{n} \beta_j(Kf) (A_j Kf - f_0) \right\|$$

$$\leq \sum_{j=1}^{n} \beta_j(Kf) \|A_j Kf - f_0\| .$$

Note that the definition of α_j implies that $\beta_j(Kf) = 0$ if $\|A_j Kf - f_0\| \geq 1$. Thus $\|\Phi(f) - f_0\| \leq \sum_{j=1}^{n} \beta_j(Kf) = 1$, so $\Phi(f) \in \bar{S}$.

For each j, the operator $A_j K$ is compact, and therefore $\overline{A_j KS}$ is compact. Thus $\bigcup_{j=1}^{n} \overline{A_j KS}$ is compact. We need Mazur's Theorem (see Dunford-Schwartz [1, p. 416]; Conway [1, p. 180]): The closed convex hull of a compact subset of a Banach space is compact. Let C denote the closed convex hull of $\bigcup_{j=1}^{n} \overline{A_j KS}$. Then C is compact, and therefore so is $\bar{S} \cap C$. Now, $\bar{S} \cap C$ is also convex, and $\bar{S} \cap C$ is not empty, since it contains $\Phi(S)$.

The Schauder Fixed Point Theorem (see Dunford-Schwartz [1, p. 456]) states that a continuous function mapping a compact convex subset of a Banach space into itself has a fixed point. Applying this to $\bar{S} \cap C$ yields an $f \in \bar{S}$ such that $\Phi(f) = f$. Since $\bar{S}$ does not contain 0, $f \neq 0$. For that f, define

$$A = \sum_{j=1}^{n} \beta_j(Kf) A_j.$$

Then $AKf = f$, so $1 \in \Pi_0(AK)$. $\qquad\qquad\qquad\qquad\square$

Remark 7.3.2.   Lomonosov's Lemma gives the original proof of Lomonosov's Theorem (Theorem 6.3.4), as follows: If K is a nonzero compact operator, let $\mathcal{J}$ denote the commutant of K. If $\mathcal{J}$ were transitive, Lomonosov's Lemma would provide an $A \in \mathcal{J}$ such that $1 \in \Pi_0(AK)$. If K itself had nonempty point spectrum, then (since eigenspaces are hyperinvariant) K would have a hyperinvariant subspace. Thus we can assume that K is quasinilpotent, by the Fredholm Alternative (Theorem 6.2.8).

But if K is quasinilpotent and $AK = KA$, then the spectral radius of AK is computed as follows (by Theorem 6.1.10):

$$\lim_{n\to\infty} \|(AK)^n\|^{1/n} = \lim_{n\to\infty} \|A^n K^n\|^{1/n} \quad (\text{since } AK = KA)$$

$$\leq \lim_{n\to\infty} \|A^n\|^{1/n} \lim_{n\to\infty} \|K^n\|^{1/n}$$

$$= r(A)\cdot 0 = 0 .$$

This contradicts $1 \in \Pi_0(AK)$.   □

It will be shown below (Corollary 7.4.12) that Lomonosov's Lemma yields a much stronger result: Every non-scalar operator that commutes with a compact operator other than 0 has a hyperinvariant subspace.

The next theorem is the infinite-dimensional extension of McCoy's Theorem (Theorem 1.3.4)

Theorem 7.3.3. *If A and B are compact, then $\{A, B\}$ is triangularizable if and only if $p(A, B)(AB - BA)$ is quasinilpotent for every (not necessarily commutative) polynomial p.*

Proof. The Spectral Mapping Theorem (Theorem 7.2.6) implies that $p(A, B)(AB - BA)$ is quasinilpotent when $\{A, B\}$ is triangularizable.

Conversely, note first that the condition of quasinilpotence is inherited by quotients: $\|\hat{A}\| \leq \|A\|$, so $\rho(\hat{A}) \leq \rho(A)$ by the spectral radius formula (Theorem 6.1.10). Thus the Triangularization Lemma (Lemma 7.1.11) implies that it suffices to show that the algebra generated by $\{A, B\}$ is not transitive.

There are two cases. If $AB = BA$, then the algebra is commutative and is certainly not transitive (Corollary 6.3.7). If $AB - BA \neq 0$ and the algebra were transitive, Lomonosov's Lemma (7.3.1) would yield an operator C in the algebra such that $1 \in \Pi_0(C(AB - BA))$ (since $AB - BA$ is compact). Such a C has the form $p(A, B)$, so this contradicts the hypothesis.   □

The following result will be strengthened in Corollary 8.4.5 below.

Theorem 7.3.4. *An algebra of compact operators is triangularizable if every pair of operators in the algebra is triangularizable.*

Proof. As in the proof of the Triangularization Lemma (Lemma 7.1.11), let $\mathcal{C}$ be a chain of subspaces that is maximal as a chain of invariant subspaces for the algebra. It must be shown that $\mathcal{C}$ is maximal as a subspace chain. Since $\mathcal{C}$ is obviously complete, it remains to be shown that the dimension of $\mathcal{M}/\mathcal{M}_-$ is at most 1 for all $\mathcal{M} \in \mathcal{C}$ (by Theorem 7.1.9).

Suppose that the dimension of $\mathcal{M}/\mathcal{M}_-$ is greater than 1. Let $\mathcal{A}$ denote the given algebra, and $\hat{\mathcal{A}}$ the corresponding algebra of quotients. Then $\hat{\mathcal{A}}$ is transitive (if $\mathcal{L}$ were a nontrivial invariant subspace of $\hat{\mathcal{A}}$, then $\mathcal{M}_-+\mathcal{L}$ could be added to $\mathcal{C}$, contradicting its maximality). Then $\hat{\mathcal{A}}$ is not commutative (by Corollary 6.3.7); choose S and T in $\mathcal{A}$ such that $\hat{S}\hat{T} - \hat{T}\hat{S} \neq 0$.

By Lomonosov's Lemma (Lemma 7.3.1) it follows (since $\hat{S}\hat{T} - \hat{T}\hat{S}$ is compact and $\hat{\mathcal{A}}$ is transitive) that there is an $A \in \mathcal{A}$ such that $1 \in \Pi_0\left(\hat{A}(\hat{S}\hat{T} - \hat{T}\hat{S})\right)$. Thus there is an $f \in \mathcal{M}$ such that $f \notin \mathcal{M}_-$ and $\left(\hat{A}(\hat{S}\hat{T} - \hat{T}\hat{S}) - I\right)(f + \mathcal{M}_-) = \mathcal{M}_-$. Let $\mathcal{N}$ denote the span of f and $\mathcal{M}_-$; then $(A(ST - TS) - I)\mathcal{N} \subseteq \mathcal{M}_-$, so $\mathcal{N} \in \mathrm{Lat}\,(A(ST - TS))$ and $1 \in \sigma\left(A(ST - TS)|_{\mathcal{N}}\right)$. Using the Fredholm Alternative (Theorem 6.2.8) again gives

$$1 \in \Pi_0\left(A(ST - TS)|_{\mathcal{N}}\right),$$

so 1 is an eigenvalue of $A(ST - TS)$.

On the other hand, $\{S,T\}$ triangularizable implies that $\sigma(ST - TS) = \{0\}$ (by Theorem 7.2.6), and then $\{A, ST - TS\}$ triangularizable implies

$$\sigma\left(A(ST - TS)\right) \subseteq \{\alpha\beta : \alpha \in \sigma(A), \beta \in \sigma(ST - TS)\}$$

$$= \{\alpha{\cdot}0 : \alpha \in \sigma(A)\}$$

$$= \{0\}\,.$$

This contradiction establishes the theorem.   □

The next two corollaries are immediate consequences of Theorems 7.3.4, 7.2.6, and 7.3.3. It will be shown below (Corollary 7.5.10) that very slight remnants of the spectral mapping theorem suffice to ensure triangularizability.

Corollary 7.3.5. *An algebra of compact operators is triangularizable if and only if the operator $p(A, B)(AB - BA)$ is quasinilpotent for all A and B in the algebra and all polynomials p.*

This corollary is strengthened below; it suffices to have $AB - BA$ quasinilpotent for all A and B in the algebra (Theorem 7.6.1). Note that Corollary 7.3.5 includes, in particular, the statement that an algebra of quasinilpotent compact operators is triangularizable. The generalization of this to semigroups is Turovskii's Theorem (Theorem 8.1.11), which is a crucial foundation for many results about semigroups presented in Chapter 8. A generalization of Corollary 7.3.5 to semigroups is Theorem 8.5.2.

Corollary 7.3.6. *An algebra of compact operators is triangularizable if and only if*

$$\sigma\left(p(A, B)\right) \subseteq p\left(\sigma(A), \sigma(B)\right)$$

for all A and B in the algebra and all polynomials p.

This corollary is substantially strengthened in Chapter 8.

As in the finite-dimensional case (Theorem 1.4.6), these corollaries can be re-interpreted to show that an algebra of compact operators is triangularizable if and only if it is commutative modulo its radical. However, we need to assume that the algebra is uniformly closed (i.e., closed in the norm topology of $\mathcal{B}(\mathcal{X})$). Also, the definition and results concerning radicals of algebras that were presented in Section 1.4 above require the algebra to have a unit. These results can be reformulated to apply to the non-unital case, but for present purposes it will be simpler just to adjoin the identity operator to any given algebra of compact operators.

Definition 7.3.7. A *translate* of the operator A is any operator of the form $A + \lambda I$ for λ a complex number.

Corollary 7.3.8. *If $\mathcal{A}$ is a uniformly closed algebra consisting of translates of compact operators, then $\mathcal{A}$ is triangularizable if and only if $\mathcal{A}/\mathrm{Rad}\,\mathcal{A}$ is commutative.*

Proof. This is merely a rephrasing of the above results, as in the finite-dimensional case (Theorem 1.4.6). The corollary asserts that $\mathcal{A}$ is triangularizable if and only if every commutator $BC - CB$ is in $\mathrm{Rad}\,\mathcal{A}$. This follows immediately from Corollaries 7.3.5 and 7.3.6.   □

It is sometimes useful to know that given chains of invariant subspaces can be extended to triangularizing chains.

Theorem 7.3.9. *If $\mathcal{J}$ is a triangularizable family of compact operators, then every chain of invariant subspaces of $\mathcal{J}$ is contained in a triangularizing chain.*

Proof. The proof is very similar to the proof of Theorem 7.3.4. Let $\mathcal{C}$ be a chain of subspaces that contains the given chain and is maximal as a chain of invariant subspaces for the family. It must be shown that $\mathcal{C}$ is maximal as a subspace chain, which comes down to showing that the dimension of $\mathcal{M}/\mathcal{M}_-$ is at most 1 for $\mathcal{M} \in \mathcal{C}$.

Let $\mathcal{A}$ denote the algebra generated by $\mathcal{J}$; then $\mathcal{A}$ is triangularizable, and $\mathcal{C}$ is a maximal chain in $\mathrm{Lat}\,\mathcal{A}$ (since obviously $\mathrm{Lat}\,\mathcal{J} = \mathrm{Lat}\,\mathcal{A}$). If the dimension of $\mathcal{M}/\mathcal{M}_-$ were greater than 1, then, as in the proof of Theorem 7.3.4, there would be A, S, and T in $\mathcal{A}$ such that $1 \in \Pi_0\left(A(ST - TS)\right)$. However, since $\mathcal{A}$ is triangularizable, the Spectral Mapping Theorem (7.2.6) implies that the spectrum of $A(ST - TS)$ is contained in

$$\{\alpha(\beta\gamma - \gamma\beta) : \alpha \in \sigma(A), \beta \in \sigma(S), \gamma \in \sigma(T)\} = \{0\},$$

contradicting $1 \in \sigma\left(A(ST - TS)\right)$.   □

7.4 Transitive Algebras

In this section we establish analogues of Burnside's Theorem (Theorem 1.2.2) for algebras of compact operators. We begin with a discussion of finite-rank operators (i.e., operators whose ranges are finite-dimensional).

Definition 7.4.1. For a given Banach space $\mathcal{X}$, the *dual space* $\mathcal{X}^*$ is the Banach space of all bounded linear functionals on $\mathcal{X}$. The space $\mathcal{X}$ is *reflexive* if whenever F is a bounded linear functional on $\mathcal{X}^*$ there is an $f \in \mathcal{X}$ such that

$$F(\phi) = \phi(f) \quad \text{for all } \phi \in \mathcal{X}^* .$$

For $f_0 \in \mathcal{X}$ and $\phi_0 \in \mathcal{X}^*$, the operator $f_0 \otimes \phi_0$ is defined by $(f_0 \otimes \phi_0)(f) = (\phi_0(f)) f_0$ for $f \in \mathcal{X}$. If neither f_0 nor ϕ_0 is 0, $f_0 \otimes \phi_0$ is an operator of rank 1. Conversely, if A is a rank-one operator on $\mathcal{X}$, choose any nonzero vector f_0 in the range of A. Then Af is a multiple of f_0 for each f; if ϕ is defined by $Af = \phi(f)f_0$, then ϕ is clearly a bounded linear functional and $A = f_0 \otimes \phi$.

The following lemma yields Burnside-type theorems when coupled with Lomonosov's Lemma.

Lemma 7.4.2. *Let $\mathcal{A}$ be a uniformly closed transitive subalgebra of $\mathcal{B}(\mathcal{X})$ that contains a finite-rank operator other than 0. Then there is a closed subspace $\mathcal{M}$ of $\mathcal{X}^*$ such that*

(i) *$f_0 \otimes \phi_0$ is in $\mathcal{A}$ whenever $\phi_0 \in \mathcal{M}$ and $f_0 \in \mathcal{X}$, and*
(ii) *$\phi(g) = 0$ for all $\phi \in \mathcal{M}$ implies $g = 0$.*

Proof. We first show that $\mathcal{A}$ contains at least one operator of rank 1. For this, let F be a finite-rank operator in $\mathcal{A}$ other than 0, and define $F\mathcal{A}$ to be $\{FA : A \in \mathcal{A}\}$. Let $\mathcal{N}$ denote the range of F. Now, $F\mathcal{A}$ is a subalgebra of $\mathcal{A}$, and it leaves $\mathcal{N}$ invariant. It is easily seen that the restriction $F\mathcal{A}\big|_{\mathcal{N}}$ is a transitive algebra of operators on the finite-dimensional space $\mathcal{N}$. Burnside's Theorem (Theorem 1.2.2) implies that

$$F\mathcal{A}\big|_{\mathcal{N}} = \mathcal{B}(\mathcal{N}).$$

Thus there are operators A in $\mathcal{A}$ such that $F\mathcal{A}\big|_{\mathcal{N}}$ has rank 1. If A_0 is any such operator, then FA_0F is a rank-one operator in $\mathcal{A}$. Therefore, $\mathcal{A}$ contains an operator of the form $f_0 \otimes \phi_0$ for some $f_0 \in \mathcal{X}$ and $\phi_0 \in \mathcal{X}^*$.

We now show that the existence of an $f_0 \neq 0$ with $f_0 \otimes \phi_0 \in \mathcal{A}$ implies that $f \otimes \phi_0 \in \mathcal{A}$ for all $f \in \mathcal{X}$. To see this, fix any $f \in \mathcal{X}$. Since $\mathcal{A}$ is transitive, there is a sequence $\{A_n\} \subseteq \mathcal{A}$ such that $\{A_n f_0\} \to f$. For each n, the operator

$$A_n(f_0 \otimes \phi_0) = (A_n f_0) \otimes \phi_0$$

is in $\mathcal{A}$, and

$$\|A_n f_0 \otimes \phi_0 - f \otimes \phi_0\| = \|(A_n f_0 - f) \otimes \phi_0\|,$$

so $f \otimes \phi_0$ is a norm limit of operators in $\mathcal{A}$ and is thus in $\mathcal{A}$.

We can now define $\mathcal{M}$ as follows:

$$\mathcal{M} = \{\phi \in \mathcal{X}^* : f \otimes \phi \in \mathcal{A} \text{ for all } f \in \mathcal{X}\}.$$

Clearly, $\mathcal{M}$ is a closed subspace of $\mathcal{X}^*$, and conclusion (i) holds.

To verify (ii), suppose that $\phi(g) = 0$ for all $\phi \in \mathcal{M}$. To show that $g = 0$, first choose f_0 and ϕ_0 such that $f_0 \otimes \phi_0 \in \mathcal{A}$ and $f_0 \otimes \phi_0 \neq 0$. For each $A \in \mathcal{A}$, the mapping $f \to \phi_0(Af)$ is a linear functional, so there is a ϕ depending on A such that $\phi(f) = \phi_0(Af)$ for all $f \in \mathcal{X}$. Each such ϕ is in $\mathcal{M}$, since $f_0 \otimes \phi = (f_0 \otimes \phi_0)A$. Therefore, $\phi(g) = 0$, so $\phi_0(Ag) = 0$ for all $A \in \mathcal{A}$. Since $\mathcal{A}$ is transitive, $\{Ag : A \in \mathcal{A}\}$ would be dense in $\mathcal{X}$ if $g \neq 0$. If this were the case, then ϕ_0 would be identically 0, which is a contradiction. Hence $g = 0$. $\square$

This lemma easily yields a Burnside's Theorem for finite-rank operators on reflexive spaces.

Corollary 7.4.3. *A uniformly closed transitive algebra of operators on a reflexive Banach space contains all finite-rank operators if it contains any finite-rank operator other than 0.*

Proof. Every finite-rank operator is a sum of a finite number of operators of rank 1, so it suffices to show that every rank-one operator is in the algebra.

Let $\mathcal{M}$ be the subspace described in Lemma 7.4.2; it must be shown that $\mathcal{M} = \mathcal{X}^*$. This is immediate, for if $\mathcal{M} \neq \mathcal{X}^*$, the Hahn-Banach Theorem implies that there is a nonzero linear functional on $\mathcal{X}^*$ that annihilates $\mathcal{M}$. Since $\mathcal{M}$ is reflexive, this functional is evaluation at some $g \in \mathcal{X}$. But then $\phi(g) = 0$ for all $\phi \in \mathcal{M}$, which contradicts property (ii) of Lemma 7.4.2. $\square$

Reflexivity is required in the above corollary, as shown by the following.

Example 7.4.4. *On the dual of every nonreflexive Banach space there is a uniformly closed transitive algebra of operators that contains many but not all finite-rank operators.*

Proof. If $\mathcal{X}$ is not reflexive, let $\mathcal{A}$ denote the uniform closure of the set of finite-rank operators on $\mathcal{X}$. The counterexample is $\mathcal{A}^*$, the algebra of all

Banach-space adjoints of operators in $\mathcal{A}$. To see this, note first that $\mathcal{A}^*$ is uniformly closed, since $\|A\| = \|A^*\|$ for all $A \in \mathcal{B}(\mathcal{X})$. We must show that $\mathcal{A}^*$ contains finite-rank operators and is transitive but does not contain all finite-rank operators on $\mathcal{X}^*$.

Note that, for $\phi_0 \in \mathcal{X}^*$, $w_0 \in \mathcal{X}^{**}$ and $f_0 \in \mathcal{X}$, the rank-one operator $\phi_0 \otimes w_0$ on $\mathcal{X}^*$ is the adjoint of $f_0 \otimes \phi_0$, where $w_0(\phi) = \phi(f_0)$ for all $\phi \in \mathcal{X}^*$. Thus $\mathcal{A}^*$ contains many operators of rank one but does not contain $\phi_0 \otimes w_0$ if w_0 is not in the range of the canonical embedding of $\mathcal{X}$ into $\mathcal{X}^{**}$.

It remains to be seen that $\mathcal{A}^*$ is transitive. However, this is easily seen: If $\phi_1 \in \mathcal{X}^*$ and $\phi_1 \neq 0$, and if ϕ_0 is any other element of $\mathcal{X}^*$, choose an $f_0 \in \mathcal{X}$ such that $\phi_1(f_0) \neq 0$ and let w_0 be evaluation at f_0. Then the operator $\phi_0 \otimes w_0$ is in $\mathcal{A}^*$, and

$$\frac{1}{\phi_1(f_0)}(\phi_0 \otimes w_0)\phi_1 = \phi_0.$$

(Hence $\mathcal{A}^*$ is *strictly transitive*; i.e., if $\phi_1 \neq 0$ and $\phi_0 \in \mathcal{X}^*$, there is an $A^* \in \mathcal{A}^*$ such that $A^*\phi_1$ is equal to ϕ_0, not just arbitrarily close to ϕ_0.) $\square$

We wish to show that transitive algebras containing nontrivial compact operators also contain nontrivial finite-rank operators; this will allow us to obtain analogues of Burnside's Theorem for algebras of compact operators. The following lemma is stated for the more general case of semigroups rather than for algebras; the general case will be used in Chapter 8.

Lemma 7.4.5. *If S is a semigroup of compact operators that is uniformly closed and is also closed under multiplication by positive real numbers, and if S contains an operator that is not quasinilpotent, then S contains a finite-rank operator other than 0 that is either idempotent or nilpotent.*

Proof. Multiplying by an appropriate positive number, we can assume that there is a $K \in S$ of spectral radius 1. Then K has a finite number of eigenvalues of modulus 1, say $\{\lambda_1, \ldots, \lambda_n\}$ (by the Fredholm Alternative (Theorem 6.2.8) and Lemma 6.2.7). Repeated application of Corollary 6.4.13 yields complementary invariant subspaces $\mathcal{N}$ and $\mathcal{R}$ of K such that $\mathcal{N}$ is finite-dimensional, $\sigma(K|_{\mathcal{N}}) = \{\lambda_1, \ldots, \lambda_n\}$, and $\rho(K|_{\mathcal{R}}) < 1$. The latter implies (by the spectral radius formula, Theorem 6.1.10) that $\left\{ \left\| \left(K|_{\mathcal{R}}\right)^n \right\| \right\} \to 0$. We show that there is a subsequence $\{n_i\}$ such that $\left\{ \left(K|_{\mathcal{N}}\right)^{n_i} \right\}$ converges to an operator different from 0; this is very similar to the proof of Lemma 3.1.6.

For this, first note that, since $K|_{\mathcal{N}}$ is an operator on a finite-dimensional space, $K|_{\mathcal{N}}$ is similar to its Jordan canonical form. Since $\sigma(K|_{\mathcal{N}})$ is contained in the unit circle, its Jordan canonical form can be written $U + N$ with U unitary, N nilpotent, and $UN = NU$. Writing U in diagonal form shows that U^m is close to the identity when m is an integer such that λ_i^m

is close to 1 for all i. Thus there is a sequence $\{m_i\}$ of natural numbers such that $\{\|U^{m_i} - I|_{\mathcal{N}}\|\} \to 0$, where $I|_{\mathcal{N}}$ is the identity on $\mathcal{N}$.

We distinguish two cases. If $N = 0$, then $\{U^{m_i}\} \to I|_{\mathcal{N}}$, so

$$\{K^{m_i}\} = \{(K|_{\mathcal{N}})^{m_i} \oplus (K|_{\mathcal{R}})^{m_i}\} \longrightarrow I|_{\mathcal{N}} \oplus 0 \, ;$$

then $I|_{\mathcal{N}} \oplus 0$ is a finite-rank idempotent operator in $\mathcal{S}$.

On the other hand, suppose that $N \neq 0$. Choose k such that $N^k \neq 0$ and $N^{k+1} = 0$. For each m_i, write the binomial expansion of

$$(U + N)^{m_i+k} = \sum_{j=0}^{m_i+k} \binom{m_i + k}{j} U^{m_i+k-j} N^j \, .$$

Since $N^{k+1} = 0$, the terms with $j > k$ are 0. The last nonzero term, $\binom{m_i+k}{k} U^{m_i} N^k$, dominates the other terms when m_i is large (since its binomial coefficient dominates and $\|U^{m_i} N^k\| = \|N^k\|$ for all m_i), so

$$\lim_{i \to \infty} \frac{(U + N)^{m_i+k}}{\binom{m_i+k}{k}} = \lim_{i \to \infty} U^{m_i} N^k = N^k \, .$$

Therefore,

$$\lim_{i \to \infty} \frac{K^{m_i+k}}{\binom{m_i+k}{k}} = N^k|_{\mathcal{N}} \oplus 0$$

is a finite-rank nilpotent operator in $\mathcal{S}$. $\square$

The finite-rank operators contained in a subalgebra of $\mathcal{B}(\mathcal{X})$ form an ideal in the algebra. This is one of the facts that make the next lemma useful.

Lemma 7.4.6. *A nonzero ideal of a transitive algebra of operators is also transitive.*

Proof. Let $\mathcal{I}$ be an ideal of the transitive subalgebra $\mathcal{A}$ of $\mathcal{B}(\mathcal{X})$. Fix any $f_0 \neq 0$ in $\mathcal{X}$ and let f be any vector in $\mathcal{X}$. Let A be any element of $\mathcal{I}$ other than 0. Then the transitivity of $\mathcal{A}$ implies that there is a $B \in \mathcal{A}$ such that Bf_0 is not in the kernel of A. Then $ABf_0 \neq 0$, so, for any $\varepsilon > 0$ there is a $C \in \mathcal{A}$ with $\|CABf_0 - f\| < \varepsilon$, since $\mathcal{A}$ is transitive. The operator CAB is in $\mathcal{I}$. $\square$

Theorem 7.4.7. *Let $\mathcal{A}$ be a uniformly closed transitive operator algebra that contains a compact operator different from 0. Then the subalgebra of $\mathcal{A}$ consisting of operators of finite rank is also transitive.*

Proof. Let $\mathcal{C}$ denote the set of compact operators in $\mathcal{A}$. Then $\mathcal{C}$ is an ideal, and hence is transitive by Lemma 7.4.6. Choose any $K \in \mathcal{C}$ such that $K \neq 0$; by Lomonosov's Lemma (7.3.1) there is an $A \in \mathcal{C}$ such that $1 \in \Pi_0(AK)$. Therefore, $\mathcal{C}$ contains an operator that is not quasinilpotent, so Lemma 7.4.5 shows that there is a finite-rank operator different from 0 in $\mathcal{C}$. The set of all finite-rank operators in $\mathcal{C}$ is an ideal of $\mathcal{C}$, so Lemma 7.4.6 finishes the proof. □

Corollary 7.4.8. *A uniformly closed transitive algebra of operators on a reflexive Banach space that contains a compact operator other than 0 must contain all finite-rank operators.*

Proof. This follows directly from the above theorem and Corollary 7.4.3. □

This corollary cannot be extended to conclude that such operator algebras must contain all the compact operators: There are Banach spaces on which there exist compact operators that are not uniform limits of finite-rank operators (as established in Enflo [1]). On any such space the uniform closure of the set of finite-rank operators is a transitive algebra that does not contain all compact operators. On many spaces, including all Hilbert spaces, compact operators are uniform limits of finite-rank operators. On such spaces there is the following version of Burnside's Theorem for compact operators.

Corollary 7.4.9. *Let $\mathcal{X}$ be a reflexive Banach space such that every compact operator in $\mathcal{B}(\mathcal{X})$ is a uniform limit of operators of finite rank. Then every uniformly closed transitive subalgebra of $\mathcal{B}(\mathcal{X})$ that contains a nonzero compact operator must contain all the compact operators on $\mathcal{X}$.*

Proof. This is an immediate consequence of the preceding Corollary.

□

There is an analogue of Burnside's Theorem that holds on all Banach spaces; this requires a different topology on $\mathcal{B}(\mathcal{X})$.

Definition 7.4.10. The *strong operator topology* on $\mathcal{B}(\mathcal{X})$ is the topology with basis consisting of all subsets of the form

$$U(A; f_1, f_2, \ldots, f_n; \varepsilon) = \{B \in \mathcal{B}(\mathcal{X}) : \|Af_i - Bf_i\| < \varepsilon \text{ for all } i\}$$

for finite subsets $\{f_1, f_2, \ldots, f_n\}$ of $\mathcal{X}$, $A \in \mathcal{B}(\mathcal{X})$, and $\varepsilon > 0$. It is easily seen that a net $\{A_\alpha\}$ converges to A in the strong operator topology if and only if $\{\|Af - A_\alpha f\|\} \to 0$ for each $f \in \mathcal{X}$ (i.e., convergence in the strong operator topology is pointwise convergence on $\mathcal{X}$. Note that the strong closure of any subset of $\mathcal{B}(\mathcal{X})$ is, in particular, uniformly closed.

Theorem 7.4.11. *The only strongly closed transitive subalgebra of $\mathcal{B}(\mathcal{X})$ that contains a compact operator different from 0 is $\mathcal{B}(\mathcal{X})$ itself.*

Proof. On reflexive spaces this follows immediately from Lemma 7.4.5 and Corollary 7.4.3: It is easily seen (see the end of this proof) that $\mathcal{B}(\mathcal{X})$ is the strong closure of the set of all finite-rank operators. The proof on arbitrary spaces is similar.

By Theorem 7.4.7, the ideal of finite-rank operators in the algebra is transitive; let $\mathcal{A}$ denote its strong closure. Consider the subspace $\mathcal{M}$ of Lemma 7.4.2 determined by $\mathcal{A}$. Property (ii) of Lemma 7.4.2 implies that $\mathcal{M}$ is weak-$*$ dense in $\mathcal{X}^*$ (the weak-$*$ topology on $\mathcal{X}^*$ is the topology induced by $\mathcal{X}$ acting as linear functionals; see Rudin [2], Conway [1]), since the image of $\mathcal{X}$ in $\mathcal{X}^{**}$ separates each element ϕ_0 in $\mathcal{X}^*$ from weak-$*$ closed subspaces not containing ϕ_0. Thus for any $\phi_0 \in \mathcal{X}^*$ there is a net $\{\phi_\alpha\} \subseteq \mathcal{M}$ such that $\{\phi_\alpha(f)\} \to \phi_0(f)$ for all $f \in \mathcal{X}$.

Suppose that $f_0 \otimes \phi_0$ is any rank-one operator on $\mathcal{X}$. Choose a net $\{\phi_\alpha\} \subseteq \mathcal{M}$ such that $\{\phi_\alpha(f)\} \to \phi_0(f)$ for all $f \in \mathcal{X}$. Then $f_0 \otimes \phi_\alpha \in \mathcal{A}$ for all α, and $\{f_0 \otimes \phi_\alpha\}$ converges to $f_0 \otimes \phi_0$ in the strong operator topology. Thus $\mathcal{A}$ contains every operator of rank one. Since every finite-rank operator is a finite sum of rank-one operators, $\mathcal{A}$ contains all finite-rank operators. The set of all finite-rank operators is strongly dense in $\mathcal{B}(\mathcal{X})$: Given any strong neighborhood of an $A_0 \in \mathcal{B}(\mathcal{X})$,

$$\mathcal{U}(A_0; f_1, f_2, \ldots, f_n; \varepsilon) = \{A \in \mathcal{B}(\mathcal{X}) : \|A_0 f_i - A f_i\| < \varepsilon \text{ for } i = 1, \ldots, n\},$$

simply choose a finite-rank operator F such that

$$F f_i = A_0 f_i \quad \text{for} \quad i = 1, \ldots, n .$$

Clearly, $F \in \mathcal{U}(A_0; f_1, f_2, \ldots, f_n; \varepsilon)$. Thus every strong neighborhood of A_0 meets $\mathcal{A}$, so A_0 is in the strong closure of $\mathcal{A}$, which is $\mathcal{A}$. $\square$

A generalization of what we have called Lomonosov's Theorem (6.3.4), also due to Lomonosov, follows as a corollary.

Corollary 7.4.12. *If A is not a multiple of the identity, and if A commutes with a compact operator other than 0, then A has a nontrivial hyperinvariant subspace.*

Proof. Let $\mathcal{A}$ denote the commutant of A. If $\mathcal{A}$ were transitive, then Theorem 7.4.11 would imply that $\mathcal{A} = \mathcal{B}(\mathcal{X})$ (since $\mathcal{A}$ contains a compact operator different from 0 and is, clearly, strongly closed). Thus A would commute with every operator; this is easily seen to imply that A is a multiple of the identity, contrary to the hypothesis. Hence $\mathcal{A}$ is not transitive. $\square$

This corollary yields the existence of invariant subspaces for a wide class of operators.

Corollary 7.4.13. *An operator has a nontrivial invariant subspace if it commutes with an operator other than a multiple of the identity that itself commutes with a compact operator other than 0.*

Proof. The operator that the given one commutes with has a hyperinvariant subspace, by Corollary 7.4.12.   □

Corollary 7.4.13 appears to have broad application. For some years after it was first established, it was not known whether there were any operators on Hilbert space that failed to satisfy its hypothesis. It has now been shown that there are operators (certain weighted shifts) that are not covered (cf. Hadwin-Nordgren-Radjavi-Rosenthal [1]), but the scope of the corollary is still not clear.

7.5 Block Triangularization and Applications

The block triangularization theorem for algebras of matrices (Theorem 1.5.1) can be generalized to algebras of compact operators, although both the statement and the proof are considerably more complex. As in the finite-dimensional case (Theorem 1.5.5), the generalization can be applied to show that even slight vestiges of the spectral mapping property imply triangularizability.

We require several properties of idempotents. Recall that an idempotent is a bounded operator P such that $P^2 = P$, and P is the projection onto its range along its kernel (see Definitions 6.4.2 and 6.4.3). The algebraic properties of idempotents are the same as in the finite-dimensional case. The following definitions and lemma are similar to those at the end of Section 3.1 above.

Definition 7.5.1. The idempotent P is *contained in* the idempotent Q, denoted $P \leq Q$, if $PQ = QP = P$.

Lemma 7.5.2. *For idempotents P and Q, $P \leq Q$ if and only if the range of P is contained in the range of Q and the kernel of P contains the kernel of Q. The relation $\leq$ is a partial ordering on the set of idempotents in $\mathcal{B}(\mathcal{X})$.*

Proof. Suppose first that $PQ = QP = P$. If $f = Pf$ is in the range of P, then $Qf = QPf = Pf = f$, so f is in the range of Q. Similarly, if $Qg = 0$, then $0 = PQg = QPg = Pg$, so g is in the kernel of P.

Conversely, suppose that the containments hold. Then, for any $f \in \mathcal{X}$, $QPf = Pf$, since the range of P is contained in the range of Q; hence $QP = P$. Also, for any $g \in \mathcal{X}$, $P(1 - Q)g = 0$, since the kernel of P contains that of Q. Hence $P = PQ$, so $P = PQ = QP$. The properties of a partial ordering for $\leq$ follow from the corresponding properties for containment of ranges and kernels. $\square$

The definition of minimality given below is equivalent to that of Definition 3.1.7.

Definition 7.5.3. An idempotent Q is a *minimal idempotent in the operator semigroup* $\mathcal{A}$ if Q is in $\mathcal{A}$, $Q \neq 0$, and the only idempotents P in $\mathcal{A}$ satisfying $P \leq Q$ are $P = 0$ and $P = Q$.

Note that it follows immediately from Lemma 7.5.2 that every idempotent of rank one is minimal. A semigroup may, however, have minimal idempotents of higher rank. We will be considering semigroups of compact operators, and every compact idempotent has finite rank (since the range of the idempotent is its eigenspace corresponding to eigenvalue 1).

Definition 7.5.4. The idempotents P and Q are *disjoint* if $PQ = QP = 0$.

Note that if P and Q are disjoint, then $P + Q$ is an idempotent:

$$(P + Q)^2 = P^2 + PQ + QP + Q^2$$
$$= P^2 + Q^2 = P + Q .$$

We wish to show that finite-rank idempotents can be written as sums of minimal idempotents. The following lemma will be needed.

Lemma 7.5.5. *If P and Q are idempotents with $P \leq Q$, then $Q - P$ is an idempotent that is disjoint from P.*

Proof. First,

$$(Q - P)^2 = Q^2 - QP - PQ + P^2$$
$$= Q^2 - P - P + P^2$$
$$= Q - P - P + P = Q - P,$$

so $Q - P$ is idempotent.

Then $P(Q - P) = PQ - P^2 = P - P = 0$ and $(Q - P)P = QP - P^2 = 0$. $\square$

Theorem 7.5.6. *If P is a finite-rank idempotent other than 0 in an operator algebra $\mathcal{A}$, then P can be written*

$$P = P_1 + P_2 + \cdots + P_k,$$

where the $\{P_i\}$ are mutually-disjoint minimal idempotents in $\mathcal{A}$ and k is at most the rank of P.

Proof. If P is minimal, there is nothing to prove. Otherwise, choose an idempotent Q_1 in $\mathcal{A}$ such that $Q_1 \leq P$ and $Q_1 \neq P$, $Q_1 \neq 0$. By Lemma 7.5.5, $P - Q_1$ is an idempotent. If Q_1 is minimal, let $P_1 = Q_1$. If Q_1 is not minimal, choose a $Q_2 \leq Q_1$ with $Q_2 \neq Q_1$, $Q_2 \neq 0$. If Q_2 is not minimal, choose a Q_3, and so on. Notice that the ranks of the $\{Q_j\}$ are strictly decreasing and are all less than the rank of P, so a finite number of steps produces a Q_j that is minimal, let P_1 denote such a Q_j. Then $P - P_1$ is an idempotent disjoint from P_1 (by Lemma 7.5.5).

Now apply the same procedure starting with $P - P_1$, obtaining a minimal idempotent P_2 such that $P_2 \leq P - P_1$. Note that the rank of $P - P_1$ is less than that of P, and the rank of $P - P_1 - P_2$ is smaller still. Continuing this process inductively produces an idempotent $P - P_1 - P_2 - \cdots - P_{k-1}$ that is minimal. If that idempotent is labeled P_k, then $P = P_1 + P_2 + \cdots + P_k$. Since the range of each of the P_i is at least one-dimensional and the P_i are mutually disjoint, the range of P has dimension at least k; i.e., k is at most the rank of P. $\qquad\square$

Lemma 7.5.7. *If $\lambda \neq 0$ is in the spectrum of the compact operator K, then there is a finite-rank idempotent P ("the Riesz projection for K corresponding to λ") in the uniformly closed algebra generated by K such that the spectrum of the restriction of K to $P\mathcal{X}$ is $\{\lambda\}$ and the spectrum of the restriction of K to $(1 - P)\mathcal{X}$ is $\sigma(K) \setminus \{\lambda\}$.*

Proof. We decompose the space with respect to the eigenvalues one at a time. Begin by assuming that $|\lambda| = \rho(K)$. Then, by multiplying by the reciprocal of λ, we can also assume that $\lambda = 1$. We first show that there is such a P in the algebra generated by K and the identity operator on $\mathcal{X}$. For this, replace K by $\frac{1}{2}(1 + K)$, so that $1 \in \sigma(K)$ and every other element in $\sigma(K)$ has modulus less than 1.

By the Riesz Decomposition Theorem (6.4.11), there is an N such that $\text{null}(K - 1)^N$ and $\text{ran}(K - 1)^N$ are complementary invariant subspaces of K. Let P denote the projection on $\text{null}(K - 1)^N$ along $\text{ran}(K - 1)^N$ (Definition 6.4.3). We show that P is in the uniformly closed algebra generated by K and I. Denote the restriction of K to $\text{null}(K - 1)^N$ by A and the restriction to $\text{ran}(K - 1)^N$ by B. Then $K = A \oplus B$ and $\rho(B) < 1$.

Since A is an operator on a finite-dimensional space and $\sigma(A) = \{1\}$, A has the form $I + F$, where $F^{k+1} = 0$ for some k (Jordan form). For each natural number n, write

$$I^n = (A - F)^n = \sum_{j=0}^{n} \binom{n}{j} A^{n-j}(-F)^j \,.$$

If $n > k$, then the terms for $j > k$ are 0, so

$$I = \sum_{j=0}^{k} \binom{n}{j} A^{n-j}(-F)^j = \sum_{j=0}^{k} \binom{n}{j} A^{n-j}(I - A)^j .$$

For each $n > k$, define the polynomial $p_n(X)$ by

$$p_n(X) = \sum_{j=0}^{k} \binom{n}{j} X^{n-j}(1 - X)^j .$$

We claim that $\lim_{n \to \infty} \|p_n(B)\| = 0$. For this, let

$$C = \max \left\{ \|(I - B)^j\| : j = 1, 2, \ldots, k \right\} .$$

Note that $\binom{n}{j} \leq n^k$ for $1 \leq j \leq k$. For each j, $\left\| \binom{n}{j} B^{n-j}(I - B)^j \right\| \leq n^k \|B^{n-j}\| C$. Now

$$\lim_{n \to \infty} \left(n^k \|B^{n-j}\| C \right)^{\frac{1}{n-j}} = \lim_{n \to \infty} \left(n^{\frac{1}{n-j}} \right)^k \lim_{n \to \infty} \|B^{n-j}\|^{\frac{1}{n-j}} \lim_{n \to \infty} C^{\frac{1}{n-j}}$$

$$= r(B) .$$

Since $r(B) < 1$, $\lim_{n \to \infty} \left(n^k \|B^{n-j}\| C \right) = 0$, so each term of $p_n(B)$ approaches 0 as $n \to \infty$. Thus $\lim_{n \to \infty} \|p_n(B)\| = 0$, and $\lim_{n \to \infty} p_n(K) = I \oplus 0 = P$.

We need to show that the identity operator is not needed. For each n, write $p_n(K) = \lambda_n I + q_n(K)$, where the polynomial q_n has constant term 0. Then modulo the compact operators, $P = 0$ and $q_n(K) = 0$, so $\{p_n(K)\} \to P$ implies that $\{\lambda_n\} \to 0$. Thus $\{q_n(K)\} \to P$.

This proves the theorem for every eigenvalue of modulus $\rho(K)$. An easy induction finishes the proof: Given any λ, let $P_1, P_2, \ldots, P_m$ be the Riesz projections corresponding to the $\lambda_j \in \sigma(K)$ with $|\lambda_j| > |\lambda|$ and let $\mathcal{M} = (1 - P_1)(1 - P_2) \cdots (1 - P_m)\mathcal{X}$. By the above, the Riesz projection P_0 for $K\big|_{\mathcal{M}}$ corresponding to λ is in the algebra generated by $K\big|_{\mathcal{M}}$. Then $P = P_0(1 - P_1)(1 - P_2) \cdots (1 - P_m)$ is the Riesz projection for K corresponding to λ, and if $p_0, p_1, p_2, \ldots, p_m$ are polynomials such that $p_0(K\big|_{\mathcal{M}})$ is close to P_0 and $p_j(K)$ is close to P_j, then

$$p_0(K) \left(1 - p_1(K)\right) \left(1 - p_2(K)\right) \cdots \left(1 - p_m(K)\right)$$

is close to P.   □

We now present the block triangularization theorem. As in the finite-dimensional case, the block triangularization is with respect to any maximal chain of invariant subspaces for the algebra. If the chain is not triangularizing, then, since it is clearly complete, there are subspaces $\mathcal{M}$ in the chain with the dimension of $\mathcal{M}/\mathcal{M}_-$ greater than 1 (by Theorem 7.1.9).

It was necessary to discuss operators on quotient spaces as part of the triangularization lemma (Definition 7.1.10 and Lemma 7.1.11). We require more specific notation.

Notation 7.5.8. Let C be a complete chain of invariant subspaces of a family $\mathcal{F}$ of operators and let $M \in C$ with $M_- \neq M$ (see Definition 7.1.6). For any $A \in \mathcal{F}$, A_M denotes the operator induced by A on the quotient space M/M_-, and for any subset S of $\mathcal{F}$, $S_M = \{A_M : A \in S\}$.

In the following, the subsets $\mathcal{E}_\omega$ correspond to the "linked" subsets J_i of Theorem 1.5.1.

Theorem 7.5.9. (The Block Triangularization Theorem) *Let $\mathcal{A}$ be a uniformly closed algebra of compact operators on a Banach space. Let C be any maximal chain of invariant subspaces of $\mathcal{A}$, and let*

$$\mathcal{E} = \{M \in C : \dim M/M_- \geq 1 \text{ and } \mathcal{A}_M \neq \{0\}\} \ .$$

Assume $\mathcal{E}$ is not empty and let $\mathcal{F}$ be the ideal generated by the minimal idempotents in $\mathcal{A}$. Then there is a partition $\mathcal{E} = \bigcup_\omega \mathcal{E}_\omega$ into finite subsets $\mathcal{E}_\omega$ such that the following properties hold:

(i) *If M/M_- is finite-dimensional, then $\mathcal{F}_M = \mathcal{B}(M/M_-)$; if M/M_- is infinite-dimensional, then $\mathcal{F}_M$ is transitive and contains subalgebras isomorphic to $\mathcal{B}(\mathbb{C}^n)$ for every n.*

(ii) *If $F \in \mathcal{F}$ and $F_M \neq 0$ for some M in a given $\mathcal{E}_\omega$, then there is a $G \in \mathcal{F}$ with $G_M = F_M$ and $G_N = 0$ for every N in $\mathcal{E} \setminus \mathcal{E}_\omega$.*

(iii) *If M and N are in the same $\mathcal{E}_\omega$, the mapping ϕ from $\mathcal{A}_M$ to $\mathcal{A}_N$ defined by $\phi(A_M) = A_N$ for $A \in \mathcal{A}$ is a well-defined algebra isomorphism such that for each idempotent P in $\mathcal{F}_M$ there is an invertible operator S mapping the range of P onto the range of $\phi(P)$ satisfying*

$$\phi(PXP) = S(PXP)S^{-1}$$

for all $X \in \mathcal{A}_M$.

Proof. Let $\mathcal{P}$ denote the set of all minimal idempotents in $\mathcal{A}$. The proof is presented in a sequence of claims. Note that the condition $\mathcal{A}_M \neq \{0\}$ in defining $\mathcal{E}$ is needed only when $\dim M/M_- = 1$; if $\dim M/M_- > 1$, then $\mathcal{A}_M \neq 0$ by the maximality of the chain.

Claim (1): For each $M \in \mathcal{E}$ there is a P in $\mathcal{P}$ such that $P_M \neq 0$.

To see this, first observe that the algebra $\mathcal{A}_M$ is transitive (for if it had a nontrivial invariant subspace $\mathcal{L}$, then $M_- + \mathcal{L}$ could be added to C, so C would not be maximal). Fix any $K \in \mathcal{A}$ such that $K_M \neq 0$. By Lomonosov's Lemma (7.3.1), there is an $A \in \mathcal{A}$ such that 1 is in the point spectrum of $A_M K_M$. Then $(AK - 1)M$ is a proper subset of M, so 1 is in the spectrum of the restriction of AK to $\mathcal{M}$, and hence is an eigenvalue (by Theorem 6.2.8) of AK.

Let P denote the Riesz projection for AK about 1 (as in Lemma 7.5.7). Then P is an idempotent in the uniformly closed algebra generated by A (Lemma 7.5.7), and

$$A_{\mathcal{M}}K_{\mathcal{M}} = P_{\mathcal{M}}A_{\mathcal{M}}K_{\mathcal{M}} + (1 - P_{\mathcal{M}})A_{\mathcal{M}}K_{\mathcal{M}}.$$

Then $1 \notin \sigma((1 - P_{\mathcal{M}})A_{\mathcal{M}}K_{\mathcal{M}})$ (since otherwise, as argued above, 1 would be in $\sigma((1 - P)AK)$), so $1 \in \sigma(P_{\mathcal{M}}A_{\mathcal{M}}K_{\mathcal{M}})$. Thus $P_{\mathcal{M}} \neq 0$. This shows that there is an idempotent P in $\mathcal{A}$ such that $P_{\mathcal{M}} \neq 0$, but P may not be minimal. However, by Lemma 7.5.6, P is a sum of minimal idempotents in $\mathcal{A}$: $P = P_1 + \cdots + P_k$. Since $P_{\mathcal{M}} \neq 0$, there is an i with $(P_i)_{\mathcal{M}} \neq 0$, establishing the first claim.

We have shown, in particular, that there are minimal idempotents in $\mathcal{A}$. Now define $\mathcal{F}$ to be the ideal of $\mathcal{A}$ generated by the minimal idempotents; i.e., $\mathcal{F}$ is the set of all sums of the form $\sum_{i=1}^{k} A_i P_i B_i$ with $P_i \in \mathcal{P}$ and $A_i, B_i \in \mathcal{A}$ for all i.

Claim (2): If $\mathcal{M} \in \mathcal{E}$ and $P \in \mathcal{P}$ with $P_{\mathcal{M}} \neq 0$, then $P_{\mathcal{M}}$ has rank one.

We prove this claim by contradiction. The restriction of $P_{\mathcal{M}}A_{\mathcal{M}}$ to the range of $P_{\mathcal{M}}$ is a transitive algebra on the range of $P_{\mathcal{M}}$ (since $A_{\mathcal{M}}$ is transitive). If the dimension of this range is at least two, Burnside's Theorem (1.2.2) would imply the existence of an $A \in \mathcal{A}$ such that the restriction of $P_{\mathcal{M}}A_{\mathcal{M}}$ to the range of $P_{\mathcal{M}}$ had spectrum $\{0,1\}$. Then $\{0,1\} \subseteq \sigma(PAP)$; let R denote the Riesz projection (Lemma 7.5.7) for PAP corresponding to 1. Since R is in the uniformly closed algebra generated by P and PAP, $PR = RP = R$, so $R \leq P$. The way R was chosen shows that $R \neq P$, so this contradicts the minimality of P, establishing the claim.

The next claim is the essence of the "linking" between different blocks.

Claim (3): If P and Q are in $\mathcal{P}$ and there is some $\mathcal{M} \in \mathcal{E}$ such that $P_{\mathcal{M}} \neq 0$ and $Q_{\mathcal{M}} \neq 0$, then

$$\{\mathcal{N} \in \mathcal{E} : P_{\mathcal{N}} = 0\} = \{\mathcal{N} \in \mathcal{E} : Q_{\mathcal{N}} = 0\}.$$

For this, assume that $P_{\mathcal{N}} = 0$. To show that $Q_{\mathcal{N}} = 0$ we will establish that $Q_{\mathcal{N}}$ is a sum of terms each of which includes $P_{\mathcal{N}}$ as a factor.

Pick any $f \in \mathcal{M}/\mathcal{M}_-$ such that $Q_{\mathcal{M}}f \neq 0$. Since $A_{\mathcal{M}}$ is transitive, and $P_{\mathcal{M}} \neq 0$, there is an $A \in \mathcal{A}$ such that $P_{\mathcal{M}}A_{\mathcal{M}}Q_{\mathcal{M}}f \neq 0$. The transitivity of $A_{\mathcal{M}}$ implies that there is a $B \in \mathcal{A}$ with $Q_{\mathcal{M}}B_{\mathcal{M}}P_{\mathcal{M}}A_{\mathcal{M}}Q_{\mathcal{M}}f \neq 0$.

By Claim (2) above, $Q_{\mathcal{M}}$ and $P_{\mathcal{M}}$ each have rank one. Hence

$$Q_{\mathcal{M}}B_{\mathcal{M}}P_{\mathcal{M}}A_{\mathcal{M}}Q_{\mathcal{M}}f = \lambda_0 Q_{\mathcal{M}}f$$

for some $\lambda_0 \neq 0$. Then λ_0 is an eigenvalue of $QBPAQ$; let R be the corresponding Riesz projection. Since R is in the uniformly closed algebra generated by $QBPAQ$, R is a subprojection of Q and $RQ = QR = R$. Since Q is minimal, $R = Q$. Thus the spectrum of the restriction of $QBPAQ$ to

$Q\mathcal{X}$ is $\{\lambda_0\}$, so this restriction is invertible. Hence there is a polynomial p with zero constant term such that $p(QBPAQ) = Q$. We can compress this equation to $\mathcal{N}/\mathcal{N}_-$, getting

$$p(Q_\mathcal{N} B_\mathcal{N} P_\mathcal{N} A_\mathcal{N} Q_\mathcal{N}) = Q_\mathcal{N} \ .$$

But $P_\mathcal{N} = 0$. Since p has zero constant term, it follows that $Q_\mathcal{N} = 0$. Thus $P_\mathcal{N}$ and $Q_\mathcal{N}$ are simultaneously 0.

Claim (4): The relation $\sim$ on $\mathcal{E}$ defined by $\mathcal{M} \sim \mathcal{N}$ whenever there is a $P \in \mathcal{P}$ such that $P_\mathcal{M}$ and $P_\mathcal{N}$ are both nonzero is an equivalence relation.

This relation is obviously symmetric and reflexive, so we need only establish transitivity. Suppose that $\mathcal{L} \sim \mathcal{M}$ and $\mathcal{M} \sim \mathcal{N}$. By definition, there is a $P \in \mathcal{P}$ such that $P_\mathcal{L}$ and $P_\mathcal{M}$ are both nonzero, and a $Q \in \mathcal{P}$ such that $Q_\mathcal{M}$ and $Q_\mathcal{N}$ are both nonzero. By Claim (3) above, $P_\mathcal{N}$ is also nonzero. Therefore $\mathcal{L} \sim \mathcal{N}$.

We now partition $\mathcal{E}$ into the equivalence classes $\{\mathcal{E}_\omega\}$ defined by the above relation. We will show that these equivalence classes satisfy the conclusions of the theorem: each equivalence class consists of "linked" quotients. We first establish the following.

Claim (5): Each $\mathcal{E}_\omega$ is finite.

Fix any $\mathcal{E}_\omega$ and any $\mathcal{M} \in \mathcal{E}_\omega$. Choose (by Claim (1)) a $P \in \mathcal{P}$ such that $P_\mathcal{M} \neq 0$. Since P is a compact idempotent, the dimension of its range (which is the dimension of its eigenspace corresponding to 1) is finite. If $\mathcal{E}_\omega$ contained an infinite number of subspaces, then, by Claim (3), $\{\mathcal{N} \in \mathcal{E}_\omega : P_\mathcal{N} \neq 0\}$ would be infinite. Then for each $\mathcal{N} \in \mathcal{E}_\omega$ we could choose an $f_\mathcal{N} \in \mathcal{N} \setminus \mathcal{N}_-$ such that $P_\mathcal{N}[f_\mathcal{N}] \neq 0$. The set $\{Pf_\mathcal{N} : \mathcal{N} \in \mathcal{E}_\omega\}$ is linearly independent, for if

$$Pf_{\mathcal{N}_{k+1}} = \alpha_1 Pf_{\mathcal{N}_1} + \alpha_2 Pf_{\mathcal{N}_2} + \cdots + \alpha_k Pf_{\mathcal{N}_k}$$

with $\mathcal{N}_1 \subseteq \mathcal{N}_2 \subseteq \cdots \subseteq \mathcal{N}_k \subseteq \mathcal{N}_{k+1}$, then each $f_{\mathcal{N}_j}$ for $j = 1, 2, \ldots, k$ would be in the predecessor of $\mathcal{N}_{k+1}$ in the chain $\mathcal{C}$. Then $\{Pf_{\mathcal{N}_j} : j = 1, 2, \ldots, k\}$ would be contained in $(\mathcal{N}_{k+1})_-$. On the other hand, $f_{\mathcal{N}_{k+1}}$ is not in $(\mathcal{N}_{k+1})_-$, and, since $P_{\mathcal{N}_{k+1}}[f_{\mathcal{N}_{k+1}}] \neq 0$, neither is $Pf_{\mathcal{N}_{k+1}}$, so $Pf_{\mathcal{N}_{k+1}}$ cannot be a linear combination of $\{Pf_{\mathcal{N}_j} : j = 1, 2, \ldots, k\}$.

We can now establish the "independence" of the $\{\mathcal{E}_\omega\}$.

Claim (6): If $F \in \mathcal{F}$ and $F_\mathcal{M} \neq 0$, for some $\mathcal{M} \in \mathcal{E}_\omega$, then there is a $G \in \mathcal{F}$ with $G_\mathcal{M} = F_\mathcal{M}$ and $G_\mathcal{N} = 0$ for all $\mathcal{N} \in \mathcal{E} \setminus \mathcal{E}_\omega$.

By the definition of the ideal $\mathcal{F}$, since $F \in \mathcal{F}$, there are idempotents $\{P_j\} \subseteq \mathcal{P}$ and operators $\{A_j\} \subseteq \mathcal{A}$ and $\{B_j\} \subseteq \mathcal{A}$ such that

$$F = \sum_{j=1}^{k} A_j P_j B_j.$$

Let $J = \{j : (P_j)_{\mathcal{M}} \neq 0\}$, and define

$$G = \sum_{j \in J} A_j P_j B_j.$$

Clearly, $G_{\mathcal{M}} = F_{\mathcal{M}}$. If $\mathcal{N} \notin \mathcal{E}_\omega$, then, by Claim (3), $(P_j)_{\mathcal{N}} = 0$ for all $j \in J$, so $G_{\mathcal{N}} = 0$.

We next show that the "linking" within each $\mathcal{E}_\omega$ extends from $\mathcal{F}$ to all of $\mathcal{A}$.

Claim (7): If $\mathcal{M}$ and $\mathcal{N}$ are in a given $\mathcal{E}_\omega$ and $A_{\mathcal{N}} = 0$ for some $A \in \mathcal{A}$, then $A_{\mathcal{M}} = 0$.

Suppose, to the contrary, that $A_{\mathcal{M}} \neq 0$. Since $\mathcal{F}_{\mathcal{M}}$ is transitive and $A_{\mathcal{M}}$ is a compact operator other than 0, Lomonosov's Lemma (7.3.1) then implies that there is some $F \in \mathcal{F}$ such that 1 is an eigenvalue of $F_{\mathcal{M}} A_{\mathcal{M}}$, and hence of FA. Let R denote the Riesz projection (Lemma 7.5.7) for FA corresponding to 1. Since R has finite rank, we can apply Theorem 7.5.6 to write $R = R_1 + R_2 + \cdots + R_k$ where the $\{R_j\}$ are mutually-disjoint minimal idempotents in $\mathcal{A}$. If $(R_j)_{\mathcal{M}} = 0$ for all j, then $R_{\mathcal{M}} = 0$, contradicting the fact that $R_{\mathcal{M}} F_{\mathcal{M}} A_{\mathcal{M}}$ has 1 in its point spectrum. Hence there must be some j such that $R_{j_{\mathcal{M}}} \neq 0$. But $A_{\mathcal{N}} = 0$ implies $F_{\mathcal{N}} A_{\mathcal{N}} = 0$, so $(R_j)_{\mathcal{N}} = 0$ for all j. This contradicts Claim (3) above.

We now establish the isomorphism asserted in part (iii) of the theorem.

Claim (8): For $\mathcal{M}$ and $\mathcal{N}$ in a given $\mathcal{E}_\omega$, the mapping ϕ defined by $\phi(A_{\mathcal{M}}) = A_{\mathcal{N}}$ is a well-defined algebra isomorphism from $\mathcal{A}_{\mathcal{M}}$ onto $\mathcal{A}_{\mathcal{N}}$.

For this, note first that Claim (7) above shows that ϕ is well-defined and injective. For $A, B \in \mathcal{A}$, $(A + B)_{\mathcal{M}} = A_{\mathcal{M}} + B_{\mathcal{M}}$ and $(AB)_{\mathcal{M}} = A_{\mathcal{M}} B_{\mathcal{M}}$, so ϕ is an algebra isomorphism.

Claim (9): If ϕ is defined by $\phi(A_{\mathcal{M}}) = A_{\mathcal{N}}$ for $A \in \mathcal{A}$, then for each idempotent P in $\mathcal{A}_{\mathcal{M}}$ there is an invertible operator S from the range of P onto the range of $\phi(P)$ such that

$$\phi(PXP)f = S(PXP)S^{-1}f$$

for all $X \in \mathcal{A}$ and all f in the range of $\phi(P)$.

This can be shown as follows. Fix any idempotent P in $\mathcal{A}_{\mathcal{M}}$. Then $\phi(P)$ is an idempotent in $\mathcal{A}_{\mathcal{N}}$, since ϕ is an algebra isomorphism. Since P and $\phi(P)$ have finite-dimensional ranges and $\mathcal{A}_{\mathcal{M}}$ and $\mathcal{A}_{\mathcal{N}}$ are transitive, the algebras $P\mathcal{A}_{\mathcal{M}}P$ and $\phi(P)\mathcal{A}_{\mathcal{N}}\phi(P)$ are transitive on the ranges of P and $\phi(P)$, respectively. Burnside's Theorem (1.2.2) implies that these algebras are the algebras of all operators on the ranges of P and $\phi(P)$, respectively.

If $\mathcal{X}$ and $\mathcal{Y}$ are finite-dimensional, then every algebra isomorphism of $\mathcal{B}(\mathcal{X})$ onto $\mathcal{B}(\mathcal{Y})$ is spatial (see Theorem 1.2.4; $\mathcal{X}$ and $\mathcal{Y}$ clearly have the same dimension and can thus be regarded as the same space). Thus there is an S such that $\phi(PXP) = S(PXP)S^{-1}$ for all X in $\mathcal{A}_{\mathcal{M}}$.

Claim (10): If $\mathcal{M} \in \mathcal{E}$ and the dimension of $\mathcal{M}/\mathcal{M}_-$ is finite, then $\mathcal{F}_{\mathcal{M}} = \mathcal{B}(\mathcal{M}/\mathcal{M}_-)$.

This follows from Burnside's Theorem (1.2.2), since, as shown above, $\mathcal{F}_{\mathcal{M}}$ is transitive.

If $\mathcal{M}/\mathcal{M}_-$ is infinite-dimensional, there is a weaker version of this claim that is still useful.

Claim (11): If $\mathcal{M}/\mathcal{M}_-$ is infinite-dimensional, then, for every natural number n, $\mathcal{F}_{\mathcal{M}}$ contains a subalgebra isomorphic to $\mathcal{B}(\mathbb{C}^n)$.

We first establish the following: If a transitive algebra $\mathcal{T}$ contains a finite-rank operator F, then it contains an idempotent with the same range as F. For this, note that the restriction of $F\mathcal{T}$ to the range $F\mathcal{X}$ of F is the algebra of all operators on $F\mathcal{X}$, by Burnside's Theorem (1.2.2). Choose any $T \in \mathcal{T}$ such that $FT|_{F\mathcal{X}} = I|_{F\mathcal{X}}$ and let $P = FT$. Then

$$P^2 = FT\,FT = FT = P,$$

so P is idempotent. Clearly $P\mathcal{X} = F\mathcal{X}$.

To establish Claim (11), then, it suffices to show that $\mathcal{F}_{\mathcal{M}}$ contains operators of arbitrarily high (finite) ranks, since $P\mathcal{F}_{\mathcal{M}}P$ is isomorphic to $\mathcal{B}(\mathbb{C}^n)$ if P is an idempotent of rank n. To show that operators of arbitrarily high ranks are in $\mathcal{F}_{\mathcal{M}}$, suppose that $P_{\mathcal{M}}$ is any idempotent in $\mathcal{F}_{\mathcal{M}}$. We show that there is an $F_{\mathcal{M}} \in \mathcal{F}_{\mathcal{M}}$ with rank greater than that of $P_{\mathcal{M}}$. For this, note first that $1 - P_{\mathcal{M}} \neq 0$ and $\mathcal{F}_{\mathcal{M}}$ is transitive, so there is a $G \in \mathcal{F}$ such that $(1 - P_{\mathcal{M}})G_{\mathcal{M}}(1 - P_{\mathcal{M}}) \neq 0$. Then let $F = (1 - P)G(1 - P) + P$; the rank of $F_{\mathcal{M}} = (1 - P_{\mathcal{M}})G_{\mathcal{M}}(1 - P_{\mathcal{M}}) + P_{\mathcal{M}}$ is greater than that of $P_{\mathcal{M}}$.

This finishes the proof of all the assertions of the theorem. $\square$

The Block Triangularization Theorem can be used to obtain many sufficient conditions that a uniformly closed algebra of compact operators be triangularizable. Any condition that precludes the algebra of all 2×2 matrices being isomorphic to a subalgebra of a compression of the algebra forces $\dim(\mathcal{M}/\mathcal{M}_-)$ to be at most 1 for all $\mathcal{M}$, so that the block triangularization is actually a triangularization. A few such conditions are stated in the next corollary.

Corollary 7.5.10. *If $\mathcal{A}$ is a uniformly closed algebra of compact operators, the following are equivalent:*

(i) *$\mathcal{A}$ is triangularizable;*

(ii) *every finite-rank nilpotent operator in $\mathcal{A}$ is in the radical of $\mathcal{A}$;*

(iii) *each sum of two finite-rank nilpotent operators in $\mathcal{A}$ is quasinilpotent;*

(iv) *each product of two finite-rank nilpotent operators in $\mathcal{A}$ is quasinilpotent;*

(v) *each pair of finite-rank operators in $\mathcal{A}$ is triangularizable;*

(vi) *the trace is permutable on the set of finite-rank operators in $\mathcal{A}$.*

Proof. It is easily seen that triangularizability implies each of the other conditions. In fact, the spectral mapping theorem (7.2.6) shows that triangularizability implies the seemingly stronger result where "finite-rank nilpotent" is replaced by "quasinilpotent" in conditions (ii), (iii), and (iv). That (i) implies (v) is trivial.

In the other direction, note that (vi) implies (v) by Theorem 2.2.1, and (v) implies (iv), (iii), and (ii) by the Spectral Mapping Theorem (7.2.6). Also, (ii) implies (iii) and (iv).

It remains to be shown, then, that each of (iii) and (iv) implies triangularizability. Choose a maximal chain of invariant subspaces of $\mathcal{A}$. If $\mathcal{A}$ is not triangularizable, there is some $\mathcal{M}$ in the chain with $\dim \mathcal{M}/\mathcal{M}_- > 1$. Then $\mathcal{A}_\mathcal{M} \neq 0$ (for otherwise the chain would not be maximal). By part (i) of the Block Triangularization Theorem, there are operators C and D in $\mathcal{F}$ such that

(a) $$C_\mathcal{M} = \begin{pmatrix} 0 & 1 \\ 0 & 0 \end{pmatrix} \oplus 0 \quad \text{and} \quad D_\mathcal{M} = \begin{pmatrix} 0 & 0 \\ 1 & 0 \end{pmatrix} \oplus 0$$

with respect to some basis for $\mathcal{M}/\mathcal{M}_-$; and
(b) $C_\mathcal{N} = D_\mathcal{N} = 0$ for all $\mathcal{N}$ not in the $\mathcal{E}_\omega$ containing $\mathcal{M}$.

It then follows from part (iii) of the Block Triangularization Theorem that $C_\mathcal{L}$ and $D_\mathcal{L}$ are, respectively, similar to $C_\mathcal{M}$ and $D_\mathcal{M}$ whenever $C_\mathcal{L} \neq 0$ or $D_\mathcal{L} \neq 0$. In particular, $C_\mathcal{K}^2 = D_\mathcal{K}^2 = 0$ for all $\mathcal{K} \in \mathcal{E}$.

It follows that $\sigma(C) = \sigma(D) = \{0\}$. One may see this by Ringrose's Theorem (7.2.3): Extend the given chain to a triangularizing chain for C and note that the diagonal coefficients of C^2 are all 0, so $\sigma(C^2) = \{0\}$. Similarly, $\sigma(D^2) = \{0\}$. Once C and D are known to be quasinilpotent, it follows that they are nilpotent, since they have finite ranks. But then

$$(C + D)_\mathcal{M} = \begin{pmatrix} 0 & 1 \\ 1 & 0 \end{pmatrix} \oplus 0,$$

from which it follows that $C + D$ is not quasinilpotent, contradicting (iii).
Similarly,

$$(CD)_\mathcal{M} = \begin{pmatrix} 1 & 0 \\ 0 & 0 \end{pmatrix} \oplus 0$$

is not quasinilpotent, so neither is CD, contradicting (iv). This establishes the corollary. $\qquad\qquad\qquad\qquad\qquad\qquad\qquad\qquad\qquad\qquad\quad\Box$

The next example shows that the sum of two simultaneously triangularizable nilpotent compact operators need not be nilpotent, although it must be quasinilpotent.

Example 7.5.11. *On the space ℓ^2, define the operators A and B by*

$$Ae_n = \begin{cases} \frac{1}{2^n}e_{n-1} & \text{if } n \text{ odd,} \\ 0 & \text{if } n \text{ even;} \end{cases}$$

$$Be_n = \begin{cases} \frac{1}{2^n}e_{n-1} & \text{if } n \text{ even and } n \geq 2, \\ 0 & \text{if } n \text{ odd or } n = 0, \end{cases}$$

with respect to the standard basis $\{e_n\}_{n=0}^{\infty}$ of ℓ^2. Then both of A and B are in upper triangular form with respect to the $\{e_n\}_{n=0}^{\infty}$, and $A^2 = B^2 = 0$. However, $A + B$ is not nilpotent (though it is, of course, quasinilpotent).□

A sufficient condition for triangularizability that is obviously not necessary is nonetheless worth mentioning; it is implicit in Corollary 7.5.10.

Corollary 7.5.12. *A uniformly closed algebra of compact operators is triangularizable if 0 is the only finite-rank operator in the algebra.*

Proof. If any of the $\mathcal{M}/\mathcal{M}_-$ arising in a maximal triangularization have dimension greater than 2, the Block Triangularization Theorem (7.5.9) provides a nontrivial ideal $\mathcal{J}$ of finite-rank operators in the algebra.   □

7.6 Approximate Commutativity

There are infinite-dimensional versions of the approximate commutativity results of Section 1.6. We begin with a result that gives a certain indication of the extent to which triangularizability generalizes commutativity: $AB - BA$ must be merely quasinilpotent rather than 0.

Theorem 7.6.1. *An algebra $\mathcal{A}$ of compact operators is triangularizable if and only if each commutator $AB - BA$ of operators A and B in $\mathcal{A}$ is quasinilpotent.*

Proof. If $\mathcal{A}$ is triangularizable, then every commutator is quasinilpotent, by the spectral mapping theorem (7.2.6).

For the converse, first note that the property of having quasinilpotent commutators is inherited by quotients, so the Triangularization Lemma (Lemma 7.1.11) implies that it suffices to show that every such $\mathcal{A}$ has nontrivial invariant subspaces. Given $\mathcal{A}$, its uniform closure $\bar{\mathcal{A}}$ also satisfies the hypothesis, since a uniform limit of compact quasinilpotent operators is quasinilpotent (Corollary 7.2.11). By Theorem 7.4.7, then, the set of finite-rank operators in $\bar{\mathcal{A}}$ is a transitive subalgebra.

Let $F \in \bar{A}$ and F have any finite rank ≥ 2. Then the restriction of $F\bar{A}$ to $F\mathcal{X}$ consists of all operators on the range of F (by Burnside's Theorem 1.2.2), and thus FAF contains operators of the form

$$C = \begin{pmatrix} 0 & 1 \\ 0 & 0 \end{pmatrix} \oplus 0 \quad \text{and} \quad D = \begin{pmatrix} 0 & 0 \\ 1 & 0 \end{pmatrix} \oplus 0 .$$

Then $CD - DC = \begin{pmatrix} 1 & 0 \\ 0 & -1 \end{pmatrix} \oplus 0$ is not quasinilpotent.   □

This result is generalized to semigroups in Theorem 8.5.2.

We can now show that a collection of compact operators is triangularizable if all its finite subsets are approximately similar to commuting operators.

Theorem 7.6.2. *Let $\mathcal{J}$ be a family of compact operators with the following property: Given any finite subset $\{A_1, A_2, \ldots, A_k\}$ of $\mathcal{J}$ there is a constant $M > 0$ such that for every $\varepsilon > 0$ there are commuting compact operators $\{D_1, D_2, \ldots, D_k\}$ with $\|D_j\| \leq M$ and there is an invertible S satisfying*

$$\|S^{-1}A_j S - D_j\| < \varepsilon$$

for every j. Then $\mathcal{J}$ is triangularizable.

Proof. By the previous theorem, it suffices to show that $AB - BA$ is quasinilpotent for all A and B in the algebra generated by $\mathcal{J}$. Given A and B, there are noncommutative polynomials p and q in some number, say k, of variables, and operators $\{A_1, A_2, \ldots, A_k\}$ in $\mathcal{J}$ such that $A = p(A_1, A_2, \ldots, A_k)$ and $B = q(A_1, A_2, \ldots, A_k)$. We show that for every $\varepsilon > 0$ there is an invertible operator S such that

$$\|S^{-1}(AB - BA)S\| < \varepsilon ;$$

this clearly implies that $AB - BA$ is quasinilpotent (since similarity preserves spectrum).

Note that every polynomial, and, in particular, the polynomial h defined by

$$h(X_1, X_2, \ldots, X_k) = p(X_1, X_2, \ldots, X_k)q(X_1, X_2, \ldots, X_k) -$$
$$q(X_1, X_2, \ldots, X_k)p(X_1, X_2, \ldots, X_k),$$

is uniformly continuous on bounded subsets of $\mathcal{B}(\mathcal{X})$. Thus, given $\varepsilon > 0$, there is a $\delta > 0$ such that

$$\|h(X_1, X_2, \ldots, X_k) - h(Y_1, Y_2, \ldots, Y_k)\| < \varepsilon$$

whenever $\|X_j - Y_j\| < \delta$ and $\|X_j\| \leq M + 1$ and $\|Y_j\| \leq M + 1$ for all j.

Choose, using the hypothesis of the theorem, commuting compact operators $\{D_1, D_2, \ldots, D_k\}$ with $\|D_j\| \leq M$ for all j and an invertible S such that

$$\|S^{-1}A_j S - D_j\| < \delta \quad \text{for all } j.$$

Note that $h(D_1, D_2, \ldots, D_k) = 0$, since the $\{D_j\}$ commute. Note also that

$$S^{-1}h(A_1, A_2, \ldots, A_k)S = h(S^{-1}A_1 S, S^{-1}A_2 S, \ldots, S^{-1}A_k S) .$$

Combining these facts with

$$\|h(S^{-1}A_1 S, S^{-1}A_2 S, \ldots, S^{-1}A_k S) - h(D_1, D_2, \ldots, D_k)\| < \varepsilon$$

gives

$$\|S^{-1}h(A_1, A_2, \ldots, A_k)S\| < \varepsilon,$$

or

$$\|S^{-1}(AB - BA)S\| < \varepsilon,$$

so $AB - BA$ has spectral radius 0. $\qquad\qquad\qquad\qquad\qquad\qquad\square$

To get a converse to this theorem, that triangularizability implies approximate similarity to a commutative set, we need to restrict to operators on Hilbert space. We require the following lemma.

Lemma 7.6.3. *If Q is a compact quasinilpotent operator on a separable Hilbert space and $\mathcal{C}$ is any triangularizing chain for Q, then, for every $\varepsilon > 0$ there exists a finite sequence*

$$0 = P_0 < P_1 < P_2 < \cdots < P_k = I$$

of orthogonal projections onto subspaces in $\mathcal{C}$ such that

$$\|(P_{i+1} - P_i)\,Q\,(P_{i+1} - P_i)\| < \varepsilon$$

for $i = 0, 1, \ldots, k - 1$.

Proof. We proceed in several steps. We identify the subspaces in $\mathcal{C}$ with the orthogonal projections onto them.

Claim: Given any $P \in \mathcal{C}$ other than I there is an $R \in \mathcal{C}$ such that $P < R$ and $\|(R - P)Q(R - P)\| < \varepsilon$.

This claim is established by considering two separate cases. First suppose that $P\mathcal{H}$ has an immediate successor in $\mathcal{C}$. In that case we simply let R be the projection onto the successor. Then $P\mathcal{H} = (R\mathcal{H})_-$, so $(R - P)Q(R - P)$ is the diagonal coefficient of Q (times a rank-one identity) and is thus 0 (since Q is quasinilpotent; see Theorem 7.2.3). Suppose that $P\mathcal{H}$ has no immediate successor. Then $P\mathcal{H} = \cap\{\mathcal{M} \in \mathcal{C} : \mathcal{M} \supsetneq P\mathcal{H}\}$.

If $P\mathcal{H} = \cap \mathcal{M}_\alpha$ with $\mathcal{M}_\alpha \in \mathcal{C}$, then $P\mathcal{H} = \bigcap_{j=1}^{\infty} \mathcal{M}_{\alpha_j}$ for some sequence $\{\mathcal{M}_{\alpha_j}\} \subseteq \mathcal{C}$ with $\mathcal{M}_{\alpha_{j+1}} \subseteq \mathcal{M}_{\alpha_j}$. (The separable Hilbert space $\mathcal{H}$ has the Lindelöf property that every open cover of a subset has a countable subcover, so $\bigcup(\mathcal{H} \setminus \mathcal{M}_\alpha) = \bigcup_{j=1}^{\infty}(\mathcal{H} \setminus \mathcal{M}_{\beta_j})$ for some sequence $\mathcal{M}_{\beta_j}$, and therefore $\bigcap \mathcal{M}_\alpha = \bigcap_{j=1}^{\infty} \mathcal{M}_{\beta_j}$. Then let $\mathcal{M}_{\alpha_j} = \bigcap_{i=1}^{j} \mathcal{M}_{\beta_i}$ for each j.) Let P_j denote the orthogonal projection onto $\mathcal{M}_{\alpha_j}$. We claim that $\|(P_j - P)Q(P_j - P)\| < \varepsilon$ for some j.

If not, then for each j there is a unit vector f_j in $\mathcal{M}_{\alpha_j}$ orthogonal to $P\mathcal{H}$ with $\|(P_j - P)Q(P_j - P)f_j\| \geq \frac{\varepsilon}{2}$. Some subsequence of the $\{f_j\}$ converges weakly to a vector f. The compactness of Q implies that $\{\|Qf_j - Qf\|\} \to 0$. Thus $\|Qf\| \geq \frac{\varepsilon}{2}$, so $f \neq 0$. On the other hand, $f \in P\mathcal{H}$ and $f \perp P\mathcal{H}$, so $f = 0$. This contradiction shows that $\|(P_j - P)Q(P_j - P)\| < \varepsilon$ for some j.

Now let $\mathcal{S}$ denote the set of all projections $P \in \mathcal{C}$ such that there is a finite chain

$$0 = P_0 < P_1 < \cdots < P_k = P$$

with

$$\|(P_j - P_{j-1})Q(P_j - P_{j-1})\| < \varepsilon \quad \text{for } j = 1, \ldots, k.$$

We must show that $I \in \mathcal{S}$. Let $\hat{P}$ denote the span of the elements of $\mathcal{S}$ (i.e., the projection onto the span of the subspaces corresponding to projections in $\mathcal{S}$). It suffices to show that $\hat{P} \in \mathcal{S}$, for then if $\hat{P} \neq I$, we could add another projection to the chain to go beyond $\hat{P}$, by the claim proven above.

We now claim that there is a $P' < \hat{P}$ in $\mathcal{C}$ such that $\|(\hat{P} - P')Q(\hat{P} - P')\| < \varepsilon$. This can be shown by imitating the proof of the earlier claim establishing the existence of a projection above a given one. Alternatively, this can be derived by applying the previous claim to Q^* and taking adjoints ($\mathcal{M}^\perp \in$ Lat Q if and only if $\mathcal{M} \in$ Lat Q^*).

Given such a P', there is a $P_n \in \mathcal{S}$ with $P_n > P'$. If $0 = P_0 \subset P_1 \subset \cdots \subset P_n$ satisfy

$$\|(P_{j+1} - P_j)Q(P_{j+1} - P_j)\| < \varepsilon$$

for $j = 0, 1, \ldots, n$, then, since

$$\|(\hat{P} - P_n)Q(\hat{P} - P_n)\| \leq \|(\hat{P} - P')Q(\hat{P} - P')\|,$$

it follows that $\hat{P} \in \mathcal{S}$, and the proof is complete.   □

We can now show that, up to similarity, simultaneous triangularizability implies approximate commutativity.

Theorem 7.6.4. *If $\{A_1, A_2, \ldots, A_k\}$ is a triangularizable family of compact operators on a Hilbert space, then there exist commuting compact normal operators $\{D_1, D_2, \ldots, D_k\}$ such that for every $\varepsilon > 0$ there is an invertible operator S satisfying*

$$\|S^{-1} A_j S - D_j\| < \varepsilon$$

for $j = 1, 2, \ldots, k$.

Proof. First, the D_j can be explicitly defined in terms of the A_j and the triangularizing chain. Let $\mathcal{C}_0$ be the collection of subspaces $\mathcal{M}$ in the chain such that $\mathcal{M}_- \neq \mathcal{M}$, and, for each j and each $\mathcal{M} \in \mathcal{C}_0$, let $\lambda_{\mathcal{M}}(A_j)$ be the diagonal coefficient (see Definition 7.2.2) of A_j corresponding to $\mathcal{M}$. Let $P_{\mathcal{M}}$ and $P_{\mathcal{M}_-}$ denote the orthogonal projections onto $\mathcal{M}$ and $\mathcal{M}_-$, respectively. Then we can define

$$D_j = \sum_{\mathcal{M} \in \mathcal{C}_0} \lambda_{\mathcal{M}}(A_j)(P_{\mathcal{M}} - P_{\mathcal{M}_-}) \quad \text{for each } j \ .$$

Since $\mathcal{C}_0$ is totally ordered, the $\{P_{\mathcal{M}} - P_{\mathcal{M}_-} : \mathcal{M} \in \mathcal{C}_0\}$ are mutually orthogonal, so $\|D_j\|$ is the supremum of the $\{|\lambda_{\mathcal{M}}(A_j)|\}$, which is the spectral radius of A_j. Each D_j is a compact normal operator, and the $\{D_j\}$ obviously commute with each other.

For each j, let $Q_j = A_j - D_j$. Then Q_j is compact, and every diagonal coefficient of Q_j with respect to the given chain is 0, since $\lambda_{\mathcal{M}}(Q_j) = \lambda_{\mathcal{M}}(A_j) - \lambda_{\mathcal{M}}(D_j) = 0$. Thus each Q_j is quasinilpotent, by Ringrose's Theorem (7.2.3).

Fix $\varepsilon > 0$. Lemma 7.6.3 can be applied to each Q_j to get a finite set of orthogonal projections onto subspaces in the given triangularizing chain so that the corresponding diagonal blocks have norms less than $\frac{\varepsilon}{2}$. It is clear that refining the partition by adding additional projections onto subspaces in the triangularizing chain improves the corresponding inequalities. Thus if $\{P_1, P_2, \ldots, P_n\}$ is the union, in increasing order, of the sets of projections arising for all the $\{Q_j\}$, and if $P_0 = 0$, $P_{n+1} = I$, then $\|(P_i - P_{i-1}) Q_j (P_i - P_{i-1})\| < \frac{\varepsilon}{2}$ for $j = 1, 2, \ldots, k$ and for $i = 1, 2, \ldots, n+1$.

Let δ be a positive number less than 1; δ will be specified below. In terms of δ, define the operator S by $S = \sum_{i=1}^{n+1} \delta^i (P_i - P_{i-1})$. We claim that, for δ sufficiently small, this S satisfies the conclusion of the theorem; i.e., that

$$\|S^{-1} A_j S - D_j\| < \varepsilon \ .$$

To see this, let $Q_j^{(\ell)}$ denote the "ℓ-th superdiagonal of Q_j"; i.e.,

$$Q_j^{(\ell)} = \sum_{i=1}^{n+1-\ell} (P_i - P_{i-1}) Q_j (P_{i+\ell} - P_{i+\ell-1}) \ .$$

In this notation,

$$S^{-1}Q_jS = Q_j^{(0)} + \delta Q_j^{(1)} + \delta^2 Q_j^{(2)} + \cdots + \delta^n Q_j^{(n+1)} \ ,$$

as can be verified by direct computation. (Alternatively, and perhaps more intuitively, write each of S and Q_j as block matrices with respect to the decomposition of the space corresponding to the $\{P_j\}$, and use block matrix multiplication to verify the above equation.)

Now $Q_j^{(0)} = \sum_{i=1}^{n+1} (P_i - P_{i-1}) Q_j (P_i - P_{i-1})$. Since

$$\|(P_i - P_{i-1}) Q_j (P_i - P_{i-1})\| < \frac{\varepsilon}{2}$$

for each i and the $\{P_i - P_{i-1}\}$ are mutually orthogonal, $\|Q_j^{(0)}\| < \frac{\varepsilon}{2}$ for every j. Also, $\|Q_j^{(\ell)}\| \le \|Q_j\|$ for every ℓ. Hence

$$\|S^{-1}Q_jS\| < \frac{\varepsilon}{2} + (\delta + \delta^2 + \cdots + \delta^n) \|Q_j\|$$

$$\le \frac{\varepsilon}{2} + \frac{\delta}{1-\delta} \|Q_j\| \ .$$

Choose δ sufficiently small that $\frac{\delta}{1-\delta}\|Q_j\| < \frac{\varepsilon}{2}$ for $j = 1, 2, \ldots, k$. Then

$$\|S^{-1}A_jS - D_j\| = \|S^{-1}A_jS - S^{-1}D_jS\|$$

(since S commutes with all the $\{D_j\}$). It follows that

$$\|S^{-1}A_jS - D_j\| = \|S^{-1}Q_jS\| < \varepsilon$$

for each j, proving the theorem.   □

This theorem can be rephrased to show that triangularizability implies a different kind of approximate commutativity.

Corollary 7.6.5. *If $\{A_1, A_2, \ldots, A_k\}$ is a triangularizable family of compact operators on a Hilbert space, and if $\varepsilon > 0$, then there exists an invertible operator T such that*

$$\|(T^{-1}A_iT)(T^{-1}A_jT) - (T^{-1}A_jT)(T^{-1}A_iT)\| < \varepsilon$$

for $i, j = 1, 2, \ldots, k$.

Proof. Let $\varepsilon > 0$, and let M denote the maximum of the norms of the $\{A_j\}$. Define ε_0 to be the smallest of 1, $\frac{\varepsilon M}{6}$, and $\frac{\varepsilon}{6}$.

Use the previous theorem to obtain an invertible T such that

$$\|T^{-1}A_jT - D_j\| < \varepsilon_0 \quad \text{for } j = 1, 2, \ldots, k.$$

For each j, let $E_j = T^{-1}A_jT - D_j$, so that $T^{-1}A_jT = D_j + E_j$ and $\|E_j\| < \varepsilon_0$. Then for any i and j,

$$\|(T^{-1}A_iT)(T^{-1}A_jT) - (T^{-1}A_jT)(T^{-1}A_iT)\|$$
$$= \|(D_i + E_i)(D_j + E_j) - (D_j + E_j)(D_i + E_i)\|$$
$$\leq \|D_iD_j - D_jD_i\| + \|D_iE_j\|$$
$$+ \|E_iD_j\| + \|E_iE_j\| + \|E_jE_i\| + \|E_jD_i\| + \|D_jE_i\|.$$

The first term is 0, and each of the other terms is less than $\frac{\varepsilon}{6}$, so

$$\|(T^{-1}A_iT)(T^{-1}A_jT) - (T^{-1}A_jT)(T^{-1}A_iT)\| < \varepsilon. \qquad \square$$

There is another corollary of Theorem 7.6.4.

Corollary 7.6.6. *If $\{A_1, A_2, \ldots, A_k\}$ is a triangularizable family of compact operators on Hilbert space, and if $\varepsilon > 0$, then there is an invertible operator S such that*

$$\|S^{-1}A_jS\| < \rho(A_j) + \varepsilon$$

for $j = 1, 2, \ldots, k$ (where $\rho(A_j)$ is the spectral radius of A_j).

Proof. In the notation of the proof of Theorem 7.6.4,

$$\|S^{-1}A_jS - D_j\| < \varepsilon$$

implies that

$$\|S^{-1}A_jS\| < \|D_j\| + \varepsilon.$$

As pointed out towards the beginning of the proof of Theorem 7.6.4, $\|D_j\|$ is the spectral radius of A_j, so the corollary follows. $\qquad \square$

7.7 Notes and Remarks

There have been several infinite-dimensional generalizations of the concept of a collection of triangular matrices. For some results about operators on Hilbert space that have upper triangular matrices with respect to an orthonormal basis (the strictest notion of triangularity), see Herrero [1] and Larson-Wogen [1].

The first systematic study of operator algebras analogous to algebras of upper triangular matrices was that of Kadison-Singer [1], which considers operator algebras on Hilbert space whose intersection with the algebra of adjoints is a maximal abelian von Neumann algebra. Shortly afterwards, Ringrose ([2], [3]) introduced "nest algebras," which are algebras of the form Alg $\mathcal{C}$ where $\mathcal{C}$ is a chain of subspaces of a Hilbert space. There has been a lot of subsequent work on nest algebras, and there is a beautiful exposition describing most of that work in Davidson [1].

Davidson [1] also discusses the (more general) CSL algebras introduced in Arveson [3]. Even more generally, an algebra of operators is said to be reflexive if $\mathcal{A} = \text{Alg Lat} \mathcal{A}$. The study of reflexivity can be said to have been initiated by Sarason [1], although the idea of undertaking a general study was introduced in Radjavi-Rosenthal [1], which was inspired by both Sarason [1] and Arveson [1]. Some early work on reflexive algebras is contained in Radjavi-Rosenthal [2, Section 9.2]. A very nice overview of all of these concepts, and of other related studies, is given in Larson [3].

Kadison and Singer [1] defined a triangular operator algebra to be *hyper-reducible* if it has a chain of invariant subspaces such that the self-adjoint projections onto the subspaces generate a maximal abelian von Neumann algebra. Kadison-Singer [1] asked whether every operator is contained in some hyperreducible triangular algebra. J.R. Ringrose raised the specific question of whether there is a similarity that the "Volterra lattice" (i.e., the continuous chain of Example 7.1.3 in the case of $\mathcal{L}^2(0,1)$) onto a chain that does not generate a maximal abelian von Neumann algebra. Larson [2] showed that there is such a similarity S, and therefore that $S^{-1}VS$ (where V is the Volterra operator defined in Example 7.1.4) is not in any hyperreducible triangular algebra. In fact, Larson [2] (using, in part, the results of Andersen [1]) showed that any chain of subspaces of a Hilbert space that is order-isomorphic to the Volterra chain is similar to it. Davidson [2] improved this to the theorem that two chains of subspaces each of which is closed under intersections and spans must be similar if there is a dimension-preserving order isomorphism of one onto the other. An excellent exposition of this theorem is presented in Davidson [1]. A study of "hypertriangularizability" is begun in Hadwin-Nordgren-Radjabalipour-Radjavi-Rosenthal [2].

Don Hadwin has suggested the study of triangularization in the Calkin algebra $(\mathcal{B}(\mathcal{H})/\mathcal{K}(\mathcal{H})$ for $\mathcal{H}$ a separable Hilbert space). Elements of the Calkin algebra have many "invariant subspaces": In fact, Voiculescu [1] proved that every closed separable subalgebra is reflexive. However, Hadwin points out, there is no analogue of the Triangularization Lemma, so it is not clear whether individual elements of the Calkin algebra are triangularizable in it, even if they are normal. Hadwin has shown that the image in the Calkin algebra of a commutative separable C^*-subalgebra of $\mathcal{B}(\mathcal{H})$ is triangularizable. A remarkable study of normal elements of the Calkin algebra is presented in Brown-Douglas-Fillmore [1]. Hadwin's question also suggests the possibility of studying triangularization in other C^*-algebras.

There is an interesting theory of "triangular models" of certain compact operators; this was developed by a number of Soviet mathematicians; see Gohberg-Krein [2] and Brodskii [1]. It is shown in Radjavi-Rosenthal [1] (see also Radjavi-Rosenthal [2, Example 9.2.6]) that, in the case $p = 2$, the algebra of Example 7.1.3 is generated as a strongly closed algebra by the Volterra operator and multiplication by x.

The definition of triangularizable given in Definition 7.1.1 was introduced in Wojtyński [1], where Engel's Theorem was generalized to the result that a Lie algebra of quasinilpotent operators in a C_p class is triangularizable. Several years later, Laurie-Nordgren-Radjavi-Rosenthal [1] made the same definition (although they were unaware of Wojtyński's paper).

The idea and notation in Definition 7.1.6 is due to Ringrose [2], which also contains Definition 7.2.2 and the beautiful Theorem 7.2.3. The proof of Theorem 7.2.3 we have given appears to be a little simpler than other proofs, although the basic ideas are similar. For other proofs of Ringrose's Theorem (7.2.3) see Ringrose [5] and, in the Hilbert space case, Davidson [1] and Radjavi-Rosenthal [2, Theorem 5.12].

The Triangularization Lemma (7.1.11) was implicit in most earlier triangularization proofs but was first explicitly formulated in Radjavi-Rosenthal [5]. The Spectral Mapping Theorem (7.2.6) is taken from Laurie-Nordgren-Radjavi-Rosenthal [1]. The version of Ringrose's Theorem including multiplicity, Theorem 7.2.9, was contained in Ringrose's original paper [2], with a different proof.

Lomonosov's Lemma 7.3.1 is the essence of his seminal paper Lomonosov [1]. The scope of Corollary 7.4.13, which was immediately recognized by those working on invariant subspaces to be a consequence of Lomonosov [1], was not at all clear for several years: It seemed possible that it might apply to every operator on Hilbert space. This was shown not to be the case in Hadwin-Nordgren-Radjavi-Rosenthal [1]. Lomonosov [2] contains a modest extension of his earlier work (also see Chevreau-Li-Pearcy [1]); the extension still does not apply to all operators (Hadwin [2]).On the other hand, Troitsky [1] shows that an operator constructed in Read [1], which has no nontrivial invariant subspaces, commutes with a nonscalar operator S satisfying $ST = TS$ for some nonscalar T commuting with an operator of rank one. Lomonosov's techniques were extended in Simonič [2] to show the existence of invariant subspaces for compact perturbations of self-adjoint operators on a real Hilbert space.

The first infinite-dimensional version of Burnside's Theorem for topological irreducibility was established by Arveson [1], who showed that the only strongly closed transitive algebra of operators that contains a maximal abelian von Neumann algebra is the algebra of all operators. Several other such theorems can be found in Radjavi-Rosenthal [2, Chapter 8]. The assertion of Theorem 7.4.11 with "finite rank" instead of "compact" was established in Nordgren-Radjavi-Rosenthal [3]; the extension to compact operators was an immediate combination of this result and Lomonosov [1]. The improvement contained in Lemma 7.4.2 and Corollary 7.4.3 was discovered by Barnes [1], whose proofs use algebraic techniques including

Wedderburn theory. The proofs given in the text were obtained in unpublished work of the authors and Eric Nordgren, as described in Rosenthal [1], which also contains Example 7.4.4.

Lemma 7.4.5, which is needed in Chapter 8 of this book for triangularization of semigroups, is a variant of a result obtained in Radjavi [1]. It is easy to show the existence of a finite-rank operator in a uniformly closed algebra containing a nonzero quasinilpotent compact operator by using the analytic functional calculus (e.g., see the proof of Theorem 8.23 in Radjavi-Rosenthal [2]).

The infinite-dimensional version of McCoy's Theorem (7.3.3), Theorem 7.3.4, and Corollaries 7.3.6 and 7.3.7 are from Laurie-Nordgren-Radjavi-Rosenthal [1]. Murphy [1] observed that Corollary 7.3.8 follows from these results. Other related results can be found in Barnes-Katavolos [1] which includes a study of operator algebras whose quasinilpotent operators form an ideal.

The material on idempotents in Section 7.5 is widely known and arises in several contexts. The Block Triangularization Theorem (7.5.9) is from Radjavi-Rosenthal [5]; some of the ideas of the proof go back to Radjavi-Rosenthal ([6], [4]) and Radjavi [2]. Theorem 7.6.1 is in Katavolos-Radjavi [1]. Lemma 7.6.3 was established by Ringrose ([3],[4]) in the course of his development of nest algebras (see also Davidson [1]). The rest of section 7.6, the results on "approximate commutativity," are all from Jafarian-Radjavi-Rosenthal-Sourour [1].

CHAPTER 8
Semigroups of Compact Operators

In this chapter we show that most of the results of Chapters 2 through 5 generalize to semigroups of compact operators on Banach spaces. In many cases we can show that the norm closure of $\mathbb{R}^+ S$ contains a finite-rank operator other than zero. This often allows us to reduce the given question to the case of operators on a finite-dimensional space and then to use the results of the first five chapters. One important case, in which finite-rank operators are conspicuously absent, is treated in the first section of this chapter, where we establish Turovskii's Theorem that a semigroup of compact quasinilpotent operators is triangularizable.

8.1 Quasinilpotent Compact Operators

The principal aim of this section is to prove Turovskii's Theorem (8.1.11).

Notation 8.1.1. If A is an operator on the Banach space $\mathcal{X}$, then L_A and R_B denote the left and right multiplication operators on $\mathcal{B}(\mathcal{X})$, respectively: $L_A(T) = AT$ and $R_B(T) = TB$ for $T \in \mathcal{B}(\mathcal{X})$. The symbols $\mathcal{F}(\mathcal{X})$ and $\mathcal{K}(\mathcal{X})$ denote, respectively, the set of all finite-rank operators and the set of all compact operators on $\mathcal{X}$.

Lemma 8.1.2. *If A and B are compact operators on $\mathcal{X}$, then $L_A R_B$ is a compact operator on $\mathcal{B}(\mathcal{X})$.*

Proof. Denote the closed unit balls of $\mathcal{X}$ and $\mathcal{B}(\mathcal{X})$ by $\mathcal{X}_1$ and $\mathcal{B}_1$, respectively. Let $\mathcal{C}$ denote the Banach space of all continuous functions mapping the compact space $\overline{B\mathcal{X}_1}$ into $\mathcal{X}$, endowed with the uniform norm. Let $\mathcal{E} = \{ AT|_{\overline{B\mathcal{X}_1}} : T \in \mathcal{B}_1 \}$. We first show that $\mathcal{E}$ is precompact (i.e., that its closure is compact) as a subset of $\mathcal{C}$. For this, we use Ascoli's Theorem (Dieudonné [1, p. 137]), which states that it suffices to show that $\mathcal{E}$ is equicontinuous (i.e., for each $f_0 \in \overline{B\mathcal{X}_1}$ and every $\varepsilon > 0$ there is a $\delta > 0$ such that $\|ATf_0 - ATf\| < \varepsilon$ whenever $AT \in \mathcal{E}$ and $\|f_0 - f\| < \delta$) and that $\{ ATf : AT \in \mathcal{E} \}$ is a precompact subset of $\mathcal{X}$ for each fixed $f \in \overline{B\mathcal{X}_1}$.

For equicontinuity, simply note that $\|ATf_0 - ATf\| \leq \|AT\| \|f_0 - f\| \leq \|A\| \|f_0 - f\|$. The second condition follows immediately from the compactness of A and the boundedness of $\{ Tf : T \in \mathcal{B}_1 \}$.

Now, to show that $L_A R_B$ is a compact operator, suppose that $\{T_n\}$ is a bounded sequence; we must show that $\{ AT_{n_i} B \}$ converges in norm for some subsequence $\{T_{n_i}\}$. We can assume $\{T_n\}$ is in $\mathcal{B}_1$. Then $\{ AT_n|_{\overline{B\mathcal{X}_1}} \}$ has a convergent subsequence $\{ AT_{n_i}|_{\overline{B\mathcal{X}_1}} \}$, so $\{ AT_{n_i} B \}$ converges uniformly on $\mathcal{X}_1$, and thus in norm. $\qquad\square$

Notation 8.1.3. *For subsets $\mathcal{E}$ and $\mathcal{F}$ of $\mathcal{B}(\mathcal{X})$, and subsets $\mathcal{D}$ of $\mathcal{X}$ and Γ of $\mathbb{C}$, let*

$$\mathcal{E}\mathcal{F} = \{EF : E \in \mathcal{E}, F \in \mathcal{F}\},$$
$$\mathcal{E}^2 = \mathcal{E}\mathcal{E},$$
$$\mathcal{E}\mathcal{D} = \{Ex : E \in \mathcal{E}, x \in \mathcal{D}\},$$
$$\Gamma\mathcal{E} = \{\gamma E : \gamma \in \Gamma, E \in \mathcal{E}\},$$

and

$$\|\mathcal{D}\| = \sup\{\|x\| : x \in \mathcal{D}\},$$
$$\|\mathcal{E}\| = \sup\{\|E\| : E \in \mathcal{E}\}.$$

Products $\mathcal{E}_1\mathcal{E}_2\cdots\mathcal{E}_n$ and $\mathcal{E}^n$ are defined by induction, and $\mathcal{E}\{x\}$ is abbreviated to $\mathcal{E}x$. Note that if $\mathcal{S}(\mathcal{E})$ is the semigroup generated by a set $\mathcal{E}$ of operators, then $\mathcal{S}(\mathcal{E}) = \bigcup_{j=1}^{\infty} \mathcal{E}^j$. Inequalities such as

$$\|\mathcal{E}\mathcal{F}\| \leq \|\mathcal{E}\| \cdot \|\mathcal{F}\|$$

are trivial to check.

Definition 8.1.4. *The joint spectral radius of a bounded set $\mathcal{E} \subseteq \mathcal{B}(\mathcal{X})$ is*

$$\rho(\mathcal{E}) = \inf\{\|\mathcal{E}^n\|^{\frac{1}{n}} : n \in \mathbb{N}\}.$$

Note that, by Theorem 6.1.10, this reduces to the ordinary spectral radius in the case where $\mathcal{E}$ consists of a single operator.

Lemma 8.1.5. *For a bounded set $\mathcal{E} \subseteq \mathcal{B}(\mathcal{X})$,*

(i) $\rho(\mathcal{E}) = \lim\limits_{n\to\infty} \|\mathcal{E}^n\|^{\frac{1}{n}}$.

(ii) $\rho(A) \leq \rho(\mathcal{E})$ for each A in $\mathcal{E}$, and

(iii) $\rho(\mathrm{co}\,\mathcal{E}) = \rho(\mathcal{E})$, where $\mathrm{co}\,\mathcal{E}$ is the convex hull of $\mathcal{E}$.

Proof. To prove (i), it suffices to show that

$$\limsup_n \|\mathcal{E}^n\|^{\frac{1}{n}} \leq \rho(\mathcal{E}).$$

Fix a positive integer m. For each n, let r_n be the remainder that n leaves upon division by m, so that $n = mq_n + r_n$ for some integer q_n. Note that

$$\lim_{n\to\infty} \frac{r_n}{n} = 0 \quad \text{and} \quad \lim_{n\to\infty} \frac{mq_n}{n} = 1.$$

Note also that, for integers j and k, we have

$$\|\mathcal{E}^{jk}\| = \|(\mathcal{E}^j)^k\| \leq \|\mathcal{E}^j\|^k \quad \text{and} \quad \|\mathcal{E}^{j+k}\| \leq \|\mathcal{E}^j\| \cdot \|\mathcal{E}^k\|.$$

It follows that

$$\|\mathcal{E}^n\|^{\frac{1}{n}} = \|\mathcal{E}^{mq_n+r_n}\|^{\frac{1}{n}} \leq (\|\mathcal{E}^m\|^{\frac{1}{m}})^{\frac{mq_n}{n}} \cdot \|\mathcal{E}\|^{\frac{r_n}{n}}.$$

Since $\lim_{n\to\infty} r_n/n = 0$ and $\lim_{n\to\infty} mq_n/n = 1$, this yields

$$\limsup_n \|\mathcal{E}^n\|^{\frac{1}{n}} \leq \|\mathcal{E}^m\|^{\frac{1}{m}}.$$

This holds for every natural number m, so the right side of this inequality can be replaced with $\rho(\mathcal{E})$, as desired.

The proof of (ii) is easy: Just note that $\|A^n\|^{\frac{1}{n}} \leq \|\mathcal{E}^n\|^{\frac{1}{n}}$ and take limits.

To prove (iii), observe that $\|\mathrm{co}\,\mathcal{F}\| \leq \|\mathcal{F}\|$ for any set $\mathcal{F}$. Also, we have $(\mathrm{co}\,\mathcal{E})^n \subseteq \mathrm{co}(\mathcal{E}^n)$ for every integer n, so that

$$\|\mathcal{E}^n\| \leq \|(\mathrm{co}\,\mathcal{E})^n\| \leq \|\mathrm{co}(\mathcal{E}^n)\| \leq \|\mathcal{E}^n\|.$$

Thus $\lim_{n\to\infty} \|\mathcal{E}^n\|^{\frac{1}{n}} = \lim_{n\to\infty} \|(\mathrm{co}\,\mathcal{E})^n\|^{\frac{1}{n}}$. $\qquad\square$

Lemma 8.1.6. *Assume that $\mathcal{S}$ is a semigroup of compact operators with $\|\mathcal{S}^n\| = 1$ for all n. Assume also that $\mathcal{S}$ is compact as a subset of $\mathcal{B}(\mathcal{X})$ with the norm topology. Then $\mathcal{S}$ contains an operator with 1 as an eigenvalue.*

Proof. Let

$$\mathcal{S}_0 = \{A \in \mathcal{S} : A = \lim_{k\to\infty} S_k \quad \text{for some} \quad S_k \in \mathcal{S}^{n_k} \text{ with } \{n_k\} \to \infty\}.$$

For each n, $\mathcal{S}^n$ is compact, so it contains a member S_n with $\|S_n\| = 1$. Then there is a subsequence $\{S_{n_k}\}$ with a limit, say A_0, in $\mathcal{S}_0$; clearly $\|A_0\| = 1$. We can split S_{n_k} into a product $R_k T_k$ with $R_k \in \mathcal{S}^{r_k}$ and $T_k \in \mathcal{S}^{t_k}$, with $\{t_k\} \to \infty$ and $\{r_k\} \to \infty$. By passing to convergent subsequences we get A_1 and B_1 in $\mathcal{S}_0$ with $A_0 = B_1 A_1$. Observe that $\|A_1\| = \|B_1\| = 1$, because $\|A_0\| = 1$.

By induction, we can find sequences $\{A_m\}$ and $\{B_m\}$ of operators in $\mathcal{S}_0$ with norm one such that $A_m = B_{m+1} A_{m+1}$. We next choose a subsequence $\{A_{m_i}\}$ converging to some A. Note that

$$A_{m_i} = T_i A_{m_{i+1}},$$

where $T_i = B_{m_i+1} \cdots B_{m_{i+1}}$. By passing to a further subsequence to make $\{T_i\}$ convergent, we obtain T in $\mathcal{S}$ with $A = TA$. Since $A \neq 0$, there are nonzero vectors in the range of A; every such vector is in the eigenspace of T corresponding to eigenvalue 1. $\qquad\square$

Lemma 8.1.7. *Let $\mathcal{F}$ be a finite set of compact operators. If $\mathcal{S}(\mathcal{F})$ (the semigroup generated by $\mathcal{F}$) is bounded, then it is precompact.*

Proof. Let $\{S_n\}$ be any sequence in $\mathcal{S} = \mathcal{S}(\mathcal{F})$. Since $\mathcal{S} = \bigcup_{k=1}^{\infty} \mathcal{F}^k$, and $\mathcal{F}$ is finite, $\mathcal{S}$ differs from $\bigcup_{k=3}^{\infty} \mathcal{F}^k$ by a finite set. Thus we can assume without loss of generality that $\{S_n\}$ is contained in $\bigcup_{k=3}^{\infty} \mathcal{F}^k$ and has a subsequence of the form $F_1 R_{n_i} F_2$, with $R_{n_i} \in \mathcal{S}$ and where F_1 and F_2 are fixed members of $\mathcal{F}$. It follows from the boundedness of $\mathcal{S}$ and Lemma 8.1.2 that some subsequence of $\{F_1 R_{n_i} F_2\}$ is convergent. $\qquad\square$

The following definition is consistent with that given in Definition 6.3.1 for hyperinvariant subspaces of a single operator.

Definition 8.1.8. If $\mathcal{E}$ is a set of operators on a Banach space $\mathcal{X}$ and $\mathcal{M}$ is a subspace of $\mathcal{X}$, we say that $\mathcal{M}$ is *hyperinvariant* for $\mathcal{E}$ if it is invariant under $\mathcal{E}$ and under its commutant $\mathcal{E}' = \{T \in \mathcal{B}(\mathcal{X}) : TS = ST : S \in \mathcal{E}\}$.

Lemma 8.1.9. *Let $\mathcal{F}$ be a bounded set of operators with $\rho(\mathcal{F}) = 1$. For each n, let $\mathcal{S}_n$ be the semigroup generated by $(1 - 1/n)\mathcal{F}$ and the identity. Then the following hold:*

(i) *$\mathcal{S}_n$ is bounded for every n, and*

$$|x|_n = \frac{\|\mathcal{S}_n x\|}{\|\mathcal{S}_n\|}$$

defines a norm on $\mathcal{X}$ equivalent to $\|x\|$. Moreover, $(1 - 1/n)|Ax|_n \leq |x|_n$ for all A in $\mathcal{F}$ and $x \in \mathcal{X}$.

(ii) *$\phi(x) = \limsup_n |x|_n$ is a continuous seminorm on $\mathcal{X}$ whose kernel $\mathcal{N}$ is hyperinvariant for $\mathcal{S}(\mathcal{F})$. Furthermore, $\phi(Ax) \leq \phi(x)$ for all A in $\mathcal{F}$ and $x \in \mathcal{X}$.*

Proof. Let $t = 1 - 1/n$. Since $\rho(t\mathcal{F}) = t < 1$, we have $\|(t\mathcal{F})^k\| < 1$ for sufficiently large k. This shows that $\mathcal{S}_n$ is bounded. The inequality $|x|_n \leq \|x\|$ is clear, and $(\|x\|/\|\mathcal{S}_n\|) \leq |x|_n$ follows from the fact that $I \in \mathcal{S}_n$, so $|\cdot|$ is equivalent to $\|\cdot\|$. Also, for A in $\mathcal{F}$ and $x \in X$, $\mathcal{S}_n(tAx) \subseteq \mathcal{S}_n x$, so that $|tAx|_n \leq |x|_n$. This proves (i).

It is easily seen that $\phi(x)$ is a seminorm on $\mathcal{X}$; its continuity follows from $|x|_n \leq \|x\|$ for all x. Thus its kernel, $\mathcal{N}$, is closed. If $T \in \mathcal{S}(\mathcal{F})$, then $T \in \mathcal{F}^k$ for some k, so that $t^k T \in \mathcal{S}_n$ and

$$\phi(Tx) = \limsup_n \frac{\|\mathcal{S}_n Tx\|}{\|\mathcal{S}_n\|} \leq \limsup_n \frac{\|\mathcal{S}_n x\|}{t^k \|\mathcal{S}_n\|} = \phi(x).$$

This shows that $T\mathcal{N} \subseteq \mathcal{N}$. If S is in the commutant of $\mathcal{S}(\mathcal{F})$, then

$$\phi(Sx) = \limsup_n \frac{\|\mathcal{S}_n Sx\|}{\|\mathcal{S}_n\|} = \limsup_n \frac{\|S\mathcal{S}_n x\|}{\|\mathcal{S}_n\|} \leq \|S\|\phi(x),$$

proving $S\mathcal{N} \subseteq \mathcal{N}$. □

Lemma 8.1.10. *Let $\mathcal{F}$ be a finite set of compact operators with $\rho(\mathcal{F}) = 1$ and let ϕ and $\mathcal{N}$ be as in the lemma above. If $\mathcal{S}(\mathcal{F})$ is unbounded, then $\mathcal{N}$ is a nontrivial subspace of $\mathcal{X}$.*

Proof. We first show that $\mathcal{N} \neq \mathcal{X}$; that is, ϕ is not identically 0. Fix any n. By choosing an $S \in \mathcal{S}_n$ such that $\|S\|$ is close to $\|\mathcal{S}_n\|$ and then a unit vector x_n such that $\|Sx_n\|$ is close to $\|S\|$, we get an x_n with $\|x_n\| = 1$

such that $|x_n|_n > t_n = 1 - 1/n$. By the definition of $|x_n|_n$, there is an S_n in $\mathcal{S}_n$ such that $\|S_n x_n\| > t_n \|S_n\|$.

Since $\|S_n\| \to \infty$ and $\mathcal{F}$ is finite, $S_n \notin \left(1 - \frac{1}{n}\right) \mathcal{F}$, for sufficiently large n. Thus we can assume $S_n = R_n T_n$ with $R_n \in \mathcal{S}_n$ and $T_n \in t_n \mathcal{F}$. Since $\mathcal{F}$ is finite, it contains a member F such that $T_n = t_n F$ for infinitely many n. Since $\mathcal{F}$ is compact and $\lim_{n \to \infty} t_n = 1$, by passing to a subsequence we can assume that $\{T_n x_n\}$ converges to some x_0 in $\mathcal{X}$ (with respect to $\|\cdot\|$). We show that $\phi(x_0) \neq 0$. For sufficiently large n in this subsequence,

$$t_n < \frac{\|R_n T_n x_n\|}{\|S_n\|} \leq \frac{\|S_n T_n x_n\|}{\|S_n\|} = |T_n x_n|_n.$$

Also, $|T_n x_n - x_0|_n \leq \|T_n x_n - x_0\|$, so that

$$|x_0|_n \geq |T_n x_n|_n - \|T_n x_n - x_0\|.$$

Since $\lim_{n \to \infty} \|T_n x_n - x_0\| = 0$, we obtain

$$\phi(x_0) \geq \limsup_{n \to \infty} |T_n x_n| \geq \limsup_{n \to \infty} t_n = 1.$$

We have shown that $\phi \neq 0$. To complete the proof we must show the existence of nonzero vectors in $\mathcal{N}$. Assume that $\mathcal{N} = \{0\}$. This means that ϕ is a norm on $\mathcal{X}$. In other words, the identity mapping from $(\mathcal{X}, \|\cdot\|)$ to $(\mathcal{X}, \phi(\cdot))$ is injective as well as surjective. Since $\phi(x) \leq \|x\|$ for all $x \in \mathcal{X}$, this mapping is a bounded operator. Thus it has a bounded inverse, by the closed graph theorem. Hence there is a positive ε such that $\varepsilon \|x\| \leq \phi(x) \leq \|x\|$ for every x.

We now construct a unit vector y_0 with $\phi(y_0) = 0$ (which implies that $\mathcal{N} \neq \{0\}$ and completes the proof). Since $\left\{\left\|\bigcup_{j=1}^n \mathcal{F}^j\right\|\right\}_{n=1}^{\infty}$ is unbounded, there is an increasing sequence n_k of integers such that

$$\|\mathcal{F}^{n_k}\| > \left\|\bigcup_{j=1}^{n_k - 1} \mathcal{F}^j\right\|.$$

Hence we can pick S_k in $\mathcal{F}^{n_k}$ with $\lim_{k \to \infty} \|S_k\| = \infty$. Using the finiteness of $\mathcal{F}$ and passing to a further subsequence of $\{n_k\}$, we can assume $S_k = F_1 R_k F_2$ for a fixed pair F_1 and F_2 in $\mathcal{F}$, where $R_k \in \mathcal{F}^{n_k - 2}$. Since the sequence $\{R_k / \|S_k\|\}$ is bounded and F_1 and F_2 are compact, it follows from Lemma 8.1.2 that a subsequence of $\{S_k / \|S_k\|\}$ is convergent, say to T. Then T is compact and of norm 1. We pick a unit vector y_k for each S_k with

$$\|S_k y_k\| > \left(1 - \frac{1}{k}\right) \|S_k\|$$

and make a final passing to a subsequence to guarantee the convergence of $\{Ty_k\}$ to a unit vector y_0. Thus $\lim_{k\to\infty}(S_k y_k/\|S_k\|) = y_0$. Since $\phi(S_k y_k) \leq \phi(y_k)$ by Lemma 8.1.9, and since $\phi(y_k) \leq 1$, we get

$$\phi(y_0) = \lim_{k\to\infty} \frac{\phi(S_k y_k)}{\|S_k\|} \leq \lim_{k\to\infty} \frac{1}{\|S_k\|} = 0.$$

$\square$

Theorem 8.1.11. (Turovskii's Theorem) *A semigroup of compact quasinilpotent operators on a Banach space is triangularizable.*

Proof. If the algebra generated by the semigroup consists of quasinilpotent operators, then it (and hence the semigroup $\mathcal{S}$) is triangularizable by Corollary 7.3.5. Without loss of generality, we can assume that the semigroup is closed under multiplication by complex scalars. Then the algebra generated by the semigroup is the same as its convex hull. Thus we can assume that there exist $\{S_j\}$ in the semigroup and positive scalars $\{\alpha_j\}$ summing to 1 such that the operator $T = \Sigma_{j=1}^n \alpha_j S_j$ has nonzero spectral radius. We show that this is impossible.

Let $\mathcal{F} = \{S_1, \ldots, S_n\}$. Note that $\mathcal{F}$ is not triangularizable, for if it were, then T would have spectral radius 0 by the Spectral Mapping Theorem (Theorem 7.2.6). Let $\mathcal{C}$ denote a maximal chain of common invariant subspaces of $\mathcal{F}$. We use the notation of Chapter 7: For $\mathcal{M} \in \mathcal{C}$ with $\mathcal{M}_- \neq \mathcal{M}$ and $A \in \text{Alg}\,\mathcal{C}$, $A_\mathcal{M}$ is the quotient operator induced by A on $\mathcal{M}/\mathcal{M}_-$.

By the block-triangular form of Ringrose's Theorem (7.2.7), $T_\mathcal{M}$ has positive spectral radius for some $\mathcal{M}$. Note that such an $\mathcal{M}$ cannot have $\mathcal{M}/\mathcal{M}_-$ one-dimensional, for if that were the case, each $(S_j)_\mathcal{M}$ would simply be the product of the diagonal coefficient of S_j corresponding to $\mathcal{M}$ and a one-dimensional identity. Thus each $(S_j)_\mathcal{M}$ would be zero, and $T_\mathcal{M}$ would therefore be zero.

For any $\mathcal{M}$ with $\rho(T_\mathcal{M}) > 0$, we prove that $\mathcal{F}_\mathcal{M}$ has a nontrivial invariant subspace, contradicting the maximality of $\mathcal{C}$. Lemma 8.1.5 implies that $\rho(\mathcal{F}_\mathcal{M}) > 0$; rescale the $\{S_j\}$ if necessary (i.e., multiply by $1/\rho(\mathcal{F}_\mathcal{M})$) so that $\rho(\mathcal{F}_\mathcal{M}) = 1$. Let $\mathcal{S}_\mathcal{M}$ denote the semigroup generated by $\mathcal{F}_\mathcal{M}$ and the identity. We claim that $\mathcal{S}_\mathcal{M}$ is unbounded. For if $\mathcal{S}_\mathcal{M}$ were bounded, we could define an equivalent norm on $\mathcal{M}/\mathcal{M}_-$ by

$$|x| = \|\mathcal{S}_\mathcal{M} x\| \quad \text{for} \quad x \in \mathcal{M}/\mathcal{M}_-.$$

Then $|\mathcal{S}_\mathcal{M}| \leq 1$. However, $\rho(\mathcal{F}_\mathcal{M}) = 1$ implies $\rho(\mathcal{S}_\mathcal{M}) \geq 1$, so $|\mathcal{S}_\mathcal{M}^n| \geq 1$ for all n. Thus $|\mathcal{S}_\mathcal{M}^n| = 1$ for all n, and Lemma 8.1.6 yields a nonquasinilpotent operator in the norm closure of $\mathcal{S}_\mathcal{M}$. Since every norm limit of compact quasinilpotent operators is quasinilpotent, this is a contradiction. Hence $\mathcal{S}_\mathcal{M}$ is unbounded, and has a nontrivial hyperinvariant subspace by Lemma 8.1.10. This contradicts the maximality of $\mathcal{E}$. $\square$

Corollary 8.1.12. *If S is a nontriangularizable semigroup of compact operators, then $\overline{\mathbb{R}^+S}$ contains finite-rank operators other than zero.*

Proof. By Turovskii's Theorem (8.1.11), $\overline{\mathbb{R}^+S}$ contains an operator that is not quasinilpotent. Thus Lemma 7.4.5 is applicable. □

Corollary 8.1.13. *If a semigroup S consists of quasinilpotent compact operators, then so does the algebra generated by S.*

Proof. By Turovskii's Theorem (8.1.11), S is triangularizable. Therefore, since the algebra generated by S consists of linear combinations of elements of S, this corollary follows from the Spectral Mapping Theorem (7.2.6). □

Corollary 8.1.14. *If A and B are compact operators such that every word in $\{A, B\}$ is quasinilpotent, then $A + B$ is quasinilpotent.*

Proof. This is a special case of the preceding corollary. □

The following lemma will be used several times.

Lemma 8.1.15. *If S is an irreducible semigroup of compact operators and r is the minimal nonzero rank of elements of $\overline{\mathbb{R}^+S}$ (r is finite, by Corollary 8.1.12), then $\overline{\mathbb{R}^+S}$ contains an idempotent of rank r.*

Proof. Replace S by $\overline{\mathbb{R}^+S}$ and let F be an element of rank r. The restriction of FS to $F\mathcal{X}$ is not $\{0\}$, for otherwise choosing any f_0 with $Ff_0 \neq 0$ would yield a nontrivial invariant subspace

$$\vee\{SFf_0 : S \in S\}$$

for S. Thus there is an $S \in S$ with $FSF \neq \{0\}$.

Let S_0 denote the semigroup consisting of $\{FSF|_{F\mathcal{X}} : S \in S\}$. Then S_0 is irreducible, for if g_0 were a nonzero vector in $F\mathcal{X}$ such that $\vee\{FSg_0 : S \in S\}$ was a proper subspace of $F\mathcal{X}$, we could choose a nonzero linear functional ϕ_0 on $F\mathcal{X}$ such that $\phi_0(FSg_0) = 0$ for all $S \in S$. Then the closed linear span in $\mathcal{X}$ of $\{Sg_0 : S \in S\}$ would be a proper invariant subspace of S, since $\phi_0(Ff) = 0$ for all f in the subspace. However, S is irreducible, so this cannot occur. Hence S_0 is irreducible.

It now follows from Lemma 3.1.6(iii) that S_0 contains the identity operator, and therefore FSF contains an idempotent. □

8.2 A General Approach

Many definitions and fundamental lemmas used in the finite-dimensional case can easily be adapted to the present case.

The definition of a semigroup ideal is unchanged (Definition 2.1.9).

Lemma 8.2.1. *A nonzero semigroup ideal of an irreducible semigroup S in $\mathcal{B}(\mathcal{X})$ is irreducible.*

Proof. The proof is the same as that of the corresponding Lemma 2.1.10, except that the span $\vee\{J\mathcal{M} : J \in \mathcal{J}\}$ given there should be interpreted as the closed linear span.   □

We restate Definition 2.1.3 to allow various subsemigroups of $\mathcal{B}(\mathcal{X})$ as domains.

Definition 8.2.2. Let S_0 be a given subsemigroup of $\mathcal{B}(\mathcal{X})$ and ϕ a function from S_0 into a set $\mathcal{E}$. We say that ϕ is *permutable* on a collection $\mathcal{F} \subseteq S_0$ if, for every integer n, members $A_1, \ldots, A_n$ of $\mathcal{F}$, and every permutation τ of $\{1, 2, \ldots, n\}$, we have

$$\phi(A_1 A_2 \cdots A_n) = \phi(A_{\tau(1)} A_{\tau(2)} \cdots A_{\tau(n)}).$$

Lemma 8.2.3. *Let ϕ be a function whose domain contains a semigroup S. Then ϕ is permutable on S if and only if*

  (i) $\phi(ST) = \phi(TS)$ *and*
  (ii) $\phi(STR) = \phi(TSR)$ *for all $R, S,$ and T in S.*

Proof. The proof of Lemma 2.1.14 applies without change.   □

Lemma 8.2.4. *Let ϕ be a nonzero continuous linear functional on the space $\mathcal{F}(\mathcal{X})$ of finite-rank operators on a reflexive Banach space $\mathcal{X}$. If $\mathcal{E}$ is a subset of $\mathcal{F}(\mathcal{X})$ on which ϕ is permutable, then $\mathcal{E}$ is reducible.*

Proof. We first note that ϕ is permutable on the algebra $\mathcal{A}$ generated by $\mathcal{E}$. (This is proved just as in Lemma 2.1.4.) By continuity, ϕ is permutable on the norm closure of $\mathcal{A}$. Assume, then, that $\mathcal{A} = \overline{\mathcal{A}}$. Now suppose that $\mathcal{E}$ is irreducible. Then $\mathcal{A}$ is transitive and thus contains all finite-rank operators, by Corollary 7.4.3. Pick A and B in $\mathcal{F}(\mathcal{X})$ with $C = AB - BA \neq 0$. Then the permutability of ϕ on $\mathcal{F}(\mathcal{X})$ implies that $\phi(SCT) = 0$ for all S and T in $\mathcal{F}(\mathcal{X})$. Thus ϕ is identically zero on the ideal $\mathcal{J}$ of $\mathcal{F}(\mathcal{X})$ generated by C. Since $\mathcal{J} \neq \{0\}$, we have $\mathcal{J} = \mathcal{F}(\mathcal{X})$, which is a contradiction.   □

Corollary 8.2.5. *Let $\mathcal{X}$ be reflexive and let ϕ be a nonzero continuous linear functional on $\mathcal{F}(\mathcal{X})$. If the restriction of ϕ to a subsemigroup S of $\mathcal{F}(\mathcal{X})$ is multiplicative, then S is reducible. In particular, S is reducible if this restriction is constant.*

Proof. Constancy implies multiplicativity, which implies permutability, so the previous corollary applies. □

Reflexivity is needed in the previous two results, as shown by the following.

Example 8.2.6. *There is a Banach space $\mathcal{X}$, a continuous, nonzero linear functional ϕ on $\mathcal{F}(\mathcal{X})$, and an irreducible subalgebra of $\mathcal{F}(\mathcal{X})$ on which ϕ is identically zero.*

Proof. Let $\mathcal{A}^*$ be the irreducible (transitive) algebra of Example 7.4.4. Recall that $\mathcal{A}^*$ is uniformly closed and there exists a finite-rank operator F that does not belong to $\mathcal{A}^*$. By the Hahn-Banach Theorem there is a continuous linear functional ϕ on the compact operators that vanishes on $\mathcal{A}^*$ and that is not 0 on F. □

For a special class of functionals we do not need the reflexivity hypothesis. The functionals introduced in the definition below generalize the "matrix-entry" functionals in finite dimensions, where, for fixed i and j, the (i,j) entry of each matrix M defines $\phi(M)$.

Definition 8.2.7. A linear functional ϕ on a linear space $\mathcal{L}$ of operators on $\mathcal{X}$ is called a *coordinate functional* if there is a nonzero x in $\mathcal{X}$ and a nonzero linear functional f on $\mathcal{X}$ such that $\phi(L) = f(Lx)$ for every L in $\mathcal{L}$.

Note that a coordinate functional on $\mathcal{L}$ is automatically nonzero if $\mathcal{L}$ separates points in $\mathcal{X}$; i.e., $\mathcal{L}x = 0$ implies $x = 0$. Also, it is easily seen that a coordinate functional ϕ is continuous on $\mathcal{B}(\mathcal{X})$ if and only if the corresponding functional f is continuous.

Lemma 8.2.8. *Let S be an arbitrary semigroup in $\mathcal{B}(\mathcal{X})$ and ϕ a continuous coordinate functional on $\mathcal{B}(\mathcal{X})$. Then S is reducible if either of the following conditions holds:*

   (i) *ϕ is constant on S.*
  (ii) *ϕ is multiplicative on S (i.e., $\phi(AB) = \phi(A)\phi(B)$ for all A and B in S).*

Proof. Let $\phi(S) = f(Sx)$ for all S, where $x \in \mathcal{X}$ and $f \in \mathcal{X}^*$.

   (i) The case where $\phi|_S = 0$ is trivial, for the closed linear span $\mathcal{M}$ of Sx is invariant under S. It is also a proper subspace of $\mathcal{X}$, because $f|_{\mathcal{M}} = 0$. Thus if $\mathcal{M} \neq \{0\}$, it is a nontrivial invariant subspace. Otherwise, the one-dimensional space spanned by x is invariant.

Assume now that $\phi(S) = \lambda \neq 0$ for all S in S. If S is a singleton $\{A\}$, then A is an idempotent in $\mathcal{B}(\mathcal{X})$ and has invariant subspaces. Thus assume that S has distinct members A and B. Let $\mathcal{A}$ be the algebra spanned by

S and let $\mathcal{J}$ be the ideal of $\mathcal{A}$ generated by $A - B$. An easy calculation shows that $\phi|_{\mathcal{J}} = 0$, so that $\mathcal{J}$ is reducible by the preceding paragraph. Since $\mathcal{J} \neq \{0\}$, the reducibility of $\mathcal{A}$ follows from Lemma 8.2.1.

(ii) The set

$$\mathcal{S}_0 = \{S \in \mathcal{S} : \phi(S) = 0\}$$

is a semigroup ideal by the multiplicativity of ϕ on $\mathcal{S}$. It is also reducible by (i). Thus, if $\mathcal{S}_0$ properly contains $\{0\}$, we are done by Lemma 8.2.1. We now assume $\mathcal{S}_0 \subseteq \{0\}$ and observe that the set

$$\mathcal{S}_1 = \left\{ \frac{S}{\phi(S)} : 0 \neq S \in \mathcal{S} \right\}$$

is a semigroup. Since ϕ takes the constant value 1 on $\mathcal{S}_1$, this semigroup is reducible by (i) above. Hence so is $\mathcal{S}$, since $\mathcal{S} \subseteq \mathbb{C}\mathcal{S}_1$. □

Corollary 8.2.9. *Let $\mathcal{E}$ be a noncommutative family of operators and ϕ a continuous coordinate functional. If ϕ is permutable on $\mathcal{E}$, then $\mathcal{E}$ is reducible.*

Proof. We can assume, as in Lemma 2.1.4, that $\mathcal{E}$ is an algebra. Pick A and B in $\mathcal{E}$ with $C = AB - BA \neq 0$, and let $\mathcal{J}$ be the ideal of $\mathcal{E}$ generated by C. Observe that $\phi|_{\mathcal{J}} = 0$. Thus $\mathcal{J}$ is reducible by the preceding lemma, and so is $\mathcal{E}$ by Lemma 8.2.1. □

We give another general, simple lemma before returning to compact operators; it is an extension of the fact that if a coordinate functional on $\mathcal{S}$ is zero, then $\mathcal{S}$ is reducible. It says that if a "corner" of $\mathcal{S}$ has a chain $\mathcal{C}$ of invariant subspaces, then $\mathcal{S}$ itself has a chain at least as long as $\mathcal{C}$.

Lemma 8.2.10. *Let P be a projection on $\mathcal{X}$ and $\mathcal{S}$ any semigroup in $\mathcal{B}(\mathcal{X})$. Let $\mathcal{C}$ be a chain of invariant subspaces for the collection $P\mathcal{S}|_{P\mathcal{X}}$ of operators on $P\mathcal{X}$. Then there is a one-to-one, order-preserving map from $\mathcal{C}$ into the lattice of invariant subspaces of $\mathcal{S}$. (Note that P is not assumed to be in $\mathcal{S}$.)*

Proof. For each $\mathcal{M} \in \mathcal{C}$, let $\mathcal{M}_1$ be the closed linear span of $\mathcal{M} \cup \mathcal{S}\mathcal{M}$. Then $\mathcal{M}_1$ is clearly an invariant subspace of $\mathcal{S}$ containing $\mathcal{M}$. Now let $\mathcal{M}$ and $\mathcal{N}$ be members of $\mathcal{C}$ with $\mathcal{M}$ properly contained in $\mathcal{N}$, so that $\mathcal{M}_1 \subseteq \mathcal{N}_1$. To see that this inclusion is proper, pick x in $\mathcal{N} \setminus \mathcal{M}$. Since x is in $\mathcal{N}_1$, we need only show that $x \in \mathcal{M}_1$ yields a contradiction. Since $Px = x$, $P\mathcal{M} = \mathcal{M}$, and $P\mathcal{S}\mathcal{M} \subseteq \mathcal{M}$ by hypothesis, $x \in \mathcal{M}_1$ implies

$$x \in P(\mathcal{M} \vee \mathcal{S}\mathcal{M}) = P\mathcal{M} = \mathcal{M}.$$

□

Lemma 8.2.11. *Let ϕ be a continuous coordinate functional. If $\mathcal{E}$ is a collection in $\mathcal{K}(\mathcal{X})$ on which ϕ is permutable, then $\mathcal{E}$ is reducible.*

Proof. We can again assume that $\mathcal{E}$ is an algebra. If $\mathcal{E}$ is commutative, we are done by Theorem 7.6.1. If not, the assertion follows from Corollary 8.2.9. □

Lemma 8.2.12. *Let $\mathcal{S}$ be an irreducible semigroup of rank-one operators on an infinite-dimensional Banach space.*

(i) *For each positive integer k, there is a k-dimensional subspace $\mathcal{M}$ of $\mathcal{X}$ and a subsemigroup $\mathcal{S}_0$ of $\mathcal{S}$ leaving $\mathcal{M}$ invariant such that $\mathcal{S}_0|_{\mathcal{M}}$ is irreducible.*

(ii) *$\mathcal{S}$ contains members A and B with independent ranges R_1 and R_2 such that the restrictions of A and B to $R_1 + R_2$ are simultaneously similar to*

$$A_0 = \begin{pmatrix} \alpha & 0 \\ \beta & 0 \end{pmatrix} \quad \text{and} \quad B_0 = \begin{pmatrix} 0 & \gamma \\ 0 & \delta \end{pmatrix}$$

with $\alpha\delta - \beta\gamma \neq 0$ and $\beta\gamma \neq 0$.

Proof. Since (ii) follows from (i) together with Lemma 4.2.4(iii), we need only show (i).

First note that if $f \in \mathcal{X}^*$, then the span of $\mathcal{S}^* f = \{S^* f : S \in \mathcal{S}\}$ is infinite-dimensional. For otherwise there would exist $S_1, \ldots, S_m$ such that $\mathcal{S}^* f \subseteq \vee\{S_i^* f : i = 1, \ldots, m\}$. Since each S_i is of rank one, and $\mathcal{X}$ is infinite-dimensional, we can pick $x \in \cap\{\ker S_i : i = 1, \ldots, m\}$. Then, for every S in $\mathcal{S}$,

$$f(Sx) = (S^* f)(x) \in (\vee S_i^* f)(x),$$

and, since $S_i^* f(x) = f(S_i x) = 0$ for $i = 1, \ldots, m$, we deduce $f(Sx) = 0$. This contradicts the irreducibility of $\mathcal{S}$ by (an easy case of) Lemma 8.2.8.

Now pick a nonzero member $K = x \otimes f$ of $\mathcal{S}$ with $x \in X$ and $f \in \mathcal{X}^*$. By the preceding paragraph, there exist k independent functionals $f_j = T_j^* f$, where $T_j \in \mathcal{S}$, $j = 1, \ldots, k$. Let $\mathcal{N} = \cap_{j=1}^k (\ker f_j)$, so that $\mathcal{N}$ has co-dimension k in $\mathcal{X}$. Since the span of Sx is dense in $\mathcal{X}$ by the irreducibility of $\mathcal{S}$, we can find k members S_i of $\mathcal{S}$ such that the span of the vectors $x_i = S_i x$ is a complement of $\mathcal{N}$. (Simply choose $S_1 x$ not in $\mathcal{N}$; then $S_2 x$ not in the span of $\mathcal{N}$ and $S_1 x$, and so on.) Let $\mathcal{M}$ denote that complement.

Let $\mathcal{S}_0$ be the subsemigroup of $\mathcal{S}$ generated by the k^2 members

$$S_i K T_j = (S_i x) \otimes (T_j^* f) = x_j \otimes f_j$$

for $i, j = 1, \ldots, k$. Since $\mathcal{M} = \vee\{\operatorname{ran} S_i : i = 1, \ldots, k\}$, $\mathcal{S}_0$ leaves $\mathcal{M}$ invariant. Also, $\{x_i \otimes f_j\}$ is a linearly independent set of operators of the form $A_{ij} \oplus 0$ relative to the decomposition $\mathcal{X} = \mathcal{M} \oplus \mathcal{N}$. Thus $\mathcal{S}_0|_{\mathcal{M}}$ has a basis for $\mathcal{B}(\mathcal{M})$ and is irreducible. □

The following result will often be used to "downsize" given reducibility problems concerning compact operators to the corresponding problems in finite dimensions.

Lemma 8.2.13. (The Downsizing Lemma) *Let $\mathcal{P}$ be a property defined for semigroups of compact operators on Banach spaces. Assume that whenever the semigroup $\mathcal{S}$ has property $\mathcal{P}$, so do*

   (i) *every subsemigroup of $\mathcal{S}$,*
   (ii) *$\mathcal{S}|_{\mathcal{X}_0}$, where $\mathcal{X}_0 = \vee\{\text{ran } S : S \in \mathcal{S}\}$, and*
   (iii) *the semigroup $\overline{\mathbb{R}^+\mathcal{S}}$.*

Let $\mathcal{S}$ be an irreducible semigroup in $\mathcal{K}(\mathcal{X})$ with property $\mathcal{P}$. Then there is an integer $k \geq 2$ and an idempotent E of rank k on $\mathcal{X}$ such that $\mathcal{S}$ contains a subsemigroup $\mathcal{S}_0$ satisfying $\mathcal{S}_0 = E\mathcal{S}_0$, where $\mathcal{S}_0|_{\text{ran } E}$ is an irreducible semigroup in $\mathcal{B}(\mathbb{C}^k)$ with property $\mathcal{P}$. Moreover, E can be chosen from $\overline{\mathbb{R}^+\mathcal{S}}$ if the minimal positive rank in $\overline{\mathbb{R}^+\mathcal{S}}$ is greater than 1.

Proof. By Turovskii's Theorem (8.1.11), $\mathcal{S}$ does not consist of quasinilpotent operators. Thus $\overline{\mathbb{R}^+\mathcal{S}}$ has nonzero operators of finite rank by Lemma 7.4.5. Without loss of generality assume $\mathcal{S} = \overline{\mathbb{R}^+\mathcal{S}}$.

First assume that $\mathcal{S}$ contains a rank-one operator. Then

$$\mathcal{S}_1 = \{S \in \mathcal{S} : \text{rank } (S) \leq 1\}$$

is an ideal in $\mathcal{S}$ and is thus irreducible by Lemma 8.2.1. Also, $\mathcal{S}_1$ has property $\mathcal{P}$ by (i). By Lemma 8.2.12, there is a 2-dimensional subspace $\mathcal{M}$ of $\mathcal{X}$ and operators A and B in $\mathcal{S}_1$ whose ranges span $\mathcal{M}$ such that, if $\mathcal{S}_0$ denotes the semigroup generated by A and B, then $\mathcal{S}_0|_{\mathcal{M}}$ is irreducible. Now, $\mathcal{S}_0$ has property $\mathcal{P}$ by (i), and if E denotes any idempotent with range $\mathcal{M}$, then $E\mathcal{S}_0 = \mathcal{S}_0$ and $\mathcal{S}_0|_{\mathcal{M}}$ has property $\mathcal{P}$ by (ii).

To complete the proof, we assume that the minimal positive rank in $\mathcal{S}$ is $k \geq 2$. Pick a member F of rank k and note that, by Lemma 8.2.10, the collection $F\mathcal{S}|_{\text{ran } F}$ is an irreducible semigroup in $\mathcal{B}(\mathbb{C}^k)$ whose nonzero members are all of rank k. (Otherwise, there would be an FS with $S \in \mathcal{S}$ such that $FS|_{\text{ran } F}$ was nonzero and had zero in its spectrum. Then FSF would have rank less than k, which is a contradiction.) It follows from Lemma 3.1.6 that $F\mathcal{S}|_{\text{ran } F} \setminus \{0\}$ is an irreducible group contained, up to similarity, in $\mathbb{R}^+\mathcal{U}$, where $\mathcal{U}$ is a unitary group acting on $\mathbb{C}^k$. In particular, $\overline{\mathbb{R}^+\mathcal{S}}$ contains an idempotent E with ran $E = $ ran F. Thus the semigroup

$$E\mathcal{S}|_{\text{ran } E} = F\mathcal{S}|_{\text{ran } E}$$

has property $\mathcal{P}$ by (i) and (ii). □

Triangularizability of certain ideals in a semigroup implies that of the semigroup, as the following extensions of Lemmas 2.2.6 and 2.2.7 demonstrate.

Lemma 8.2.14. *Let S be a semigroup of compact operators on a Banach space and k a natural number. If the ideal S^k is triangularizable, so is S itself.*

Proof. Let S^k be triangularizable. Then S is reducible, by Turovskii's Theorem (8.1.11) if $S^k = \{0\}$ and by Lemma 8.2.1 if $S^k \neq \{0\}$.

Now, triangularizability of any family of compact operators is inherited by quotients, by Theorem 7.3.9. This (applied to S^k) together with the Triangularization Lemma (7.1.11) implies that S is triangularizable.   □

Lemma 8.2.15. *Let S be a semigroup of compact operators. If an ideal $\mathcal{J}$ of S has a unique triangularizing chain of subspaces, then S is triangularizable (and has a unique triangularizing chain).*

Proof. The proof can be obtained from that of Lemma 3.4.9 by replacing Lemmas 2.1.10, 1.5.3, and 1.1.4 with their appropriate infinite-dimensional versions: 8.2.1, 7.3.9, and 7.1.11.   □

8.3 Permutability and Submultiplicativity of Spectra

Recall that spectrum is permutable on a semigroup S of compact operators if and only if $\sigma(ABC) = \sigma(BAC)$ for all A, B, and C in S. (This follows from Lemma 8.2.3, since $\sigma(ST) = \sigma(TS)$ for any compact S and T.) Note that σ is a function into sets without regard to multiplicity: The multiplicity of an element λ in $\sigma(ABC)$ is allowed to differ from that in $\sigma(BAC)$.

Theorem 8.3.1. *Every semigroup of compact operators with permutable spectrum is reducible.*

Proof. In view of Theorem 3.2.2., we can assume that the underlying Banach space is infinite-dimensional. The property $\mathcal{P}$ of permutability of spectrum clearly satisfies condition (i) of the Downsizing Lemma (8.2.13); it also satisfies condition (iii) by homogeneity and continuity of spectrum for compact operators.

To verify condition (ii), let $\mathcal{X}_0 = \vee\{\operatorname{ran} S : S \in \mathcal{S}\}$ and pick any three members A, B, and C in the semigroup. Observe that

$$\sigma(S) = \sigma(S|_{\mathcal{X}_0}) \cup \{0\}$$

for every $S \in \mathcal{S}$. If $\mathcal{X}_0$ is infinite-dimensional, $\sigma(S) = \sigma(S|_{\mathcal{X}_0})$ for all $S \in \mathcal{S}$, since compact operators are not invertible. If $\mathcal{X}_0$ is finite-dimensional but at least one of the restrictions is not invertible, then the equality

$$\sigma\left[(A|_{\mathcal{X}_0})(B|_{\mathcal{X}_0})(C|_{\mathcal{X}_0})\right] = \sigma\left[(B|_{\mathcal{X}_0})(A|_{\mathcal{X}_0})(C|_{\mathcal{X}_0})\right]$$

follows from the corresponding $\sigma(ABC) = \sigma(BAC)$. On the other hand, if all of the restrictions are invertible, then

$$\sigma\left[(A|_{\mathcal{X}_0})(B|_{\mathcal{X}_0})(C|_{\mathcal{X}_0})\right] = \sigma(ABC) \setminus \{0\}$$
$$= \sigma(BAC) \setminus \{0\}$$
$$= \sigma\left[(B|_{\mathcal{X}_0})(A|_{\mathcal{X}_0})(C|_{\mathcal{X}_0})\right].$$

Thus, in all cases, the restrictions have permutable spectrum. If $\mathcal{S}$ were irreducible, the Downsizing Lemma (8.2.13) would yield an irreducible semigroup $\mathcal{S}_0|_{\operatorname{ran} E}$ with permutable spectrum on a finite-dimensional space of dimension at least 2. This would contradict Theorem 3.2.2, so it follows that $\mathcal{S}$ is reducible.   □

Corollary 8.3.2. *A semigroup of bounded operators on which spectrum is permutable is reducible if it contains a compact operator other than 0.*

Proof. The ideal of compact operators in the semigroup is reducible, by Theorem 8.3.1, and thus so is the semigroup (by Lemma 8.2.1).   □

As we have seen (Example 2.1.11), permutability of spectra does not imply triangularizability except under additional hypotheses.

Theorem 8.3.3. *A semigroup of operators of rank at most one with permutable spectrum is triangularizable.*

Proof. The proof is analogous to that of the finite-dimensional case (Theorem 3.2.7). By the Triangularization Lemma (7.1.11) and Theorem 8.3.1, it suffices to show the following: If S_1, S_2, and S_3 are members of the semigroup leaving subspaces $\mathcal{M}$ and $\mathcal{N}$ of the underlying space $\mathcal{X}$ invariant and $\mathcal{M} \supset \mathcal{N}$, then the quotient operators $\hat{S}_j$ on $\mathcal{M}/\mathcal{N}$ satisfy $\sigma(\hat{S}_1\hat{S}_2\hat{S}_3) = \sigma(\hat{S}_2\hat{S}_1\hat{S}_3)$. This is obviously the case if either $\mathcal{M}/\mathcal{N}$ has dimension one or at least one of the quotient operators $\hat{S}_j$ is zero. So assume that the dimension of $\mathcal{M}/\mathcal{N}$ is greater than one, and that every $\hat{S}_j$ is nonzero. Since each S_j has rank 1, it follows that $S_j|_{\mathcal{N}}$ and the quotient operator induced by S_j on $\mathcal{X}/\mathcal{M}$ are both zero.

Note that, for any compact operator K leaving $\mathcal{M}$ and $\mathcal{N}$ invariant, $\sigma(\hat{K}) \subseteq \sigma(K)$. Thus if $K|_{\mathcal{N}}$ and the quotient operator induced by K on

$\mathcal{X}/\mathcal{M}$ are both zero, then $\sigma(K) = \sigma(\hat{K}) \cup \{0\}$. In particular, if $0 \in \sigma(\hat{K})$, then $\sigma(\hat{K})$ equals $\sigma(K)$. Hence

$$\sigma(\hat{S}_1 \hat{S}_2 \hat{S}_3) = \sigma(S_1 S_2 S_3) = \sigma(S_1 S_2 S_3) = \sigma(\hat{S}_1 \hat{S}_2 \hat{S}_3). \qquad \square$$

Depending on the minimal rank in a semigroup, we may obtain more than one invariant subspace from permutability of the spectrum. The following result extends the assertion (Theorem 3.2.1) that groups with permutable spectrum are triangularizable.

Theorem 8.3.4. *Let S be a semigroup of compact operators with permutable spectrum. Denote by r the minimal nonzero rank in $\overline{\mathbb{R}^+ S}$. If $\overline{\mathbb{R}^+ S}$ contains an idempotent of rank r, then S has a chain of distinct invariant subspaces*

$$0 = \mathcal{M}_0 \subset \mathcal{M}_1 \subset \cdots \subset \mathcal{M}_r.$$

Proof. Assume $S = \overline{\mathbb{R}^+ S}$ without loss of generality, and let E be an idempotent of rank r in S. By the minimality of r, the closed semigroup $\mathcal{G} = ESE|_{E\mathcal{X}} \setminus \{0\}$ is a group (by Lemma 3.1.6).

It is easily seen that spectrum is permutable on $\mathcal{G}$ (by the permutability hypothesis on S together with the invertibility of members of $\mathcal{G}$). Hence $\mathcal{G}$ is triangularizable, by Theorem 3.2.1. This means that $ES|_{E\mathcal{X}}$ has a chain of $r + 1$ distinct invariant subspaces (including $\{0\}$ and $E\mathcal{X}$). The conclusion then follows from Lemma 8.2.10. $\qquad \square$

The definition of submultiplicativity for spectra (Definition 3.3.1) is unchanged for the infinite-dimensional case. Recall that, in finite dimensions, this condition does not guarantee reducibility (Example 3.3.7). The following result is not surprising in light of Theorem 3.3.2.

Theorem 8.3.5. *Let S be a semigroup of compact operators on an infinite-dimensional Banach space. If spectrum is submultiplicative on S, then S is reducible.*

Proof. We can assume, by homogeneity and continuity of spectrum for compact operators, that $S = \overline{\mathbb{R}^+ S}$.

Suppose that S is irreducible. Then S contains operators of finite rank (by Corollary 8.1.12). Let m be the minimal nonzero rank in S. By Lemma 8.2.1, we can pass to an ideal and assume that S consists of operators of rank m or zero. By irreducibility, S contains an idempotent E of rank m (Lemma 8.1.15). At this point it is possible to repeat the proof given in the finite-dimensional case. However, since we have not yet dealt with trace conditions in infinite dimensions, we give a slightly different proof, which is also useful in other situations.

Observe that $\mathcal{G} = E\mathcal{S}E \setminus \{0\}$ is a group with identity E whose restriction to $E\mathcal{X}$ is similar to a unitary group (Lemma 3.1.6). We first claim that the map $S \mapsto ESE$ is multiplicative: $ESTE = ESETE$ for all S and T in $\mathcal{S}$. If either ES or TE is zero, there is nothing to prove. Otherwise, $ESE \neq 0$ and $ETE \neq 0$. (If $ESE = 0$, then ES is nilpotent and thus equals zero. A similar argument applies to TE.) Let $S' = ES'E$ and $T' = ET'E$ be the respective inverses of ESE and ETE in $\mathcal{G}$.

Now, $S'S$ and TT' are idempotents, because

$$(S'S)^2 = S' \cdot ESES' \cdot S = S'ES = S'S$$

and

$$(TT')^2 = T \cdot ET'ET \cdot T' = TET' = TT'.$$

Hence $\sigma(S'S \cdot TT') \subseteq \{0,1\} \cdot \{0,1\} = \{0,1\}$, by hypothesis. Since $S'STT'$ is in $\mathcal{G}$ and its restriction to $E\mathcal{X}$ is similar to a unitary operator, we deduce that $S'STT' = E$. It follows that

$$ESETE = ES \cdot S'STT' \cdot TE$$
$$= ESES' \cdot ST \cdot T'ETE$$
$$= ESTE$$

as claimed. Thus $ES(I - E)TE = 0$ for all S and T in $\mathcal{S}$.

The closed linear span of $\mathcal{S}(E\mathcal{X})$ is clearly invariant under $\mathcal{S}$, and hence equals $\mathcal{X}$ by irreducibility. Thus the equation just proved implies $ES(I - E) = 0$; i.e., $(I - E)\mathcal{X}$ is an invariant subspace for $\mathcal{S}$.   □

Corollary 8.3.6. *Let $\mathcal{S}$ be a semigroup of bounded operators on an infinite-dimensional Banach space. If $\mathcal{S}$ has submultiplicative spectrum and contains a nonzero compact operator, then $\mathcal{S}$ is reducible.*

Proof. The ideal of compact operators in the semigroup is reducible, by Theorem 8.3.5, so the result follows from Lemma 8.2.1.   □

Infinite-dimensionality is necessary in the results above, as demonstrated by Example 3.3.7.

For rank-one operators we have the following stronger result.

Theorem 8.3.7. *A semigroup of operators of rank at most one with submultiplicative spectrum is triangularizable.*

Proof. The proof is almost the same as that of Theorem 8.3.3. Just use Theorem 8.3.5 instead of 8.3.1, and show that if S_1 and S_2 are members of the semigroup leaving $\mathcal{M}$ and $\mathcal{N}$ invariant and $\dim \mathcal{M}/\mathcal{N} \geq 2$, then the quotient operators $\hat{S}_j$ satisfy

$$\sigma(\hat{S}_1\hat{S}_2) = \sigma(S_1 S_2) \subseteq \sigma(S_1)\sigma(S_2) = \sigma(\hat{S}_1)\sigma(\hat{S}_2).$$ □

Theorem 8.3.8. *Let S be a semigroup of compact operators with $\sigma(S) \subseteq \{0,1\}$ for every S in S. If for A and B in S, $1 \in \sigma(AB)$ implies $1 \in \sigma(A) \cap \sigma(B)$, then S is reducible.*

Proof. The hypothesis implies submultiplicativity of spectra on S.   □

The definition of band is unchanged (Definition 2.3.4).

Theorem 8.3.9. *Every band in $\mathcal{K}(\mathcal{X})$ is triangularizable.*

Proof. Since quotients of idempotents are also idempotents, it suffices, by the Triangularization Lemma (7.1.11), to show reducibility. This follows from the preceding theorem in the same way as Theorem 2.3.5 followed from Theorem 2.3.1.   □

8.4 Subadditivity and Sublinearity of Spectra

The definitions of subadditive and sublinear spectra are as in the finite-dimensional case (see Definition 4.2.1). We need an analogue of Lemma 4.2.2. This is necessarily different from (but is also applicable to) the finite-dimensional case.

Lemma 8.4.1. *Let A and B be compact operators on $\mathcal{X}$ with a common invariant subspace $\mathcal{M}$. If spectrum is sublinear on $\{A, B\}$, then it is also sublinear on the pair $\{A|_\mathcal{M}, B|_\mathcal{M}\}$. The same assertion is true for subadditivity if A and B have rank at most one.*

Proof. Assume that spectrum is sublinear on A and B, and let $A_0 = A|_\mathcal{M}$ and $B_0 = B|_\mathcal{M}$. For each fixed ordered pair $\langle \alpha, \beta \rangle$ in

$$[\sigma(A) \times \sigma(B)] \setminus [\sigma(A_0) \times \sigma(B_0)]$$

define

$$\mathcal{E}_{\alpha,\beta} = \{\lambda \in \mathbb{C} : \alpha + \lambda\beta \in \sigma(A_0 + \lambda B_0)\}.$$

The set $\mathcal{E}_{\alpha,\beta}$ is easily seen to be closed. We claim that it is nowhere dense. Suppose otherwise. Then $(A_0 - \alpha I) + \lambda(B_0 - \beta I)$ is non-invertible for uncountably many values of λ. Since $\langle \alpha, \beta \rangle$ is not in $\sigma(A_0) \times \sigma(B_0)$, at least one of the operators $A_0 - \alpha I$ and $B_0 - \beta I$ is invertible. If $A_0 - \alpha I$ is invertible, then the operator $I + \lambda(B_0 - \beta I)(A_0 - \alpha I)^{-1}$ is non-invertible for uncountably many values of λ. In other words, the operator $I + (B_0 - \beta I)(A_0 - \alpha I)^{-1}$ has uncountable spectrum, which is a contradiction, because it is a translate of a compact operator. Similarly, a contradiction is obtained if $B_0 - \beta I$ is assumed invertible. Hence $\mathcal{E}_{\alpha,\beta}$ is nowhere dense.

Since

$$[\sigma(A) \times \sigma(B)] \setminus [\sigma(A_0) \times \sigma(B_0)]$$

is countable, it follows from the Baire Category Theorem that $\mathcal{E} = \mathbb{C} \setminus \cup_{\alpha,\beta} \mathcal{E}_{\alpha,\beta}$ is a dense G_δ set. Observe that

$$\sigma(A_0 + \lambda B_0) \subseteq \sigma(A_0) + \lambda \sigma(B_0)$$

for all $\lambda \in \mathcal{E}$, for the sublinearity hypothesis for $\{A, B\}$ implies that any element of $\sigma(A_0 + \lambda B_0)$ has the form $\alpha + \lambda\beta$ with $\alpha \in \sigma(A)$ and $\beta \in \sigma(B)$. If $(\alpha, \beta) \in \sigma(A_0) \times \sigma(B_0)$, we are done. Otherwise, $\lambda \notin \mathcal{E}_{\alpha,\beta}$ implies $\alpha + \lambda\beta \notin \sigma(A_0 + \lambda B_0)$, which is a contradiction. Since $\mathcal{E}$ is dense, to show $\mathcal{E} = \mathbb{C}$ it suffices to show that $\mathcal{E}$ is closed. This follows from the continuity of spectrum for compact operators (Theorem 7.2.13): Just observe that if $\{\lambda_n\}$ is a sequence in $\mathcal{E}$ converging to λ and if γ is any member of $\sigma(A_0 + \lambda B_0)$, then there is a sequence $\{\gamma_n\}$ in $\mathbb{C}$ such that $\gamma_n \in \sigma(A_0 + \lambda_n B_0)$ and $\lim_{n\to\infty} \gamma_n = \gamma$. For each n,

$$\gamma_n = \alpha_n + \lambda_n \beta_n$$

for some $\alpha_n \in \sigma(A_0)$ and some $\beta_n \in \sigma(B_0)$. Choosing suitable subsequences, we can assume that $\lim_{n\to\infty} \alpha_n = \alpha$ and $\lim_{n\to\infty} \beta_n = \beta$, so that $\gamma = \alpha + \lambda\beta$ and hence $\gamma \in \sigma(A_0) + \lambda\sigma(B_0)$. Thus $\mathcal{E} = \mathbb{C}$.

The proof of the assertion on subadditivity is exactly the same as that of the rank-one case of Lemma 4.2.2. $\qquad\square$

Corollary 8.4.2. *Sublinearity of spectrum for compact operators is inherited by quotients. Subadditivity is inherited for operators of rank ≤ 1.*

Proof. In view of the preceding lemma, we need only show that if a subspace $\mathcal{M}$ of $\mathcal{X}$ is invariant under compact A and B, then the property (of sublinearity or subadditivity) also holds for the induced operators $\{\hat{A}, \hat{B}\}$ on $\mathcal{X}/\mathcal{M}$. Let $\mathcal{M}^\perp$ denote the annihilator of $\mathcal{M}$ (the set of all elements of $\mathcal{X}^*$ that vanish on $\mathcal{M}$). Then there is an isometric isomorphism from $\mathcal{M}^\perp$ onto $(\mathcal{X}/\mathcal{M})^*$ (see Rudin [2, p. 96]). If K is any compact operator on $\mathcal{X}$ leaving $\mathcal{M}$ invariant (so that $\mathcal{M}^\perp$ is invariant under K^*), then the operators $K^*|_{\mathcal{M}^\perp}$ and $\hat{K}^*$ can be identified through that isomorphism, and $\sigma(K^*) = \sigma(K)$ and $\sigma(\hat{K}^*) = \sigma(\hat{K})$. Therefore the property assumed for $\{A, B\}$ holds for $\{A^*|_{\mathcal{M}^\perp}, B^*|_{\mathcal{M}^\perp}\}$ by the lemma above, and thus also for $\{\hat{A}, \hat{B}\}$. $\qquad\square$

Theorem 8.4.3. *A semigroup of compact operators on which spectrum is sublinear is triangularizable.*

Proof. By the Triangularization Lemma (7.1.11) and Corollary 8.4.2, it suffices to show reducibility. The property of sublinearity implies reducibility in finite dimensions (Theorem 4.2.11). Thus we need only verify that sublinearity satisfies the three conditions of the Downsizing Lemma (8.2.13). Note that (i) is clear, (ii) follows from Lemma 8.4.1, and (iii) is a consequence of continuity and homogeneity of spectra of compact operators.

$\square$

The next result employs the following natural extension of properly L to pairs of compact operators. The sequences $\{\alpha_i\}$ and $\{\beta_i\}$, given in Definition 4.2.1, are made into infinite sequences by supplying zero terms if the nonzero eigenvalues form a finite set. (For a quasinilpotent operator, for example, every term of the sequence is zero.)

Corollary 8.4.4. *A semigroup of compact operators satisfying property* L *is triangularizable.*

Proof. This is a special case of the above theorem. $\square$

One consequence of the preceding theorem is the result that local triangularizability implies global triangularizability, as in the finite-dimensional case.

Corollary 8.4.5. *Let S be a semigroup of compact operators on a Banach space $\mathcal{X}$. If every pair in S is triangularizable, then S is triangularizable.*

Proof. This follows directly from the theorem above, as Corollary 4.2.14 did from Theorem 4.2.11. $\square$

For self-adjoint semigroups on Hilbert spaces, sublinearity of spectrum yields diagonalizability (as in finite dimensions).

Corollary 8.4.6. *A self-adjoint semigroup S of compact operators on a Hilbert space with sublinear spectrum is abelian.*

Proof. Let $\mathcal{C}$ be a triangularizing chain for S (such exists by Theorem 8.4.3). If $\mathcal{M}$ is in $\mathcal{C}$ and $\mathcal{M}_- \neq \mathcal{M}$, then the one-dimensional subspace $\mathcal{M} \ominus \mathcal{M}_-$ and its orthogonal complement are invariant under S, by self-adjointness. Let $\mathcal{H}_0$ be the direct sum of all such gaps of dimension one obtained from $\mathcal{C}$. Then the underlying Hilbert space decomposes into $\mathcal{H}_0 \oplus \mathcal{H}_1$ with $\mathcal{H}_0$ and $\mathcal{H}_1$ invariant under S. We need only observe that $S|_{\mathcal{H}_0}$ is diagonal and $S|_{\mathcal{H}_1} = \{0\}$. (The latter statement follows from the fact that $S|_{\mathcal{H}_1}$ is quasinilpotent for every S in S, by Ringrose's Theorem (7.2.3), so S^*S is also quasinilpotent, which implies $S = 0$.) $\square$

Note that the sublinearity condition was not used for all scalars; in fact, if it is assumed for a sufficiently large set of scalars, triangularizability follows. One such set is $\mathbb{R}$, as we now show. (Also see Theorem 4.2.15.)

Theorem 8.4.7. *For a semigroup S of compact operators on a Banach space, the following conditions are mutually equivalent:*

   (i) S *is triangularizable;*
   (ii) *spectrum is sublinear on S;*
   (iii) *spectrum is real-sublinear on S; i.e., for every real λ and A, B in S,*

$$\sigma(A + \lambda B) \subseteq \sigma(A) + \lambda \sigma(B);$$

   (iv) *for every integer k, members $\{A_1, \ldots, A_k\}$ of S, and complex scalars $\{\lambda_1, \ldots, \lambda_k\}$,*

$$\sigma(\lambda_1 A_1 + \cdots + \lambda_k A_k) \subseteq \lambda_1 \sigma(A_1) + \cdots + \lambda_k \sigma(A_k).$$

Proof. That (i) implies (iv) is a consequence of the Spectral Mapping Theorem (7.2.6). Clearly, (iv) implies (ii) and (ii) implies (iii), so that we need only show that (iii) implies triangularizability.

The Baire-Category argument used in the proof of Lemma 8.4.1 can be applied to the present situation, proving as in Corollary 8.4.2 that real-sublinearity is inherited by quotients. Thus by the Triangularization Lemma (7.1.11), it suffices to prove that S is reducible. This is an easy consequence of the Downsizing Lemma (8.2.13) together with Corollary 4.2.15.   □

The sublinearity hypothesis can sometimes be weakened to subadditivity.

Theorem 8.4.8. *A semigroup of operators of rank at most one is triangularizable if it has subadditive spectrum.*

Proof. By the Triangularization Lemma (7.1.11) and Corollary 8.4.2, it suffices to show reducibility of such a semigroup S. The Downsizing Lemma (8.2.13) cannot be used here, but the reduction to finite dimensions can be achieved via Lemma 8.2.12 directly. By that lemma, if S were irreducible, then it would contain members $A = A_0 \oplus 0$ and $B = B_0 \oplus 0$ with

$$A_0 = \begin{pmatrix} \alpha & 0 \\ \beta & 0 \end{pmatrix}, \quad B_0 = \begin{pmatrix} 0 & \gamma \\ 0 & \delta \end{pmatrix},$$

and $(\alpha\delta - \beta\gamma)\beta\gamma \neq 0$. Thus spectrum would be subadditive on A_0 and B_0, yielding a contradiction as in the proof of Theorem 4.2.5.   □

Definition 8.4.9. An operator T on a Banach space is called a *strong quasiaffinity* if $\overline{TM} = M$ for every invariant subspace M of T.

Note that if T is a strong quasiaffinity, then it is a quasiaffinity; i.e., it is injective and has dense range.

Corollary 8.4.10. *Let S be a semigroup of bounded operators with sublinear spectrum. If S contains a nonzero compact K, then it is reducible. If K is a strong quasiaffinity, then S is triangularizable.*

Proof. It is not hard to verify that if T is a strong quasiaffinity, then so are its quotients. Since sublinearity is also inherited by quotients, it suffices, by the Triangularization Lemma (7.1.11), to prove reducibility. The ideal of compact operators in S is reducible by Theorem 8.4.3, so the result follows from Lemma 8.2.1. □

8.5 Polynomial Conditions on Spectra

Many of the results of Section 4.4 involving polynomial conditions have appropriate extensions to infinite dimensions. We treat only those with less obvious proofs.

If g is a polynomial in one variable, we use the notation f_g introduced at the beginning of Section 4.4 for the noncommutative polynomial in two variables obtained from g. The appropriate version of Definition 4.4.1 follows.

Definition 8.5.1. Let f be any polynomial in two variables. We say that f *is quasinilpotent on a semigroup S* if $f(A, B)$ is quasinilpotent for all A and B in S.

We first dispose of the case $f = f_g$ where $g(x) = x - 1$.

Theorem 8.5.2. *Let S be a semigroup of compact operators such that $AB - BA$ is quasinilpotent for all A and B in S. Then S is triangularizable.*

Proof. The property is inherited by quotients. Thus we need only show reducibility (by the Triangularization Lemma 7.1.11). In view of Theorem 4.4.12, it suffices to verify that the property satisfies the conditions of the Downsizing Lemma (8.2.13), which is very easy. □

As was shown in Section 4.4., we do not always obtain triangularizability for nonlinear polynomials g. The counterexamples 4.4.5 and 4.4.6 can easily be extended to infinite dimensions.

For an extension of Theorem 4.4.9, we need the following result.

Proposition 8.5.3. *Let $S = \mathbb{C}S$ be an irreducible semigroup of operators of rank ≤ 1 on $\mathcal{X}$, and let $f(x, y) = \Sigma_{i=0}^m a_i x^i y x^{m-i}$ with $a_0 a_m \neq 0$. If f is quasinilpotent (equivalently, nilpotent) on S, then:*

   (i) *the idempotents in S form an abelian band $\mathcal{E}$;*
   (ii) *the closed linear span of $\{E\mathcal{X} : E \in \mathcal{E}\}$ is $\mathcal{X}$; and*

(iii) *for each nonzero A in S, there are unique E and F in $\mathcal{E}$ such that EAF is nonzero; moreover, $A = EAF$.*

Proof. Let $\mathcal{E}$ be the set of idempotents in S. By irreducibility, for each nonzero A in S there is an E in $\mathcal{E}$ with $EA = A$. (Just observe that AS has a nonzero idempotent; otherwise, the ideal generated by A would consist of nilpotent operators, which contradicts irreducibility. If $AB = E$ is such an idempotent with $B \in S$, it is easy to see that $EA = ABA = A$, since ABA has rank one.) This shows that $\mathcal{E}S = S$. Similarly, $S = S\mathcal{E}$. Since the span of the ranges of members of S is invariant under S, we have proved (ii) and the existence part of (iii).

To prove (i), pick distinct members E and F of $\mathcal{E}$. We must show that $EF = 0$. Assume that $F \neq 0$, $E \neq 0$. We first prove that $E\mathcal{X} \neq F\mathcal{X}$. Suppose otherwise. Then $E = x_0 \otimes \phi$ and $F = x_0 \otimes \psi$ for some x_0 in $\mathcal{X}$ and some independent ϕ and ψ in $\mathcal{X}^*$ satisfying $\phi(x_0) = \psi(x_0) = 1$. Let $\mathcal{N} = \ker \phi \cap \ker \psi$. Then $\mathcal{N} \oplus \mathbb{C}x_0$ has codimension one in $\mathcal{X}$, and, by irreducibility, S contains a member $T = x_1 \otimes \theta$ whose range, $\mathbb{C}x_1$, is not contained in $\mathcal{N} \oplus \mathbb{C}x_0$. Replacing θ with some $S^*\theta$, we can also assume that $\theta(x_0) \neq 0$. (If this replacement were impossible, $\theta(Sx_0)$ would be zero, contradicting irreducibility.) Now, the span of $\{x_1, x_2\}$ is invariant under E, F, and T, whose respective restrictions to that span have the following matrices relative to the basis $\{x_0, x_1\}$:

$$A = \begin{pmatrix} 1 & \alpha \\ 0 & 0 \end{pmatrix}, \quad B = \begin{pmatrix} 1 & \beta \\ 0 & 0 \end{pmatrix}, \quad C = \begin{pmatrix} 0 & 0 \\ \gamma & \delta \end{pmatrix}.$$

Note that $\alpha \neq \beta$, because $E = A \oplus 0$, $F = B \oplus 0$, and $E \neq F$. Also, $\gamma \neq 0$, because $\theta(x_0) \neq 0$. It is easy to verify that A, B, and C generate an irreducible semigroup S_0 in $\mathcal{M}_2(\mathbb{C})$. Since f is nilpotent on S_0 (and since $m \geq 2$ by Theorem 8.5.2), this yields $AB = 0$ by Proposition 4.4.7, which is a contradiction. Thus, $E\mathcal{X} \neq F\mathcal{X}$.

To complete the proof of (i), let $E = x \otimes \phi$, $F = y \otimes \psi$, and denote the span of x and y by $\mathcal{M}$. Then $E|_{\mathcal{M}}$ and $F|_{\mathcal{M}}$ have matrices of the form

$$A = \begin{pmatrix} 1 & \alpha \\ 0 & 0 \end{pmatrix}, \quad B = \begin{pmatrix} 0 & 0 \\ \beta & 1 \end{pmatrix}$$

relative to the basis $\{x, y\}$. By irreducibility, S has a member S such that the kernel of $S^*\phi$ does not contain $\alpha x - y$. Thus $T = (x \otimes \phi)S = x \otimes S^*\phi$ has a restriction

$$C = \begin{pmatrix} \gamma & \delta \\ 0 & 0 \end{pmatrix}$$

to $\mathcal{M}$ that is independent of A. The argument now concludes exactly as in the second paragraph of the proof of Proposition 4.4.7.

We must prove (iii). The existence of E and F in $\mathcal{E}$ with $EAF = A$ was shown above (and used only the irreducibility hypothesis). Now assume $E_1AF_1 \neq 0$. Suppose that $E_1 \neq E$ or $F_1 \neq F$. Then, by what we have proved, $E_1AF_1 = E_1EAFF_1 = 0$, which is a contradiction.   □

The following result extends Theorem 4.4.9 to infinite dimensions.

Theorem 8.5.4. *Let $g(x) = a_0 + \cdots + a_m x^m$ with $a_0 a_m \neq 0$, and assume that g is not divisible by $x^p - 1$ for any prime p. If S is a semigroup of compact strong quasiaffinities on which f_g is quasinilpotent, then S is triangularizable.*

Proof. If T is a strong quasiaffinity, then so are its quotients. Since quasinilpotence of f_g is also inherited by quotients, we must only show reducibility.

Now, f_g is quasinilpotent on $\overline{CS}$ by continuity of spectrum for compact operators, so if $\overline{CS}$ does not contain any operators of rank one, then it is reducible by the Downsizing Lemma together with Theorem 4.4.9. (All the conditions of the Downsizing Lemma are easily seen to be satisfied.) Thus we assume that there are rank-one operators in $\overline{CS}$.

Suppose that S is irreducible, so that the ideal $\mathcal{J}$ of rank-one operators in $\overline{CS}$ is also irreducible (by Lemma 8.2.1). Let $\mathcal{E}$ be the abelian band in $\mathcal{J}$ given by Proposition 8.5.3 and observe that $S\mathcal{E} = \mathcal{E}S = \mathcal{J}$. Let $\mathcal{E}_1$ be the set of nonzero elements in $\mathcal{E}$, and pick E_0 in $\mathcal{E}_1$. By irreducibility, there is a (strong quasiaffinity) T in S such that $TE_0 \neq E_0 TE_0$. Then TE_0 is nonzero by injectivity of T, and there is an E_1 in $\mathcal{E}_1$ with $TE_0 = E_1 TE_0$. Similarly, there is a member E_2 of $\mathcal{E}_1$ such that $TE_1 = E_2 TE_1 \neq 0$. Continuing in this fashion, we obtain a sequence $\{E_n\}$ in $\mathcal{E}_1$ with $TE_n = E_{n+1} TE_n \neq 0$ for each n. We now distinguish two cases.

(1) The sequence $\{E_n\}$ has distinct members. In this case, we use irreducibility to pick R in $\mathcal{J}$ with $R = E_0 R E_m \neq 0$ (where m is the degree of g). Since $T^j R T^{m-j}$ is nonzero, it is of rank one for $j = 0, \ldots, m$. Since

$$T^j E_0 = E_j T^j E_0 \neq 0 \quad \text{and} \quad T^{m-j} E_j = E_m T^{m-j} E_j \neq 0,$$

we deduce that

$$E_j T^j R T^{m-j} E_j = E_j T^j E_0 R E_m T^{m-j} E_j \neq 0.$$

It follows from (iii) of Proposition 8.5.3 that

$$T^j R T^{m-j} = E_j (T^j R T^{m-j}) E_j,$$

and thus $T^j R T^{m-j} = \mu_j E_j$ for some $\mu_j \neq 0$. The hypothesis on f_g now implies that the diagonalizable operator

$$\sum_{j=0}^{m} a_j \mu_j E_j = \sum_{j=0}^{m} a_j T^j R T^{m-j}$$

is nilpotent. Thus $a_j = 0$ for every j, which is a contradiction.

(2) There are distinct integers $i < j$ with $E_i = E_j$. We can assume, by passing to a power of T, that $j = i + p$ with p prime. (Note that the equality $j = i + 1$ is impossible, because it would imply that $i \neq 0$ by the choice of E_0, and then $TE_i = E_i TE_i$ and $TE_{i-1} = E_i TE_{i-1}$ would be linearly dependent, which contradicts the injectivity of T.)

With no loss of generality, assume $i = 0$ and $j = p$. Let $\mathcal{M}$ be the span of the ranges of the E_i for $i = 0, \dots, p-1$. Clearly, $\mathcal{M}$ is invariant under T. Let $\mathcal{S}_0$ be the semigroup generated by $T|_\mathcal{M}$ and $E_i|_\mathcal{M}$ for $0 \leq i \leq p-1$. Then $\mathcal{S}_0$ is an irreducible semigroup in $\mathcal{M}_p(\mathbb{C})$ on which f_g is nilpotent. After a diagonal similarity applied to $\mathcal{S}_0$, we can assume that $A = T|_\mathcal{M}$ is a cyclic permutation matrix and $B = E_0|_\mathcal{M} = \text{diag}(1, 0, \dots, 0)$. The rest of the proof is exactly as in the end of the proof of Theorem 4.4.9: Since $f_g(A, BA^{-m})$ is nilpotent, we conclude that g is divisible by $x^p - 1$, which yields the final contradiction. $\square$

Corollary 8.5.5. *Let g be as in Theorem 8.5.4 and let S be a semigroup of bounded strong quasiaffinities on which f_g is quasinilpotent. If S contains a compact operator, then S is triangularizable.*

Proof. By the Triangularization Lemma (7.1.11), it suffices to show reducibility. The ideal of compact operators in S is reducible by Theorem 8.5.4, and so is S by Lemma 8.2.1. $\square$

Theorem 8.5.6. *Let g be as in Theorem 8.5.4, and let $S = \overline{\mathbb{R}^+ S}$ be any semigroup of bounded operators on which f_g is quasinilpotent. If S contains a finite-rank idempotent E whose rank is*

$$r = \min\{\text{rank } S : 0 \neq S \in \mathcal{S}\},$$

then S has distinct invariant subspaces

$$\{0\} = \mathcal{M}_0 \subset \mathcal{M}_1 \subset \cdots \subset \mathcal{M}_r.$$

Proof. The semigroup $ESE|_{E\mathcal{X}} \setminus \{0\}$ is a group, by minimality of r together with Lemma 3.1.6. Now, f_g is nilpotent on this group, so that the group is triangularizable by Theorem 4.4.9. The assertion now follows from Lemma 8.2.10. $\square$

8.6 Conditions on Spectral Radius and Trace

As was the case in finite dimensions, permutability and submultiplicativity of spectral radius are equivalent for semigroups of compact operators. Neither is, of course, sufficient for reducibility unless additional hypotheses are made.

Theorem 8.6.1. *Let S be an irreducible semigroup of compact operators on a Banach space $\mathcal{X}$. If spectral radius is submultiplicative on S, then it is multiplicative.*

Proof. Assume without loss of generality that $S = \overline{\mathbb{R}^+ S}$. The proof is similar to that given for Theorem 3.4.3. First observe that S has no nilpotent elements, by Turovskii's Theorem (8.1.11) and Lemma 8.2.1. Thus it does not have divisors of zero, because if A and B are nonzero members of S with $AB = 0$, then BSA consists of nilpotent elements and hence $BSA = 0$, implying that the closure of $S A \mathcal{X}$ is a nontrivial invariant subspace for S.

Next observe that if A is a member of S with $\rho(A) = 1$, then applying Lemma 7.4.5 to the semigroup generated by A yields a nonzero idempotent (of finite rank) E: Some subsequence of $\{A^n\}$ approaches E. The rest of the proof is exactly as in Theorem 3.4.3.   □

The following result is the infinite-dimensional analogue of Theorem 3.4.7.

Theorem 8.6.2. *Let S be an irreducible semigroup of compact operators on $\mathcal{X}$ and $\mathcal{J}$ a nonzero ideal of S. If $\mathcal{J}$ has submultiplicative spectral radius, then so does S.*

Proof. As in the proof of Theorem 3.4.7, we can assume $S = \overline{\mathbb{C} S}$ and verify that if $S \in S$ and $J \in \mathcal{J}$, then $\rho(SJ) \leq \rho(S)\rho(J)$. The rest of the proof is a little different.

First note that, by Corollary 8.1.12, there are nonzero operators of finite rank in S. Let F be such a member of minimal rank. Since $\mathcal{J}$ is irreducible by Lemma 8.2.1, we have $\mathcal{J}F \neq \{0\}$. Thus $\mathcal{J}$ contains the ideal $\mathcal{F}$ of all minimal-rank members of S, and we can assume with no loss that $\mathcal{J} = \mathcal{F}$. We next claim that S has no quasinilpotent member other than zero. Otherwise, let A be such a member and note that $\mathcal{J}A \neq \{0\}$ by irreducibility of $\mathcal{J}$. Also, JA is nilpotent for every $J \in \mathcal{J}$ by the preceding paragraph. This contradicts Turovskii's Theorem (8.1.11) applied to $\mathcal{J}$.

To show submultiplicativity of ρ on S, let S and T be any members of S and assume that $\rho(S) = \rho(T) = 1$. We must prove $\rho(ST) \leq 1$. We can assume $\rho(ST) \neq 0$. Then Lemma 7.4.5, applied to the semigroup generated by $ST/\rho(ST)$, yields a finite-rank idempotent P in S that commutes with ST and satisfies

$$\rho(PSTP) = \rho(ST).$$

We now consider the semigroup $PSP|_{P\mathcal{X}}$ and its nonzero ideal $P\mathcal{J}P|_{P\mathcal{X}}$, both irreducible. (The ideal is nonzero because $P\mathcal{J}P = \{0\}$ implies that $\mathcal{J}P$ consists of nilpotents and thus $\mathcal{J}P = 0$, contradicting irreducibility of $\mathcal{J}$.)

By Theorems 3.4.3 and 3.4.7, ρ is multiplicative on $PSP|_{PX}$, and thus on PSP. Let r be the rank of P. Then PJP contains r^2 independent members PJ_iP with spectral radius 1, and there are scalars α_i such that

$$P = \sum_{i=1}^{r^2} \alpha_i PJ_iP.$$

Thus $(PSTP)^k = P(ST)^kP = (ST)^kP = \Sigma\alpha_i(ST)^kPJ_iP$ for every $k \geq 1$, so that

$$|\operatorname{tr}(PSTP)^k| \leq \Sigma|\alpha_i| \cdot |\operatorname{tr}(ST)^kPJ_iP|$$
$$\leq r\Sigma|\alpha_i|\rho((ST)^kPJ_iP).$$

Since $\rho(AJ) \leq \rho(A)\rho(J)$ for all A in S and J in $\mathcal{J}$, we have

$$\rho((ST)^kPJ_iP) = \rho(ST(ST)^{k-1}PJ_iP) \leq \rho(S)\rho(T(ST)^{k-1}PJ_iP)$$
$$= \rho(T(ST)^{k-1}PJ_iP) \leq \rho(T)\rho((ST)^{k-1}PJ_iP)$$
$$= \rho((ST)^{k-1}PJ_iP).$$

By induction, $\rho((ST)^kPJ_iP) \leq \rho(PJ_iP) \leq 1$ for every k. This implies that, for all k,

$$|\operatorname{tr}(PSTP)^k| \leq r\sum_{i=1}^{r^2}\alpha_i$$

and $\rho(PSTP) \leq 1$ by Lemma 3.4.6. Since $\rho(PSTP) = \rho(ST)$, it follows that $\rho(ST) \leq 1$. $\qquad\square$

The following result does not require irreducibility (just as in finite dimensions).

Theorem 8.6.3. *Let S be a semigroup of compact operators on a Banach space X. Spectral radius is submultiplicative on S if and only if it is permutable on S.*

Proof. If ρ is permutable on S, then the short proof given for the corresponding statement in Theorem 3.4.8 applies word for word, and submultiplicativity follows. The converse needs a little more care in infinite dimensions.

Assume that ρ is submultiplicative on S. If S consists of quasinilpotent elements, there is nothing to prove. So assume otherwise, and let C be a maximal chain of invariant subspaces for S. Using the notation of Section 7.1, let

$$\mathcal{E} = \{\mathcal{M} \in C : \mathcal{M} \neq \mathcal{M}_-\},$$

and let $A_{\mathcal{M}}$ denote the quotient operator on $\mathcal{M}/\mathcal{M}_-$ for each A in $\mathcal{S}$. Denote the Banach space $\mathcal{M}/\mathcal{M}_-$ by $\mathcal{X}_{\mathcal{M}}$ and form a new Banach space

$$\mathcal{Y} = \underset{\mathcal{M} \in \mathcal{E}}{\oplus} \mathcal{X}_{\mathcal{M}}$$

using the supremum norm: If $x_{\mathcal{M}}$ is in $\mathcal{X}_{\mathcal{M}}$ for each $\mathcal{M}$, define the norm of $x = \{x_{\mathcal{M}}\}$ by $\|x\| = \sup_{\mathcal{M} \in \mathcal{E}} \|x_{\mathcal{M}}\|$.

Let $\Phi : \mathcal{S} \to \mathcal{B}(\mathcal{Y})$ be defined by $\Phi(A) = \oplus_{\mathcal{M} \in \mathcal{E}} A_{\mathcal{M}}$ (with $A_{\mathcal{M}}$ acting on $\mathcal{X}_{\mathcal{M}}$). It is easily verified that $\Phi(A)$ is compact for each A in $\mathcal{S}$. (Use the fact that $A_{\mathcal{M}}$ is compact and that given $\varepsilon > 0$ there is only a finite number of indices $\mathcal{M}$ for which $\|A_{\mathcal{M}}\| > \varepsilon$.) Also,

$$\sigma(\Phi(A)) \cup \{0\} = \bigcup_{\mathcal{M} \in \mathcal{E}} \sigma(A_{\mathcal{M}}) \cup \{0\} = \sigma(A).$$

Since $\Phi(AB) = \Phi(A)\Phi(B)$ for all A and B in $\mathcal{S}$, we can assume, with no loss of generality, that $\mathcal{X} = \mathcal{Y}$, so that $\mathcal{S}|_{\mathcal{X}_{\mathcal{M}}}$ is irreducible, $\mathcal{S}$ is contained in the direct sum of the $\mathcal{S}|_{\mathcal{X}_{\mathcal{M}}}$, and $\rho(A) = \max_{\mathcal{M} \in \mathcal{E}} \rho(A_{\mathcal{M}})$.

It is convenient to reduce the cardinality of $\mathcal{E}$. Call a subset $\mathcal{E}_1$ of $\mathcal{E}$ admissible if, for all $A \in \mathcal{S}$,

$$\rho(A) = \rho(\underset{\mathcal{M} \in \mathcal{E}_1}{\oplus} A_{\mathcal{M}}).$$

If $\mathcal{T}$ is a collection of admissible subsets of $\mathcal{E}$ that is totally ordered by inclusion, then $\mathcal{E}_0 = \cap\{\mathcal{F} \subseteq \mathcal{E} : \mathcal{F} \in \mathcal{T}\}$ is also admissible. For if A is an arbitrary nonquasinilpotent member of $\mathcal{S}$, then $\rho(A)$ is achieved on a finite number of members $\mathcal{M}_i$ of $\mathcal{E}$, $i = 1, \ldots, m$. At least one of the $\mathcal{M}_i$ has to belong to every set in the chain $\mathcal{T}$ and thus to $\mathcal{E}_0$; for quasinilpotent members of $\mathcal{S}$, the conclusion is trivial. By Zorn's Lemma, there exists a minimal admissible subset of $\mathcal{E}$. With no loss of generality, we assume that $\mathcal{E}$ itself is minimal.

We make the harmless assumption $\mathcal{S} = \overline{\mathbb{R}^+ \mathcal{S}}$. By minimality of $\mathcal{E}$, for each $\mathcal{M}$ in $\mathcal{E}$ there is at least one element $A \in \mathcal{S}$ with $\rho(A) = \rho(A_{\mathcal{M}}) = 1$ and

$$\max_{\substack{N \in \mathcal{E} \\ N \neq \mathcal{M}}} \rho(A_N) < 1.$$

Then the finite-rank operator F obtained by applying Lemma 7.4.5 to the semigroup generated by A is easily seen to satisfy $F_{\mathcal{M}} \neq 0$ and $F_N = 0$ for $\mathcal{M} \neq N \in \mathcal{E}$. The ideal $\mathcal{J}_{\mathcal{M}}$ of $\mathcal{S}|_{\mathcal{X}_{\mathcal{M}}}$ generated by $F_{\mathcal{M}}$ is nonzero and irreducible. It now follows from the hypothesis that ρ is submultiplicative on $\mathcal{J}_{\mathcal{M}}$ and thus on $\mathcal{S}|_{\mathcal{X}_{\mathcal{M}}}$ (by Theorem 8.6.2). Hence it is multiplicative on $\mathcal{S}|\mathcal{M}$, by Theorem 8.6.1. We have shown that, for every $\mathcal{M} \in \mathcal{E}$ and A, B, C in $\mathcal{S}$,

$$\rho(A_{\mathcal{M}} B_{\mathcal{M}} C_{\mathcal{M}}) = \rho(A_{\mathcal{M}})\rho(B_{\mathcal{M}})\rho(C_{\mathcal{M}}) = \rho(B_{\mathcal{M}} A_{\mathcal{M}} C_{\mathcal{M}}).$$

Then, as in the finite-dimensional case,

$$\rho(ABC) = \max_{M \in \mathcal{E}} \rho((ABC)_M) = \max_{M \in \mathcal{E}} \rho(A_M B_M C_M)$$

$$= \max_{M \in \mathcal{E}} \rho(B_M A_M C_M) = \rho(BAC). \qquad \square$$

Corollary 8.6.4. *A semigroup $\mathcal{S}$ of compact operators with submultiplicative spectral radius is triangularizable if it contains a set of quasinilpotent operators with a unique triangularizing chain.*

Proof. The set of all quasinilpotent elements of $\mathcal{S}$ is an ideal by submultiplicativity of ρ, and is triangularizable by Turovskii's Theorem (8.1.11). Since its triangularizing chain is unique by hypothesis, Lemma 8.2.15 is applicable. $\qquad \square$

Corollary 8.6.5. *A semigroup of compact operators with submultiplicative spectral radius is triangularizable if either of the following conditions holds.*

(i) *$\mathcal{S}$ acts on $\mathcal{L}^2(0,1)$ and contains the Volterra integration operator V:*

$$(Vf)(x) = \int_0^x f(y)dy.$$

(ii) *$\mathcal{S}$ acts on ℓ^2 and contains the weighted Donoghue shift S:*

$$Se_n = w_n e_{n+1},$$

where $\{e_n\}_{n=1}^\infty$ is an orthonormal basis for ℓ^2, and $\{w_n\}$ is a monotone sequence of nonzero complex numbers in ℓ^2.

Proof. Both V and S are unicellular operators, i.e., operators with a unique triangularizing chain (see Radjavi-Rosenthal [2, pp. 66–68]). $\qquad \square$

We have seen (Theorem 8.3.9) that bands of compact operators are triangularizable. The following simple example shows that a semigroup consisting of scalar multiples of idempotents can be irreducible. (See also Example 3.4.13.)

Example 8.6.6. *There exists an irreducible semigroup $\mathcal{S}$ of rank-one operators on ℓ^2 such that for every $A \in \mathcal{S}$, either A or $-A$ is an idempotent. (It follows that ρ is multiplicative, in fact constant, on $\mathcal{S}$.)*

Proof. Let $\mathcal{S}_1$ be the semigroup consisting of the eight matrices

$$\pm \begin{pmatrix} 1 & 0 \\ 0 & 0 \end{pmatrix}, \quad \pm \begin{pmatrix} 1 & 1 \\ 0 & 0 \end{pmatrix}, \quad \pm \begin{pmatrix} 1 & 0 \\ -2 & 0 \end{pmatrix}, \quad \pm \begin{pmatrix} -1 & -1 \\ 2 & 2 \end{pmatrix}.$$

Then S_1 is irreducible and consists of idempotents and their negatives. For each positive integer of the form 2^k, let S_k be the tensor product of k copies of S_1, i.e.,

$$S_k = \{E_1 \otimes \cdots \otimes E_k : E_i \in S_1, \quad i = 1, \ldots, k\}.$$

Pick an orthonormal basis $\{e_n\}_{n=1}^{\infty}$ for ℓ^2 and let S_k act on the span of the first 2^k basis vectors. Define a semigroup $\hat{S}_k$ acting on ℓ^2 by $\hat{S}_k = S_k \oplus 0$. Observe that, by construction, $\hat{S}_k \subseteq \hat{S}_{k+1}$ for each k, so that

$$S = \bigcup_{k=1}^{\infty} \hat{S}_k$$

is a semigroup and has the stated properties.   □

The following result cannot be improved, in light of the example above. It extends the result on bands (as in Theorem 3.5.2).

Theorem 8.6.7. *Let S be a semigroup of compact operators with sub-multiplicative spectral radius. If every member of S is a nonnegative scalar multiple of an idempotent, then S is triangularizable.*

Proof. The proof is the same as that of Theorem 3.5.2, except that Theorem 8.6.1 should be quoted instead of Theorem 2.3.5, and Lemma 7.1.11 instead of Lemma 1.1.4.   □

As in finite dimensions, any semigroup as in the above theorem is essentially a band.

Corollary 8.6.8. *Let S be as in Theorem 8.6.7. Then there is a band S_1 such that $S \subseteq \mathbb{R}^+ S_1$.*

Proof. This follows from Theorem 8.6.7 in the same way that Corollary 3.5.3 followed from Theorem 3.5.2.   □

The assumption of submultiplicativity in the above results is essential, as demonstrated in the finite-dimensional case.

We now turn to the trace condition. We shall confine ourselves to the case of operators on a Hilbert space; trace is defined for trace-class operators in Definition 6.5.12.

Theorem 8.6.9. *Let S be a family of trace-class operators on a Hilbert space. Then S is triangularizable if and only if trace is permutable on S.*

Proof. If S is triangularizable, then it has permutable trace by Corollary 7.2.15. We now prove the sufficiency of the condition. Since trace is permutable on the algebra generated by S, we can assume that S is an algebra.

By Theorem 8.5.2, it suffices to show that $AB - BA$ is quasinilpotent for all $\{A, B\}$ in $\mathcal{S}$. This is equivalent to the assertion that $(AB - BA)^n$ has trace zero for all natural numbers n. (If $\{\lambda_j\}$ is the sequence of eigenvalues of $AB - BA$, then, by Lidskii's Theorem (7.2.14), $\operatorname{tr}((AB - BA)^n) = \sum_j \lambda_j^n$ for each n. It is easy to verify that $\sum_j \lambda_j^n = 0$ for all n implies that every λ_j is zero; see, for example, Radjavi [1].) Now, this assertion is clearly true for $n = 1$. For $n \geq 2$, let $C = AB - BA$ and apply the permutability hypothesis to A, B, and C^{n-1}:

$$\operatorname{tr}(AB - BA)^n = \operatorname{tr} ABC^{n-1} - \operatorname{tr} BAC^{n-1} = 0. \qquad \square$$

For semigroups, the statement is simpler.

Corollary 8.6.10. *A semigroup $\mathcal{S}$ of trace-class operators is triangularizable if and only if*

$$\operatorname{tr}(ABC) = \operatorname{tr}(BAC)$$

for all choices of A, B, C in $\mathcal{S}$.

Proof. The given condition is equivalent to permutability, by Lemma 8.2.3.

Other corollaries of Theorem 8.6.9, corresponding to those in finite dimensions, are easily proved. $\qquad \square$

Corollary 8.6.11. *Let $\mathcal{F}$ be a family of trace-class operators. If trace is permutable on $\mathcal{F}^k$ for some k, then it is permutable on $\mathcal{F}$.*

Proof. This is a direct consequence of Theorem 8.6.9 and Lemma 8.2.14. $\qquad \square$

Corollary 8.6.12. *A self-adjoint family of trace-class operators is commutative if and only if it has permutable trace.*

Proof. We need only show that permutability of trace on a self-adjoint semigroup implies commutativity.

By Theorem 8.6.9, the family is triangularizable. Thus so is the semigroup it generates, and this semigroup is also self-adjoint. As shown in the proof of Corollary 8.4.6, a self-adjoint triangularizable semigroup of compact operators is commutative. $\qquad \square$

Obvious consequences of Theorem 8.6.9 include the results that if trace is multiplicative or constant on a semigroup $\mathcal{S}$ (and in particular if it is zero), then $\mathcal{S}$ is triangularizable. This can be extended to translates of trace-class operators.

Theorem 8.6.13. *If every member of a semigroup is of the form* $\lambda + T$, *where* T *is a trace-class operator and* $\operatorname{tr} T = 0$, *then the semigroup is triangularizable.*

Proof. Since quotients of operators of this form are also translates of operators with trace zero, the Triangularization Lemma (7.1.11) is applicable once we show reducibility.

Let $\mathcal{A}$ be the algebra generated by the semigroup and the identity operator. Then it is easily seen that $\mathcal{A}$ also consists of operators of the form $\lambda + T$ with $\operatorname{tr} T = 0$. Consider the Banach algebra of trace-class operators with the trace norm and adjoin the identity operator to form a Banach algebra $\mathcal{C}$. Then the closure of $\mathcal{A}$ in $\mathcal{C}$ still consists of translates of operators of trace zero, by continuity of trace. Since $\overline{\mathcal{A}}$ is a proper subalgebra of $\mathcal{C}$, the reducibility of $\mathcal{A}$ follows from an analogue of Corollary 7.4.9 for the trace class (Radjavi-Rosenthal [3]). $\qquad\square$

8.7 Nonnegative Operators

In this section we shall give extensions of the results of Chapter 5 to an infinite-dimensional setting. The most general situation to which our finite-dimensional results can be extended is that of nonnegative compact operators on a Banach lattice, but, to keep the exposition brief and self-contained, we confine ourselves to $\mathcal{L}^p$ spaces (including the discrete cases of l^p) with $p \geq 1$.

In the rest of this chapter, $\mathcal{X}$ will represent a separable complex space $\mathcal{L}^p(X, \mu)$ with $1 \leq p < \infty$, where X is a Hausdorff-Lindelöf space and μ a σ-finite, regular Borel measure on X. The necessary adjustments for the case of real function spaces are easily made.

Definition 8.7.1. A member f of $\mathcal{X} = \mathcal{L}^p(X, \mu)$ is called *positive* if (a representative of) f has positive values almost everywhere on X. The term *nonnegative* is defined analogously. The set $\mathcal{X}^+$ of all nonnegative members of $\mathcal{X}$ is called the *nonnegative cone* of $\mathcal{X}$. We write $f \geq 0$ if $f \in \mathcal{X}^+$. Less frequently, we write $f > 0$ to mean that f is positive. The inequality $f \geq g$ means $f - g \in \mathcal{X}^+$.

Definition 8.7.2. A subspace $\mathcal{M}$ of $\mathcal{X} = \mathcal{L}^p(X, \mu)$ is called *standard* if it consists of all $f \in \mathcal{X}$ vanishing outside a Borel subset Y of X; $\mathcal{M}$ is naturally identified with $\mathcal{L}^p(Y, \mu|Y)$, which we shall denote by $\mathcal{L}^p(Y, \mu)$ or simply by $\mathcal{L}^p(Y)$.

(The terms "ideal" or "lattice ideal" are used more frequently than "standard subspace," but to avoid confusion with other uses of "ideal," we prefer the latter adjective. For the same reason, we do not use the term "band" in this context.)

If $X_1, \ldots, X_r$ are mutually disjoint Borel sets with union X, then

$$\mathcal{L}^p(X) = \mathcal{L}^p(X_1) \oplus \cdots \oplus \mathcal{L}^p(X_r).$$

Relative to this direct sum, every operator on $\mathcal{L}^p(X)$ has a uniquely determined *operator matrix*

$$\begin{bmatrix} A_{11} & \cdots & A_{1r} \\ \vdots & & \\ A_{r1} & \cdots & A_{rr} \end{bmatrix},$$

where A_{ij} is an operator from $\mathcal{L}^p(X_j)$ into $\mathcal{L}^p(X_i)$.

Definition 8.7.3. Let $\mathcal{X} = \mathcal{L}^p(X, \mu)$ and $\mathcal{Y} = \mathcal{L}^p(Y, \nu)$. A bounded operator $A : \mathcal{X} \to \mathcal{Y}$ is called *nonnegative* if it takes $\mathcal{X}^+$ into $\mathcal{Y}^+$; A is called *strongly positive* if Af is positive whenever $f \geq 0$ and $f \neq 0$. We use the notation $A \leq B$ to mean that $B - A$ is nonnegative.

Note that an $n \times n$ matrix that has all of its entries positive induces a strongly positive operator on $\mathbb{C}^n$.

Definition 8.7.4. A collection $\mathcal{A}$ of bounded operators on $\mathcal{X} = \mathcal{L}^p(X, \mu)$ is called *decomposable* if it has a nontrivial standard invariant subspace and *indecomposable* otherwise; it is called *completely decomposable* if it has a triangularizing chain of standard invariant subspaces.

(The terms "reducible" for decomposable $\mathcal{A}$ and "ideal-triangularizable" for completely decomposable $\mathcal{A}$ are also used in the literature.)

The simplest example of a single indecomposable operator is a strongly positive operator. A different example can be obtained by considering a direct-sum decomposition

$$\mathcal{L}^p(X_1) \oplus \cdots \oplus \mathcal{L}^p(X_r)$$

for $\mathcal{L}^p(X)$ as in Definition 8.7.2, and strongly positive operators

$$A_i : \mathcal{L}^p(X_i) \to \mathcal{L}^p(X_{i+1})$$

for $i < r$ and $A_r : \mathcal{L}^p(X_r) \to \mathcal{L}^p(X_1)$. Then the matrix

$$\begin{bmatrix} 0 & 0 & \cdots & 0 & A_r \\ A_1 & 0 & \cdots & 0 & 0 \\ 0 & A_2 & \cdots & 0 & 0 \\ \vdots & & & & \\ 0 & 0 & \cdots & A_{r-1} & 0 \end{bmatrix}$$

is easily seen to represent an indecomposable operator on $\mathcal{L}^p(X)$.

Lemma 8.7.5. *Let $\mathcal{F}$ be a countable subset of $\mathcal{L}^p(X, \mu)$ consisting of nonnegative functions. Fix a representative for each member and assume that the set*

$$\{x : f(x) = 0 \quad \text{for all} \quad f \in \mathcal{F}\}$$

has measure zero. If A is a nonnegative operator from $\mathcal{L}^p(X, \mu)$ into $\mathcal{L}^p(Y, \nu)$ such that $Af = 0$ for all $f \in \mathcal{F}$, then $A = 0$.

Proof. We first treat the case of a singleton $\mathcal{F} = \{f\}$, where f is a positive function. Express X as the union of the sets

$$X_n = \left\{ x \in X : \frac{1}{n} \le f(x) \right\}$$

for positive integers n. Let χ_n be the characteristic function of X_n. Then $\chi_n \le nf$ implies $0 \le A\chi_n \le nAf = 0$. If E is any measurable subset of X of finite measure, then $\chi_E \chi_n \le \chi_n$ implies $A\chi_E \chi_n = 0$. Since this holds for all n, it follows that $A\chi_E = 0$. Thus $Ah = 0$ for all $h \in \mathcal{L}^p(X, \mu)$.

The general case is reduced to the special case by considering the single positive function $f = \sum_{n=1}^{\infty} f_n / (n^2 \|f_n\|)$, where $\{f_n\}$ is an enumeration of $\mathcal{F}$. $\qquad\square$

Lemma 8.7.6. *For an arbitrary semigroup of bounded nonnegative operators on $\mathcal{X} = \mathcal{L}^p(X, \mu)$, the following are mutually equivalent:*

   (i) *$\mathcal{S}$ is decomposable;*

  (ii) *$A\mathcal{S}B = \{0\}$ for some nonnegative, nonzero operators A and B;*

 (iii) *there are nonnegative, nonzero members f of $\mathcal{X}$ and ϕ of $\mathcal{X}^*$ such that $\phi(Sf) = 0$ for all S in $\mathcal{S}$;*

 (iv) *some nonzero ideal of $\mathcal{S}$ is decomposable.*

Proof. The proof is similar to the finite-dimensional version (Lemma 5.1.5). The proof that (i) implies (ii) is exactly the same. To show that (ii) implies (iii), assume (ii). Pick a nonnegative g such that $f = Bg \ne 0$. If $\mathcal{S}f = \{0\}$, then any nonnegative ϕ in $\mathcal{X}^*$ satisfies $\phi(Sf) = 0$. Thus we can assume $\mathcal{S}f \ne \{0\}$. By separability of $\mathcal{X}$, there exists a dense countable subset $\mathcal{F}$ in $\mathcal{S}f$. Fixing a representative for each member of $\mathcal{F}$, we deduce from Lemma 8.7.5, together with the hypothesis $A\mathcal{S}f = \{0\}$, that the set

$$\{x : h(x) = 0 \quad \text{for all} \quad h \in \mathcal{F}\}$$

has positive measure. Its complement X_0 in X yields a nontrivial standard subspace $\mathcal{L}^p(X_0)$ containing $\mathcal{S}f$. Now pick a nonnegative $\phi \in \mathcal{X}^*$ such that $\phi \ne 0$ and $\phi|_{\mathcal{L}^p(X_0)} = 0$. Then $\phi(Sf) = 0$. Thus (ii) implies (iii).

Next assume (iii). To show (i), assume that $\mathcal{S} \neq \{0\}$. If $\mathcal{S}f = \{0\}$, then Lemma 8.7.5 implies, after fixing a representative function for f, that the set

$$X_1 = X \setminus \{x : f(x) = 0\}$$

gives rise to a nontrivial standard subspace $\mathcal{L}^p(X_1)$ contained in the kernel of $\mathcal{S}$, and thus invariant under $\mathcal{S}$. Hence we can assume $\mathcal{S}f \neq \{0\}$. Pick a nonzero nonnegative g in $\mathcal{L}^p(X)$ and define the rank-one operator A by $Ah = \phi(h)g$, so that the hypothesis $\phi(\mathcal{S}f) = 0$ implies $A(\mathcal{S}f) = 0$. Then the nontrivial standard subspace $\mathcal{L}^p(X_0)$, as constructed in the preceding paragraph, is invariant under $\mathcal{S}$. Thus (iii) implies (i).

To complete the proof, we need only show the equivalence of (i) and (iv). This is done exactly as in the last paragraph of the proof of Lemma 5.1.5.

□

The following includes the assertion that a nonnegative compact quasi-nilpotent operator is decomposable. It will be shown (see Theorem 8.7.9) that every semigroup of such operators is completely decomposable.

Theorem 8.7.7. *Let T be a nonnegative quasinilpotent operator such that $T \geq K \geq 0$ for some compact operator K other than 0. Then T has a nontrivial standard invariant subspace. Furthermore, such a subspace can be chosen to be invariant under every nonnegative operator A that satisfies $A \leq B$ for some operator B commuting with T.*

Proof. Let the underlying space be $\mathcal{X} = \mathcal{L}^p(X)$, and let

$$\mathcal{E} = \{A \in \mathcal{B}(\mathcal{X}) : \text{there is a } B \text{ satisfying } 0 \leq A \leq B \text{ and } BT = TB\}.$$

Note that $T \in \mathcal{E}$. Also, if $A_1 \leq B_1$ and $A_2 \leq B_2$, then $A_1 A_2 \leq A_1 B_2 \leq B_1 B_2$, so $\mathcal{E}^2 \subseteq \mathcal{E}$. Thus, for every nonnegative f in $\mathcal{X}$, the set $\mathcal{E}f$ is invariant under every member of $\mathcal{E}$, and its closed linear span is an invariant subspace for $\mathcal{E}$. We claim that this span is a standard invariant subspace and $\overline{\mathcal{E}f}$ consists of all its nonnegative members. Pick representative functions $\{f_n\}_{n=1}^{\infty}$ from a countable dense subset of $\mathcal{E}f$ and let

$$F = \{t \in X : f_n(t) = 0 \text{ for } n = 1, 2, \dots\} .$$

Then every function in $\mathcal{E}f$ vanishes almost everywhere on F. Let $X_0 = X \setminus F$. We show that every nonnegative function in $\mathcal{L}^p(X_0)$ is in the closure of $\mathcal{E}f$, which establishes the claim, since every function in $\mathcal{L}^p(X_0)$ is a linear combination of nonnegative ones.

Fix any $g_0 \geq 0$ in $\mathcal{L}^p(X_0)$ and let $\varepsilon > 0$. There is a Borel set Y of finite measure in X_0 such that $\int_{X_0 \setminus Y} g_0^p d\mu < \varepsilon$. Also, we can choose $\delta > 0$ such that $\int_E g_0^p d\mu < \varepsilon$ whenever E is a Borel set of measure less than δ. For each

f_n as in the preceeding paragraph, choose $A_n \in \mathcal{E}$ such that $A_n f = f_n$. By the definition of X_0, there is an N such that

$$\left\{ t \in Y : \left(\sum_{n=1}^{N} A_n f \right)(t) = 0 \right\}$$

has measure less than δ. Let $A_0 = \sum_{n=1}^{N} A_n$; then A_0 is clearly in $\mathcal{E}$. Since the measure of $\{t \in Y_0 : (A_0 f)(t) = 0\}$ is less than δ, there exist positive integers L and M such that the measure of

$$E = \left\{ t \in Y : (A_0 f)(t) < \frac{1}{L} \right\} \cup \{ t \in Y : g_0(t) > M \}$$

is also less than δ.

Now define the function h by

$$h(t) = \begin{cases} \frac{g_0(t)}{(A_0 f)(t)} & \text{for } t \in Y \setminus E, \\ 0 & \text{otherwise.} \end{cases}$$

Then $h(t) \leq ML$ for all t, so the multiplication operator H defined by $Hg = hg$ for $g \in \mathcal{L}^p(X, \mu)$ is a bounded operator. It is obviously nonnegative. Moreover, H is in $\mathcal{E}$ because $H \leq (ML)I$. It follows that the function $HA_0 f$ belongs to $\mathcal{E}f$, and

$$\|g_0 - HA_0 f\|^p = \int_{Y \setminus E} |g_0 - HA_0 f|^p d\mu + \int_E |g_0 - HA_0 f|^p d\mu +$$

$$\int_{X_0 \setminus Y} |g_0 - HA_0 f|^p d\mu = 0 + \int_E |g_0|^p d\mu + \int_{X_0 \setminus Y} |g_0|^p d\mu < 2\varepsilon.$$

Thus g_0 is in the closure of $\mathcal{E}f$.

Note that we have not yet used the fact that K is compact. This is needed to prove that, for some nonnegative $f \neq 0$, the invariant subspace obtained above is proper; equivalently, $\overline{\mathcal{E}f}$ is a proper subset of the nonnegative cone $\mathcal{X}^+$ of $\mathcal{L}^p(X)$. The proof of this is similar to the proof of the existence of hyperinvariant subspaces for compact operators (Theorem 6.3.4). Thus assume that for every nonzero $f \geq 0$, $\overline{\mathcal{E}f}$ coincides with $\mathcal{X}^+$.

Fix an $f_0 \geq 0$ such that $Kf_0 \neq 0$. Let $\mathcal{B}$ be an open ball containing f_0 such that $\overline{K\mathcal{B}}$ does not contain zero. Let $\mathcal{B}^+ = \mathcal{B} \cap \mathcal{X}^+$ (so that $\mathcal{B}^+$ is a relatively open neighborhood of f_0 in $\mathcal{X}^+$). By what we just assumed, $\mathcal{B}^+ \cap \mathcal{E}f$ is nonempty for every nonzero f in $\mathcal{X}^+$. Thus the family of open sets $\{A^{-1}(\mathcal{B}) : A \in \mathcal{E}\}$ covers the compact set $\overline{K\mathcal{B}^+}$. It follows that, for some finite subset $\mathcal{F}$ of $\mathcal{E}$,

$$\bigcup \{ A^{-1}(\mathcal{B}^+) : A \in \mathcal{F} \} \supseteq \overline{K\mathcal{B}^+}.$$

For each $A \in \mathcal{F}$, let B_A be an operator (which exists, by the definition of $\mathcal{E}$) such that $A \leq B_A$ and $B_A T = T B_A$.

Starting with $g_0 = Kf_0$ in $K\mathcal{B}^+$, we can find A_1 in $\mathcal{F}$ with $A_1 Kf_0 \in \mathcal{B}^+$ such that $g_1 = KA_1 Kf_0 \in K\mathcal{B}^+$; then A_2 in $\mathcal{F}$ with $A_2 KA_1 Kf_0 \in \mathcal{B}^+$. By induction, we can find a sequence A_n in $\mathcal{F}$ with

$$g_m = KA_m KA_{m-1} K \cdots KA_1 Kf_0 \in K\mathcal{B}^+$$

for every integer m. Denoting B_{A_i} by B_i for each i, we have

$$g_m \leq KB_m KB_{m-1} K \cdots KB_1 Kf_0 \leq TB_m TB_{m-1} T \cdots TB_1 Tf_0$$
$$= T^{m+1} B_m B_{m-1} \cdots B_1 f_0 \ .$$

Hence $\|g_m\| \leq \|T^{m+1}\| \cdot M^m$, where $M = \max\{\|B_A\| : A \in \mathcal{F}\}$. Since $\lim_{m \to \infty} \|T^{m+1}\|^{1/m} = 0$, we get $\lim_{m \to \infty} \|g_m\| = 0$, implying $0 \in \overline{K\mathcal{B}^+}$, which is a contradiction. $\qquad\square$

The following lemma shows that maximal chains of standard invariant subspaces are triangularizing.

Lemma 8.7.8. *Let $\mathcal{C}$ be a maximal chain of standard subspaces of $\mathcal{X} = \mathcal{L}^p(X, \mu)$. Then $\mathcal{C}$ is a maximal subspace chain.*

Proof. Since $\{0\}$ and $\mathcal{X}$ are in $\mathcal{C}$, to prove the assertion using Theorem 7.1.9 we must show that $\mathcal{C}$ is complete and that all its gaps are one-dimensional.

If $\{\mathcal{M}_w\}$ is a subchain of $\mathcal{C}$, then it follows from the separability of $\mathcal{X}$ that $\cap_w \mathcal{M}_w$ is the intersection of a countable subchain $\{\mathcal{M}_i : i \in \mathbb{N}\}$ of $\{\mathcal{M}_w\}$. If $\mathcal{M}_i = \mathcal{L}^p(X_i)$, then

$$\bigcap_i \mathcal{M}_i = \mathcal{L}^p \left(\bigcap_i X_i \right)$$

and $\bigcap_i X_i$ is a Borel set. Thus $\bigcap_w \{\mathcal{M}_w\}$ is a standard subspace. Similarly, it can be verified that the closed span of $\{\mathcal{M}_w\}$ is standard. (Denoting the complement of a Borel set E in X by E^C, observe that $\bigcap_w \mathcal{L}^p(X_w^C)$ is the intersection of a countable number of X_i^C, as shown above, so that

$$\vee_w \mathcal{L}^p(X_w) = \mathcal{L}^p(\textstyle\bigcup_i X_i),$$

and $\bigcup_i X_i$ is a Borel set.)

Now let $\mathcal{M} = \mathcal{L}^p(X_1) \in \mathcal{C}$ and let $\mathcal{M}_- = \mathcal{L}^p(X_2)$; we can assume $X_2 \subseteq X_1$. To complete the proof, we must verify that if $\mathcal{M} \neq \mathcal{M}_-$ (that is, if $\mu(X_1 \setminus X_2) \neq 0$), then $\mathcal{L}^p(X_1 \setminus X_2)$ is one-dimensional. If this dimension were greater, then there would exist disjoint Borel subsets Y_1 and Y_2 of $X_1 \setminus X_2$ with nonzero measure, resulting in strict inclusions

$$\mathcal{M}_- = \mathcal{L}^p(X_2) \subset \mathcal{L}^p(X_2 \cup Y_1) \subset \mathcal{L}^2(X_1) = \mathcal{M},$$

contradicting maximality of $\mathcal{C}$. $\qquad\square$

We are now ready to establish the infinite-dimensional analogue of Theorem 5.1.2.

Theorem 8.7.9. *A semigroup of nonnegative compact quasinilpotent operators is completely decomposable.*

Proof. Let S be such a semigroup on $\mathcal{L}^p(X)$ and assume, with no loss of generality, that $S = \overline{\mathbb{R}^+ S}$. We can also assume, by Corollary 8.1.13, that S is convex.

We first show that S is decomposable. This is particularly easy if $p > 1$: Since the set of all compact operators on $\mathcal{L}^p(X)$ is separable, so is S. Pick a countable dense set $\{S_n : n \in \mathbb{N}\}$ in S, and let

$$T = \sum_{n=1}^{\infty} \frac{S_n}{2^n \|S_n\|}.$$

Now, $T \in S$ and $S_n \leq 2^n \|S_n\| T$ for every n. By Theorem 8.7.7, since $2^n \|S_n\| T$ commutes with T, there is a nontrivial standard subspace of $\mathcal{L}^p(X)$ invariant under all operators S_n, and hence under S.

When $\mathcal{L}^1(X)$ is infinite-dimensional, the space of compact operators on $\mathcal{L}^1(X)$ is not separable. The following proof applies to $\mathcal{L}^p(X)$ for all p, including $p = 1$. The set

$$\mathcal{T} = \{T : 0 \leq T \leq S \text{ for some } S \in S\}$$

is easily seen to be a semigroup of quasinilpotent operators leaving invariant every standard invariant subspace of S. (Observe that $T \leq S$ implies $T^n \leq S^n$, which implies $\|T^n\| \leq \|S^n\|$.)

Now, the ideal of compact operators in $\mathcal{T}$ is reducible by Turovskii's Theorem (8.1.11). It follows from Lemma 8.2.1 that $\mathcal{T}$ has a nontrivial invariant subspace $\mathcal{Y}$. Note that the (closed linear) span $\mathcal{Y}_0$ of $\mathcal{T}\mathcal{Y}$ is also invariant under $\mathcal{T}$ and, assuming $\mathcal{T} \neq \{0\}$ without loss of generality, we have $0 \neq \mathcal{Y}_0 \neq \mathcal{L}^p(\mathcal{Y}_0)$. To complete the proof, we shall show that $\mathcal{Y}_0$ is a standard subspace.

First note that $\mathcal{Y}_0$ is invariant under all multiplication operators M_ϕ for $\phi \in \mathcal{L}^\infty(X)$ (see Definition 7.1.4). To see this, simply observe that, for every $S \in \mathcal{T}$ and every nonnegative function $\phi \in \mathcal{L}^\infty(X)$, $M_\phi S \leq \|\phi\|_\infty S$. Therefore, $\mathcal{T}$ contains $M_\phi S$ for each such ϕ and S. Hence $M_\phi \mathcal{Y}_0 \subseteq \mathcal{Y}_0$ for all nonnegative ϕ in $\mathcal{L}^\infty(X)$. Since every member of $\mathcal{L}^\infty(X)$ is a linear combination of four nonnegative ones, $\mathcal{Y}_0$ is invariant under all multiplication operators.

Now, the only invariant subspaces of the algebra of all multiplication operators are the standard subspaces, as we proceed to show. If f is a fixed representative of any element of such a subspace $\mathcal{Y}_0$ and n and k are positive integers, let

$$\phi_{n,k}(x) = \begin{cases} \frac{1}{f(x)} & \text{if } \frac{1}{n} \leq |f(x)| \leq k, \\ 0 & \text{otherwise.} \end{cases}$$

Then $M_{\phi_{n,k}}f$ is the characteristic function of $\{x \in X : \frac{1}{n} \leq |f(x)| \leq k\}$ so $\mathcal{Y}_0$ contains all such characteristic functions and hence also their limit in $\mathcal{L}^p(X)$ as n and k approach infinity, which is the characteristic function χ of $X_0 = \{x \in X : f(x) \neq 0\}$. Thus $\mathcal{Y}_0$ contains $M_\phi \chi$ for all $\phi \in \mathcal{L}^\infty(X)$, and it follows that $\mathcal{Y}_0 \supseteq \mathcal{L}^p(X_0)$. The (closed linear) span of any collection of standard subspaces is standard (by σ-finiteness), so $\mathcal{Y}_0$ is standard.

To prove complete decomposability, let $\mathcal{C}$ be a maximal chain of standard invariant subspaces for $\mathcal{T}$. By Lemma 8.7.8, we need only show that if

$$\mathcal{M}_1 = \mathcal{L}^p(X_1) \quad \text{and} \quad \mathcal{M}_2 = \mathcal{L}^p(X_2)$$

are invariant subspaces for $\mathcal{T}$ with $X_1 \subset X_2$ and $\mu(X_2 \setminus X_1) \neq 0$, then $\mathcal{L}^p(X_2 \setminus X_1)$ is one-dimensional. Suppose otherwise. Observe that the quotient semigroup of $\mathcal{T}$ on $\mathcal{M}_2/\mathcal{M}_1$ is a semigroup of nonnegative compact quasinilpotent operators on $\mathcal{L}^p(X_2 \setminus X_1)$, and thus it has a nontrivial invariant subspace of the form $\mathcal{L}^p(Y)$ with $Y \subset X_2 \setminus X_1$. Then $\mathcal{L}^p(X_1 \cup Y)$ is invariant under $\mathcal{T}$, so that the strict inclusions

$$\mathcal{M}_1 = \mathcal{L}^p(X_1) \subset \mathcal{L}^p(X_1 \cup Y) \subset \mathcal{L}^p(X_2)$$

contradict the maximality of $\mathcal{C}$ as a chain of standard invariant subspaces. $\qquad \square$

Corollary 8.7.10. *Let $\mathcal{S}$ be a semigroup of nonnegative quasinilpotent operators. If $\mathcal{S}$ contains a nonzero compact operator, then it is decomposable.*

Proof. The ideal of $\mathcal{S}$ generated by its compact members is decomposable by Theorem 8.7.9. Thus so is $\mathcal{S}$ itself, by Lemma 8.7.6. $\qquad \square$

We need the following extensions of finite-dimensional results on the structure of idempotents.

Lemma 8.7.11. *If E is a nonnegative idempotent of rank one on $\mathcal{L}^p(X)$, then E is indecomposable if and only if E is strongly positive or, equivalently, $E = f \otimes \phi$ with strictly positive f and ϕ in $\mathcal{X}$ and $\mathcal{X}^*$ respectively; i.e., $Eg = \phi(g)f$ for all g in $\mathcal{X}$, f is positive almost everywhere, and $\phi(g) > 0$ for every nonzero $g \geq 0$ in $\mathcal{X}$. (The assertion on ϕ is equivalent to stating that ϕ is strictly positive as a member of the dual space $\mathcal{L}^q(X)$.)*

Proof. Suppose that E is indecomposable. Pick a nonzero nonnegative vector f in the range of E. Then $E = f \otimes \phi$ for some ϕ in $\mathcal{X}^*$, which is nonnegative by hypothesis. Now fix a representative of f and note that

$$X_0 = X \setminus \{x : f(x) = 0\}$$

gives rise to the standard subspace $\mathcal{L}^p(X_0)$ invariant under E. Also, for any $h \geq 0$ such that $\phi(h) = 0$, let

$$Y_h = X \setminus \{x : h(x) = 0\}.$$

As in the proof of Lemma 8.7.5, $Eg = 0$ for all g in the standard subspace $\mathcal{L}^p(Y_h)$.

Now, $\mathcal{L}^p(X_0)$ and $\mathcal{L}^p(Y_h)$ are invariant under E, so, since E is indecomposable, $\mathcal{L}^p(X_0) = \mathcal{L}^p(X)$ and $\mathcal{L}^p(Y_h) = \{0\}$ for every $h \geq 0$. In other words, $f > 0$ almost everywhere, and $\phi(h) > 0$ for all nonzero $h \geq 0$.

The converse is also easy. Assume that $E = f \otimes \phi$ has a nontrivial standard invariant subspace $\mathcal{L}^p(Z)$. If f is strictly positive, we must show that ϕ is not. But in such a case $E\mathcal{L}^p(Z)$ cannot contain f, since f is nonzero almost everywhere on Z. Thus $E\mathcal{L}^p(Z) = \{0\}$. Let h be any nonzero nonnegative vector in $\mathcal{L}^p(Z)$. Then $\phi(h) = 0$, so ϕ is not strictly positive. □

Lemma 8.7.12. *Let E be a nonnegative idempotent of finite rank r on $\mathcal{X} = \mathcal{L}^p(X)$.*

(i) *Assume that $Eh = 0$ with $0 \leq h \in \mathcal{X}$ implies $h = 0$ and that $\phi|_{E\mathcal{X}} = 0$ with $0 \leq \phi \in \mathcal{X}^*$ implies $\phi = 0$. Then there exist r disjoint Borel subsets X_j whose union is X such that E is of the form*

$$E_1 \oplus \cdots \oplus E_r,$$

where each E_j is an indecomposable nonnegative idempotent of rank one on $\mathcal{L}^p(X_j)$.

(ii) *In general, there are three disjoint Borel sets Y_1, Y_2, and Y_3 with union X such that E has an operator matrix*

$$\begin{bmatrix} 0 & AF & AFB \\ 0 & F & FB \\ 0 & 0 & 0 \end{bmatrix}$$

relative to the direct sum $\mathcal{X} = \mathcal{L}^p(Y_1) \oplus \mathcal{L}^p(Y_2) \oplus \mathcal{L}^p(Y_3)$, where F is of the form $E_1 \oplus \cdots \oplus E_r$ as in (i) above and A and B are nonnegative.

Proof. (i) If $r = 1$, the assertion follows from Lemma 8.7.11. Assume $r > 1$. It suffices to show that E has a nontrivial pair of "orthogonal" invariant subspaces of the form $\mathcal{L}^p(Y)$ and $\mathcal{L}^p(X \setminus Y)$. For then $E = F \oplus G$, where F and G are nonnegative idempotents of rank $< r$ each satisfying the hypothesis of (i), so the proof can be completed by induction and Lemma 8.7.11.

Pick linearly independent nonnegative functions f_0 and g_0 in the range of E. Replacing g_0 by $f_0 + g_0$ if necessary (and fixing representations of f_0 and g_0), we can assume $f_0 \leq g_0$. Let

$$t_0 = \sup\{t \in \mathbb{R} : t \geq 0, t g_0 \leq f_0\}.$$

Then $t_0 \geq 0$ and $t_0 g_0 \leq f_0 \leq g_0$. Also, $t_0 < 1$, since f_0 and g_0 are distinct. Pick t_1 with $t_0 < t_1 < 1$ and let $h_0 = f_0 - t_1 g_0$. Note that h_0 is neither nonnegative nor nonpositive. Consider the usual decomposition $h_0 = h_0^+ - h_0^-$, where $h_0^+ = (|h_0| + h_0)/2$ and $h_0^- = (|h_0| - h_0)/2$ so that h_0^+ and h_0^- are both nonzero.

Now, $Eh_0 = h_0$, $h_0 = Eh_0^+ - Eh_0^-$, and therefore $h_0^+ \leq Eh_0^+$ (since $Eh_0^- \geq 0$). Thus $Eh_0^+ - h_0^+ \geq 0$ and, since $E(Eh_0^+ - h_0^+) = 0$, we conclude from the hypothesis that $Eh_0^+ = h_0^+$. Let

$$Y = X \setminus \{x : h_0^+(x) = 0\}.$$

Then $\mathcal{L}^p(Y)$ is nontrivial. It is also invariant under E, because $g \leq n h_0^+$ for any $n \in \mathbb{N}$ implies

$$Eg \leq nEh_0^+ = nh_0^+ \in \mathcal{L}^p(Y),$$

and thus $Eg \in \mathcal{L}^p(Y)$. But the set of such functions g includes characteristic functions of any Borel subset W of Y that is contained in some set $\{x : h_0^+(x) \geq n\}$ with $n \in \mathbb{N}$. Thus $Ef \in \mathcal{L}^p(Y)$ for all $f \in \mathcal{L}^p(Y)$.

We will be done if we show that $\mathcal{L}^p(X \setminus Y)$ is also invariant. If we write

$$E = \begin{bmatrix} F & R \\ 0 & G \end{bmatrix}$$

relative to the direct sum $\mathcal{L}^p(X) = \mathcal{L}^p(Y) \oplus \mathcal{L}^p(X \setminus Y)$, we must show that $R = 0$. The proof is similar to the finite-dimensional case: $E = E^2$ implies that

$$F(FR + RG)G = FRG,$$

and thus $FRG = 0$. It follows that $RG = 0$, because otherwise $Fh = 0$ for some nonzero $h \geq 0$ in $\mathcal{L}^p(Y)$. Then $Eh = 0$, which is a contradiction. Similarly, it follows from $RG = 0$ that $R = 0$. For suppose otherwise. Then it would follow from Lemma 8.7.5, together with the separability of $\mathcal{L}^p(X \setminus Y)$, that there exists a nontrivial Borel subset of X_1 of $X \setminus Y$ such that $G\mathcal{L}^p(X \setminus Y) \subseteq \mathcal{L}^p(X_1)$. Hence some nonzero nonnegative functional ϕ_1 on $\mathcal{L}^p(X \setminus Y)$ annihilates $\mathcal{L}^p(X_1)$, so that

$$\phi_1|_{G\mathcal{L}^p(X \setminus Y)} = 0.$$

Extending ϕ_1 to ϕ on $\mathcal{L}^p(X)$, by defining $\phi(f) = 0$ for all $f \in \mathcal{L}^p(Y)$, we shall verify that the restriction of ϕ to the range of E is zero (a final contradiction, which will prove (i)). Just write $h \in \mathcal{L}^p(X)$ as $h = h_1 \oplus h_2$ relative to the direct sum $\mathcal{L}^p(Y) \oplus \mathcal{L}^p(X \setminus Y)$ and note that

$$\phi(Eh) = \phi((Fh_1 + Rh_2) \oplus Gh_2)$$
$$= \phi(Fh_1 + Rh_2) + \phi(Gh_2) = 0.$$

(ii) It is easily seen that there is a maximal standard subspace $\mathcal{L}^p(Y_1)$ annihilated by E. (A maximal chain of standard invariant subspaces annihilated by E has the same span as some countable subchain, by separability.) Similarly, there is a minimal standard subspace containing $\mathcal{L}^p(Y_1)$ and containing the range of E, say $\mathcal{L}^p(Y_1 \cup Y_2)$ with Y_1 and Y_2 disjoint. Letting Y_3 be the complement of $Y_1 \cup Y_2$ in X, we write the operator matrix

$$\begin{bmatrix} 0 & A & C \\ 0 & F & B \\ 0 & 0 & 0 \end{bmatrix}$$

of E relative to $\mathcal{L}^p(Y_1) \oplus \mathcal{L}^p(Y_2) \oplus \mathcal{L}^p(Y_3)$. Now $E^2 = E$ implies that

$$F^2 = F, \quad A = AF, \quad B = FB, \quad \text{and} \quad C = AB$$

as in finite dimensions. Clearly, F is a nonnegative idempotent of rank r on $\mathcal{L}^p(Y_2)$. To complete the proof we must verify that F satisfies the hypotheses given for E in the statement of (i). But if $h \geq 0$ is in $\mathcal{L}^p(Y_2)$ with $Fh = 0$, then $Ah = AFh = 0$, so that $Eh = 0$. This contradicts the maximality of Y_1. Similarly, if $\phi \geq 0$ is in the dual of $\mathcal{L}^p(Y_2)$ with $\phi|_{F\mathcal{L}^p(Y_2)} = 0$, then ϕ can be extended to a nonnegative functional on $\mathcal{L}^p(X)$ with zero restriction to the range of E. This contradicts the minimality property of $Y_1 \cup Y_2$. $\qquad \square$

Corollary 8.7.13. *Let E be a nonnegative idempotent of finite rank r on $\mathcal{X} = \mathcal{L}^p(X)$. Then the range of E contains r nonnegative vectors $f_1, \ldots, f_r$ whose nonnegative linear combinations include all $f \geq 0$ in $E\mathcal{X}$.*

Proof. This is an easy consequence of Lemma 8.7.12. $\qquad \square$

Lemma 8.7.14. *Let A be a nonzero nonnegative operator on $\mathcal{X} = \mathcal{L}^p(X)$ and assume that either $\ker A$ or $\ker A^*$ contains a nonzero nonnegative vector. Then the semigroup*

$$\mathcal{S} = \{S \in \mathcal{B}(\mathcal{X}) : S \geq 0, \quad SA = AS\}$$

is decomposable.

Proof. First let nonzero $f \geq 0$ be such that $Af = 0$. Then $ASf = SAf = 0$ for all $S \in \mathcal{S}$. Let $E \geq 0$ be a rank-one idempotent with f in its range. Then $ASE = 0$, and the decomposability of $\mathcal{S}$ follows from Lemma 8.7.6.

Next assume $A^*\phi = 0$ with $0 \neq \phi \geq 0$. Then $\phi(Ag) = 0$ for all g, and hence

$$\phi(SAh) = \phi(ASh) = 0$$

for all S in $\mathcal{S}$ and all h. Since $A \neq 0$, there is a nonzero $h \geq 0$ such that $f = Ah \neq 0$. Thus $\phi(Sf) = 0$ for all $S \in \mathcal{S}$, so that $\mathcal{S}$ is decomposable by Lemma 8.7.6. $\square$

Theorem 8.7.15. *If spectral radius is submultiplicative but not multiplicative on a semigroup $\mathcal{S}$ of nonnegative compact operators, then $\mathcal{S}$ is decomposable.*

Proof. We can assume that $\mathcal{S} = \overline{\mathbb{R}^+\mathcal{S}}$ by continuity of spectral radius. By hypothesis, there are members A and B of $\mathcal{S}$ with $\rho(AB) < \rho(A)\rho(B)$. We can assume that $\rho(A) = \rho(B) = 1$. The set of quasinilpotent members of $\mathcal{S}$ is an ideal by hypothesis. Thus it can be assumed to be trivial, by Theorem 8.7.9 and Lemma 8.7.6 (iv). This implies that, for some sequences $\{n_j\}$ and $\{m_j\}$ of integers,

$$\lim_{j\to\infty} A^{n_j} = E = E^2 \quad \text{and} \quad \lim_{j\to\infty} B^{m_j} = F = F^2.$$

Let $k_j = \min(n_j, m_j)$. By Theorem 8.6.3,

$$\rho(A^{n_j}B^{m_j}) = \rho((AB)^{k_j} \cdot A^{n_j-k_j} \cdot B^{m_j-k_j})$$
$$\leq \rho(AB)^{k_j} \cdot \rho(A)^{n_j-k_j} \cdot \rho(B)^{m_j-k_j} = \rho(AB)^{k_j}.$$

Thus $\lim_{j\to\infty} \rho(A^{n_j}B^{m_j}) = 0$, implying $\rho(EF) = 0$. Then $EF = 0$, because there are no quasinilpotents in $\mathcal{S}$ other than zero. It follows that $(FSE)^2 = \{0\}$ and hence $FSE = 0$. Lemma 8.7.6 (ii) now completes the proof. $\square$

Lemma 8.7.16. *Let $\mathcal{S}$ be a semigroup of nonnegative compact operators on $\mathcal{L}^p(X)$ and E a nonnegative idempotent of finite rank $r > 0$ (not necessarily in $\mathcal{S}$). Then the conclusions of Lemma 5.2.1 hold (where $\mathcal{V} = \mathcal{L}^p(X)$.)*

Proof. The proof of the first conclusion is the same as that in the corresponding part of Lemma 5.2.1; we just use Corollary 8.7.13 instead of Corollary 5.1.10. To prove the second conclusion, assume $ESE|_{E\mathcal{X}}$ is decomposable. Then by Lemma 8.7.6 there are nonnegative, nonzero members f and φ of $E\mathcal{X}$ and $(E\mathcal{X})^*$, respectively, such that $\varphi(ESEf) = 0$ for all $S \in \mathcal{S}$. Define the functional ψ by $\psi(T) = \varphi(ET)$. Note that ψ is nonnegative and $\psi(Sf) = 0$ for all $S \in \mathcal{S}$. Thus $\mathcal{S}$ is decomposable by Lemma 8.7.6.

$\square$

Lemma 8.7.17. *Let* $S = \overline{\mathbb{R}^+ S}$ *be an indecomposable semigroup of non-negative compact operators on* $\mathcal{X} = \mathcal{L}^p(X)$. *Then* S *contains nonzero operators of finite rank, and all the conclusions of Lemma 5.2.2 hold, where (iii) has the following interpretation:*

(iii)′ *For each nonzero* $f \geq 0$ *in* $\mathcal{X}$, *there is a minimal idempotent* E *in* S *with* $Ef \neq 0$. *Also, for each nonzero* $\phi \geq 0$ *in* $\mathcal{X}^*$, *there is a minimal idempotent* F *in* S *with* $\phi|_{F\mathcal{X}} \neq 0$.

Proof. By Theorem 8.7.9, S contains operators that are not quasinilpotent, so S contains finite-rank operators by Lemma 7.4.5. Let r denote the minimal positive rank of elements of S. The proof of all the conclusions except (iii)′ is as in Lemma 5.2.2 (where we use the appropriate infinite-dimensional analogues, which we have just proved, of the results that were used there).

Note that the "right-handed" version of (ii) is also true; i.e., given A of rank r in S, there exists a minimal idempotent E in S with $AE = A$. The proof is similar: the idempotent F with $FA = A$ was found by first picking a B in S such that $\rho(AB) = 1$; a subsequence of $\{(AB)^n\}$ was then shown to converge to F. It is easy to see that the corresponding subsequence of $\{(BA)^n\}$ converges to $E = E^2$ with $AE = A$.

Now to show (iii)′, observe that the indecomposable ideal

$$S_r = \{S \in S : \text{ rank}(S) = r\} \cup \{0\}$$

has a member A with $\phi(Af) \neq 0$, by Lemma 8.7.6. Letting E and F be minimal idempotents with $AE = A = FA$, we deduce that $Ef \neq 0$ and $\phi(FAf) \neq 0$. $\qquad\qquad\square$

Lemma 8.7.18. *Let* $S = \overline{\mathbb{R}^+ S}$ *be an indecomposable semigroup of nonnegative compact operators. Then the conclusions of Lemmas 5.2.3 and 5.2.4 hold.*

Proof. The proofs are exactly the same as in the finite-dimensional case except that Lemmas 8.7.6 and 8.7.17 should be used in place of Lemmas 5.1.5 and 5.2.2. $\qquad\qquad\square$

Lemma 8.7.19. *Let* $S = \overline{\mathbb{R}^+ S}$ *be an indecomposable semigroup of nonnegative compact operators on* $\mathcal{X} = \mathcal{L}^p(X)$ *satisfying any one, and thus all, of the conditions of Lemma 5.2.4. Let* $\mathcal{R}$ *be the common range of the minimal idempotents in* S. *Then*

(i) *the only quasinilpotent element of* S *is zero,*

(ii) *spectral radius is multiplicative on* S, *so that*

$$S_0 = \{S/\rho(S) : 0 \neq S \in S\}$$

is a semigroup on which ρ is identically one, and

(iii) *there is a nonnegative idempotent P with range R in the closed convex hull of S such that*

$$S|_\mathcal{R} = PSP|_\mathcal{R}$$

and with the following property: There is no nonzero $f \geq 0$ in $\mathcal{X}$ with $Pf = 0$, and there is no nonzero $\phi \geq 0$ in $\mathcal{X}^*$ with $\phi|_\mathcal{R} = 0$.

Proof. Since $\mathcal{R}$ is invariant under $\mathcal{S}$, $A|_\mathcal{R}$ is nilpotent for all quasinilpotent operators A in $\mathcal{S}$. Thus the proof can proceed as in Lemma 5.2.5, once we have verified (iii). To this end, let $\mathcal{E}$ be the set of all minimal idempotents in $\mathcal{S}$ and pick a countable subset $\{E_n\}$ of $\mathcal{E}$ such that the only nonnegative vector annihilated by all the E_n is zero. (This is possible by Lemma 8.7.17, together with the fact that the measure is σ-finite.) Let $\{a_n\}$ be a sequence of positive numbers with $\sum_{n=1}^{\infty} a_n = 1$ and $\sum_{n=1}^{\infty} a_n \|E_n\| < \infty$, and set $P = \sum_{n=1}^{\infty} a_n E_n$. Then $P = P^2$ and $P\mathcal{X} = \mathcal{R}$. Now, if $Pf = 0$ and $f \geq 0$, then $E_n f = 0$ for all n, so that $f = 0$. Similarly, if $\varphi|_\mathcal{R} = 0$ and $\varphi \geq 0$, then $\varphi = 0$ by Lemma 8.7.17. □

We are now ready to state the infinite-dimensional analogue of Theorem 5.2.6, extending the Perron-Frobenius Theorem to semigroups.

Theorem 8.7.20. *Let $\mathcal{S}$ be an indecomposable semigroup of nonnegative compact operators on $\mathcal{L}^p(X, \mu)$ such that $\overline{\mathbb{R}^+\mathcal{S}}$ has a unique minimal right ideal. Denote the minimal positive rank in $\mathcal{S}$ by r. Then the following hold:*

(i) *There is an f in $\mathcal{L}^p(X)$, almost everywhere positive, such that*

$$Sf = \rho(S)f$$

for every S in $\mathcal{S}$. The vector f is unique up to scalar multiples.

(ii) *Every S in $\mathcal{S}$ has at least r eigenvalues of modulus $\rho(S)$, counting multiplicities; these are all of the form $\rho(S)\alpha$, where $\alpha^{r!} = 1$.*

(iii) *There are r mutually disjoint Borel subsets X_i of X with union X such that the $r \times r$ operator matrix of every nonzero $S \in \mathcal{S}$ relative to the partition*

$$\mathcal{L}^p(X_1) \oplus \cdots \oplus \mathcal{L}^p(X_r)$$

has exactly one nonzero block in each block row and each block column.

(iv) *If the block matrix of any S in $\mathcal{S}$ has a cyclic pattern, then $\sigma(S)$ is invariant under rotation about the origin by the angle $2\pi/r$.*

(v) *$r = 1$ if and only if given a Borel subset X_0 of X with $\mu(X_0) < \infty$, a positive ε, and any nonzero $g \geq 0$ in $\mathcal{L}^p(X)$, there exists S in $\mathcal{S}$*

such that the function Sg is positive on X_0 except for a subset of measure less than ε.

Proof. The proofs of the first four assertions are almost exactly those in Theorem 5.2.6. The proof of (v) is a little different.

First assume $r = 1$, and let ε, X_0, and g be as given above. Pick a minimal idempotent $E = f \otimes \phi$ in S with $t = \phi(g) \neq 0$. Since $E = \lim_{n \to \infty} t_n S_n$ for some scalars t_n and $S_n \in S$, we have

$$\lim_{n \to \infty} \int_X |t_n S_n g - tf|^p = 0.$$

Recall that (a representative of) f is positive. Thus there is a Borel subset Y of X_0 with $\mu(Y) > \mu(X_0) - \varepsilon/2$ on which tf is bounded below. It follows that, for sufficiently large n, the function $t_n S_n g$ is positive on Y except for a subset Z of measure less than $\varepsilon/2$. Thus $t_n S_n g$ is positive on $Y \setminus Z$ and $\mu(Y \setminus Z) > \mu(X_0) - \varepsilon$.

The case $r > 1$ is easier. Pick X_0 of the form $Y_1 \cup Y_2$, where Y_1 and Y_2 are respective subsets of X_1 and X_2 from (iii) above, such that

$$0 < \mu(Y_1) = \mu(Y_2) < \infty.$$

It follows from the block partition form of the operator matrix of each S in S that, for every $g \geq 0$, the function Sg is either zero on Y_1 or on Y_2, and thus

$$\mu\{t \in X_0 : (Sg)(t) = 0\} \geq \mu(Y_1). \qquad \square$$

Lemma 8.7.21. *Let $S = \overline{\mathbb{R}^+ S}$ be an indecomposable semigroup of non-negative compact operators on $\mathcal{L}^p(X)$. Denote the center of S by $\mathcal{Z}$ and the set of minimal idempotents of S by $\mathcal{E}$. Then the following are mutually equivalent:*

(i) *$\mathcal{E}$ is a singleton;*
(ii) *$\mathcal{E} \cap \mathcal{Z} \neq 0$;*
(iii) *$\mathcal{E} \subseteq \mathcal{Z}$;*
(iv) *$SE = ES$ for some E in $\mathcal{E}$;*
(v) *$SE = ES$ for all E in $\mathcal{E}$.*

Proof. If E is the unique element in $\mathcal{E}$, then it follows from Lemma 8.7.17 that $E(SE) = SE$ for all S in S. Also, as in the proof of Lemma 8.7.17, $(ES)E = ES$, so that $ES = SE$. This shows that (i) implies (iii).

Assume (ii). It follows from Lemma 8.7.14, together with indecomposability of S, that if $E \in \mathcal{E} \cap \mathcal{Z}$, then $\ker E \cup \ker E^*$ does not contain a nonzero $f \geq 0$. Hence E is a direct sum of positive (rank-one) idempotents, by Lemma 8.7.12. Thus $EF \neq 0$ for $F \in \mathcal{E}$. Since $E \in \mathcal{Z}$, the operator

$EF = FE$ is an idempotent, and we have $E = F$, proving that (ii) implies (i). We need only show that (iv) implies (ii), which will complete the cycle via the obvious deductions of (ii) and (v) from (iii) and of (iv) from (v).

Assuming (iv), we deduce $SE = ESE$ for all S in $\mathcal{S}$, since the range of E is invariant for $\mathcal{S}$. Also, for each S, there is a T such that $ES = TE$. Thus

$$ES = E(ES) = ETE = E(TE)E = ESE = SE$$

for every S. This implies (ii). $\square$

Corollary 8.7.22. *Let $\mathcal{S}$ be an indecomposable semigroup of nonnegative compact operators on $\mathcal{X} = \mathcal{L}^p(X, \mu)$ such that $\overline{\mathbb{R}^+\mathcal{S}}$ has a unique minimal idempotent E. Denote the rank of E by r. Then all the conclusions of Theorem 8.7.20 hold. Furthermore:*

(i) *If $S \in \mathcal{S}$, $g \geq 0$ is nonzero in $\mathcal{X}$, and $Sg = 0$, then $S = 0$. Also, if $\psi \geq 0$ is nonzero in $\mathcal{X}^*$ and $S^*\psi = 0$, then $S = 0$.*

(ii) *The semigroup $\mathcal{S}^*$ on $\mathcal{X}^*$ also has a common positive eigenvector ϕ, unique up to scalar multiples, such that $S^*\phi = \rho(S)\phi$ for all $S \in \mathcal{S}$.*

(iii) *The minimal rank r is 1 if and only if, given a Borel subset X_0 of X with $\mu(X_0) < \infty$, a positive ε, and any finite set $\mathcal{F}$ of nonzero nonnegative functions in $\mathcal{L}^p(X)$, there exists $S \in \mathcal{S}$ such that every function in $S\mathcal{F}$ is positive on X_0 except for a subset of measure less than ε.*

Proof. We adapt the proof of Corollary 5.2.8 with minor necessary changes. Clearly, Theorem 8.7.20 applies, because ES is the unique minimal right ideal. Next note that, since E is in the center of $\mathcal{S}$, the kernels of E and E^* cannot contain nonzero nonnegative vectors (by Lemma 8.7.14). Thus, by Lemma 8.7.12,

$$E = E_1 \oplus \cdots \oplus E_r,$$

where each E_i is of the form $f_i \otimes \phi_i$ with positive f_i in $\mathcal{L}^p(X_i)$ and positive ϕ_i in $\mathcal{L}^q(X_i)$ ($q = \infty$ if $p = 1$). Now, $Sg = 0$ with nonzero $g \geq 0$ implies, since $ES^{r!}E = \rho(S)^{r!}E$ by Theorem 8.7.20 and since E commutes with S by Lemma 8.7.21, that

$$\rho(S)^{r!}Eg = ES^{r!}Eg = ES^{r!}g = 0.$$

Hence $\rho(S)^{r!}\phi_j(g) = 0$ for $j = 1, \ldots, r$, implying $\rho(S) = 0$. Since there are no nonzero quasinilpotents in $\mathcal{S}$ (by Lemma 8.7.19), we have $S = 0$. This proves the first part of (i). The second part is proved similarly.

To prove (ii), recall that the common eigenvector f for $\mathcal{S}$ given by Theorem 8.7.20 can be assumed to be in the form $f_1 + \cdots + f_r$, where $\{f_i\}$ are as in the preceding paragraph. Now assume, with no loss of generality, that $\rho(S) = 1$ for a given S in $\mathcal{S}$, so that $Sf = f$ and $Sf_i = f_{\tau(i)}$, where $\tau(i)$

is a permutation of $\{1, \ldots, r\}$ (depending on S). Noting that $\phi_j(f_i) = \delta_{ij}$ (the Kronecker δ), we deduce that

$$(S^*\phi_j)(f_i) = \phi_j(Sf_i) = \phi_j(f_{\tau(i)})$$
$$= \delta_{\tau(i),j} = \delta_{i,\tau^{-1}(j)} = \phi_{\tau^{-1}(j)}(f_i)$$

for all i and j. It follows from the positivity of f_i and ϕ_j (almost everywhere on the sets X_i and X_j, respectively) that $S^*\phi_j = \phi_{\tau^{-1}(j)}$ for each j, so that $S^*\phi = \phi$ with $\phi = \phi_1 + \cdots + \phi_r$. To complete the proof of (ii) we must show the uniqueness of ϕ. Thus let ψ be any nonzero vector in $\mathcal{X}^*$, not necessarily assumed nonnegative, such that $S^*\psi = \psi$ for all $S \in \mathcal{S}$ with $\rho(S) = 1$. Then, in particular, $E^*\psi = \psi$. This implies, since $\{\phi_1, \ldots, \phi_r\}$ is a basis for the range of E^*, that $\psi = a_1\phi_1 + \cdots + a_r\phi_r$ for some scalars a_i. Now it follows from the equation $S^*\phi_j = \phi_{\tau^{-1}(j)}$ proven above, together with the transitivity of the group

$$\mathcal{G} = \{ESE : 0 \neq S \in \mathcal{S}, \ \rho(S) = 1\},$$

that for every j and i in $\{1, \ldots, r\}$ there is a τ with $\tau^{-1}(j) = i$; in other words, there is an $S \in \mathcal{S}$ with $S^*\phi_j = \phi_i$. Since ψ is a fixed point for $\mathcal{G}$, we conclude that $a_i = a_j$ for all i and j. Hence $\psi = a_1\phi$.

The proof of (iii) is similar to that of (v) in Theorem 8.7.20. It suffices to observe that, in the present case, $r = 1$ implies that $Eg \neq 0$ for every $g \in \mathcal{F}$.   □

Corollary 8.7.23. *Let $\mathcal{S}$ be a commutative indecomposable semigroup of nonnegative compact operators on $\mathcal{L}^p(X, \mu)$. Then $\overline{\mathbb{R}^+\mathcal{S}}$ has a unique minimal idempotent E, and all the conclusions of Theorem 8.7.20 and Corollary 8.7.22 hold. In addition, the four assertions of Corollary 5.2.12 are true.*

Proof. The proof is the same as that of Corollary 5.2.12 if references are made to Theorem 8.7.20 and Corollary 8.7.22 instead of Theorem 5.2.6 and Corollary 5.2.8.   □

The following is an analogue of the Perron-Frobenius Theorem (Corollary 5.2.13).

Corollary 8.7.24. *Let A be an indecomposable nonnegative compact operator on $\mathcal{L}^p(X, \mu)$. Then $\rho(A) \neq 0$, so that if S is the semigroup generated by A, then $\overline{\mathbb{R}^+\mathcal{S}}$ contains a unique minimal idempotent E, whose rank we denote by r. We scale A to get $\rho(A) = 1$. Then the following hold:*

(i) *The sequence $\{A^{rj}\}_{j=1}^{\infty}$ converges to an idempotent E of rank r.*

(ii) *There are mutually disjoint Borel subsets $X_1, \ldots, X_r$ of X such that, relative to the direct sum representation*

$$\mathcal{L}^p(X) = \mathcal{L}^p(X_1) \oplus \cdots \oplus \mathcal{L}^p(X_r),$$

A has an operator matrix

$$\begin{bmatrix} 0 & 0 & \cdots & 0 & A_r \\ A_1 & 0 & \cdots & 0 & 0 \\ 0 & A_2 & \cdots & 0 & 0 \\ \vdots & & & & \\ 0 & 0 & \cdots & A_{r-1} & 0 \end{bmatrix}.$$

(iii) *There is an $f \in \mathcal{L}^p(X)$, positive almost everywhere, such that $Af = f$.*

(iv) *The set $\{\lambda \in \sigma(A) : |\lambda| = 1\}$ consists precisely of all the r-th roots of unity; each member of the set is a simple eigenvalue.*

(v) *The spectrum of A is invariant under rotation about the origin by the angle $2\pi/r$.*

(vi) *The element 1 is dominant in $\sigma(A)$ if and only if $\{A^j\}$ is convergent. This occurs if and only if, given any Borel subset X_0 of X with $\mu(X_0) < \infty$ and any finite set $\mathcal{F}$ of nonzero nonnegative functions in $\mathcal{L}^p(X)$, there is an integer n and a Borel subset of X_0, whose measure is arbitrarily close to $\mu(X_0)$, on which every member of $A^n \mathcal{F}$ is positive almost everywhere.*

Proof. All of the assertions follow from Corollary 8.7.23 in the same way that the finite-dimensional version follows from Corollary 5.2.12. □

The following simple example shows that the last assertion in the result above, and the corresponding assertions in Theorem 8.7.20 and its other corollaries, cannot be strengthened to match the finite-dimensional case (Theorem 5.2.6 and its corollaries), even if the measure is atomic (i.e., $\mathcal{X} = l^p$).

Example 8.7.25. *Let $\{\alpha_n\}$ be the standard basis for l^p and pick sequences $\{a_n\}$, $\{b_n\}$, and $\{c_n\}$ of positive numbers each converging to zero, so that the "tridiagonal" matrix defined below represents a compact operator on l^p:*

$$T\alpha_1 = a_1\alpha_1 + b_1\alpha_2$$

and

$$T\alpha_n = c_{n-1}\alpha_{n-1} + a_n\alpha_n + b_n\alpha_{n+1} \quad \text{for} \quad n \geq 2.$$

It is easily seen that the semigroup S generated by T (i.e., $S = \{T^n : n \in \mathbb{N}\}$) is indecomposable; in fact, for any pair (i,j) of integers, sufficiently high powers of T have their (i,j) entry positive. On the other hand, no member of S has a positive column.

Of course, if S is as in the above example, the semigroup $\overline{\mathbb{R}^+ S}$ does have members with positive columns.

The next theorem extends the results of Section 5.3 on decomposability and reducibility.

Theorem 8.7.26. *All the conclusions of Corollary 5.3.1, Theorem 5.3.2, Corollary 5.3.3, and Theorem 5.3.4 are valid if "matrices" is replaced with "compact operators" in their hypotheses.*

Proof. The proofs are similar to the finite-dimensional versions.   □

We conclude this section with a decomposability result on nonnegative bands.

Theorem 8.7.27. *Let S be a band of nonnegative finite-rank operators on $\mathcal{X} = \mathcal{L}^p(X, \mu)$, and denote the minimal rank of members of S by r. If $r > 1$, then S is decomposable. In fact, it has a chain of standard invariant subspaces of the form*

$$\{0\} = \mathcal{X}_0 \subset \mathcal{X}_1 \subset \cdots \subset \mathcal{X}_r = \mathcal{X}.$$

Proof. The proof is very similar to that of Theorem 5.1.13: Instead of Lemmas 5.1.15 and 5.1.9, use Lemmas 8.7.6 and 8.7.12.   □

We have seen that the result above is not valid if $r = 1$: There are indecomposable singleton bands. If zero is an essential member of a band S (that is, if S has zero divisors), then it is decomposable. For if $EF = 0$ with nonzero E and F in S, then the set FSE consists of matrices whose squares are zero; since they are also idempotents, we get $FSE = \{0\}$, implying decomposability by Lemma 8.7.6 (ii).

8.8 Notes and Remarks

The compactness result in Lemma 8.1.2 is from Vala [1]; some related material can be found in Bonsall [1]. The formula for the joint spectral radius (Lemma 8.1.5) is due to Rota-Strang [1]. Another joint spectral radius was defined in Berger-Wang [1]; some relations between the two definitions are explored in Rosenthal-Soltysiak [1].

The remaining results in Section 8.1 are from Turovskii [2]. The problem solved by Turovskii (Theorem 8.1.11) had been open since 1984, when it was raised in Nordgren-Radjavi-Rosenthal [2] and, independently, in Shulman [1]. Special cases of this result (e.g., when the operators are in a class C_p), were established in Nordgren-Radjavi-Rosenthal [1]. A monetary prize for settling the question had been announced a year before Turovskii solved it; he was not aware of the existence of the prize! For further consequences of this remarkable result see Turovskii [2] and the exposition by Yahaghi

[1]; the latter also discusses Shulman [1] and proves that a triangularizable family of compact operators containing a nonzero quasinilpotent operator has a nontrivial hyperinvariant subspace. The special case of Corollary 8.1.14 where the operators come from a Schatten class C_p was proved in Nordgren-Radjavi-Rosenthal [2] as a corollary of the corresponding special case of Turovskii's Theorem. Guinand [1] gives a pair of (noncompact) operators A and B such that every word in A and B has cube zero, but $A + B$ is not quasinilpotent.

Wojtyński [1] showed that Engel's Theorem (Corollary 1.7.6) generalizes to the theorem that a Lie algebra of quasinilpotent operators in a C_p class is triangularizable. Wojtyński [1] asked if this can be extended to an algebra of compact operators; Shulman-Turovskii [1] contains an affirmative answer to this question and many other interesting related results.

Some of the elementary results of Section 8.2 are folklore, and some appear for the first time in their present form (e.g., Lemma 8.2.8 to Lemma 8.2.11). Variants of the Downsizing Lemma and the result preceding it (Lemmas 8.2.12 and 8.2.13) have been implicitly used in Radjavi [4] and [6], but their current streamlined form is new.

Most of the results in Section 8.3 are from Lambrou-Longstaff-Radjavi [1]. Exceptions are Theorem 8.3.3 and Theorem 8.3.4 (new), Theorem 8.3.7 (from Radjavi [6]), and Theorem 8.3.9 (from Radjavi [2]).

The special case of Corollary 8.4.5 where the semigroup is contained in a Schatten class was proved in Radjavi-Rosenthal [5]; the extension to compact operators is an immediate consequence of that proof together with Turovskii's Theorem. Theorem 8.5.2 is given in Radjavi-Rosenthal-Shulman [1] with a different proof. The remaining results of Sections 8.4 and 8.5 are from Radjavi [6]; they have been extended here to the Banach-space setting. Certain more general results along these lines are in Jahandideh ([1], [2]). Kaplansky [1] studies pairs of compact normal operators satisfying property L. For a discussion of the finite-dimensional background, see Section 4.5.

Turovskii [1] includes the result that $\mathcal{S} \cup \mathcal{T}$ is triangularizable if each of $\mathcal{S}$ and $\mathcal{T}$ is a triangularizable collection of compact operators and $ST = TS$ for all $S \in \mathcal{S}$ and $T \in \mathcal{T}$. The results of Section 8.6 concerning spectral radius are mostly from Lambrou-Longstaff-Radjavi [1], adapted here to the Banach-space setting. The following open problem is posed in that paper: Does submultiplicativity imply permutability of spectra on arbitrary semigroups in Banach algebras ? Example 8.6.6, Theorem 8.6.7, and Corollary 8.6.8 are from Radjavi [4]. The results concerning trace are from Radjavi [2] with one exception: Theorem 8.6.13 is from Nordgren-Radjavi-Rosenthal [2].

In Section 5.4 we discussed the history of nonnegative operators on finite-dimensional spaces. Most of the results of Section 8.7 hold in the more general setting of Banach lattices. (For extensive general results on Banach lattices and positive operators, see the books Aliprantis-Burkinshaw

[1], Zaanen [1] and Schaefer [1].) Generalizing a theorem of Ando [1], de Pagter [1] proved that a compact, quasinilpotent, nonnegative operator on a Banach lattice is decomposable. In a sequence of papers, Abramovich, Aliprantis, and Burkinshaw have extended this result to operators that have some affinity to compact operators. A good exposition of their results is in Abramovich-Aliprantis-Burkinshaw [1] (where many other references to their work can be found). Using Turovskii's Theorem, Drnovšek [3] generalized de Pagter's Theorem to semigroups. Theorem 8.7.7 and Theorem 8.7.9 are special cases of these results, adapted to our setting of $\mathcal{L}^p$ spaces. The proof of Theorem 8.7.7 follows the lines of the original proof of de Pagter [1] and those of Abramovich-Aliprantis-Burkinshaw [1]. Lemma 8.7.8 is from Jahandideh [1]. The argument given for the general case in the proof of Theorem 8.7.9 was devised in collaboration with G. MacDonald, L. Livshits, and B. Mathes.

The structure result for nonnegative idempotents (Lemma 8.7.12) is essentially in Zhong [1], as is its proof. Theorem 8.7.15 is new. The remaining results preceding Corollary 8.7.24 are adaptations of results and arguments in Radjavi [7] to the compact case. The analogue of the Perron-Frobenius Theorem (Corollary 8.7.24) is an almost immediate corollary of the Krein-Rutman [1] Theorem (which explicitly includes (iii) of Corollary 8.7.24). For Theorem 8.7.26, the corresponding remarks from Section 5.4 apply. Theorem 8.7.27 on bands is from Marwaha [2].

A characterization of strong quasiaffinities is given in Holbrook-Nordgren-Radjavi-Rosenthal [1].

CHAPTER 9
Bounded Operators

We have shown that most of the theorems on triangularizability of operators on finite-dimensional spaces have satisfactory extensions to collections of compact operators. However, we shall see that there are few generalizations to collections of arbitrary bounded operators: There are counterexamples to most reasonable conjectures. In particular, there are commutative sets of operators that are irreducible, and there are irreducible algebras of nilpotent operators, and irreducible semigroups consisting of nilpotent operators of index two. There are also some affirmative results. There are many cases in which pairs of operators whose commutator has rank one are triangularizable, although there are counterexamples to the general extension of Laffey's Theorem. There is a similar situation with respect to nonnegative operators and for bands.

9.1 Collections of Nilpotent Operators

It is discouraging, though not surprising, that there exist bounded linear operators that have only the trivial invariant subspaces. This was first shown by Per Enflo [1], who constructed the Banach space as he constructed the operator: The space is the closure, in a certain norm, of the set of polynomials, and does not appear to be a familiar Banach space. Several years later, C. J. Read [1] (see also Read [2]) constructed an irreducible operator on the very familiar space ℓ^1, the space of all summable sequences. Thus there are irreducible semigroups of bounded operators that are singly generated. Subsequently, Read [3] constructed a quasinilpotent operator on ℓ^1 that is irreducible. These constructions of irreducible operators are quite ingenious and quite complex; rather than attempting a description, we merely refer the interested reader to the original papers (the most easily understood treatment is probably Read [3]).

It is still not known whether such a construction is possible on Hilbert space. The *invariant subspace problem* is the question: Does every bounded linear operator on a Hilbert space have a nontrivial invariant subspace? There are many affirmative results known for particular classes of operators (see the discussion in Section 9.5 below), but the existence of counterexamples on ℓ^1 might suggest that a negative answer is likely in general. On the other hand, there is also a dearth of counterexamples. The *hyperinvariant subspace problem* is the question of whether every operator on Hilbert space other than a multiple of the identity has a nontrivial subspace that is invariant under all the operators that commute with the given one. This problem is also unsolved. The *transitive algebra problem* is the question: Is every transitive algebra of bounded linear operators on Hilbert space strongly dense in the algebra of all bounded linear operators on the space?

An affirmative result would be a powerful strengthening of Burnside's Theorem (1.2.2) and of the generalizations to algebras of compact operators (e.g., Corollary 7.4.8). Since the commutant of an operator other than a multiple of the identity is not strongly dense, such a theorem would, in particular, give an affirmative answer to the hyperinvariant subspace problem. Nonetheless, no counterexample is yet known.

There is a version of Burnside's Theorem on the space ℓ^∞ of bounded sequences: The only weak* operator closed transitive subalgebra of $\mathcal{B}(\ell^\infty)$ is $\mathcal{B}(\ell^\infty)$ (Honor [1]; the weak* operator topology has quite unusual properties). Honor [1] also establishes (Corollary 6) that subalgebras of $\mathcal{B}(\ell^1)$ whose set of adjoints form transitive subalgebras of $\mathcal{B}(\ell^\infty)$ are strongly dense in $\mathcal{B}(\ell^1)$.

Even on Hilbert space, there are known counterexamples to extensions of McCoy's Theorem (7.3.3) and Turovskii's Theorem (8.1.11). There is a basic construction that produces counterexamples to these and other natural conjectures. We begin with the observation that the block matrix $\begin{pmatrix} A & -A \\ A & -A \end{pmatrix}$ has square 0 (i.e., is nilpotent of index 2) for every matrix A. Our examples are constructed from such block matrices, as follows.

Theorem 9.1.1. *On a separable Hilbert space, there is an irreducible semigroup of nilpotent operators of index 2.*

Proof. Let $\mathcal{H}$ be a Hilbert space with a fixed orthonormal basis $\{e_1, e_2, \dots \}$. Fix any natural number n. For each $2^{n-1} \times 2^{n-1}$ matrix A, define $N(A) = \begin{pmatrix} A & -A \\ A & -A \end{pmatrix}$, so $N(A)$ is a $2^n \times 2^n$ matrix. Let $N(A)$ act naturally as a matrix with respect to the basis $\{e_1, e_2, \dots, e_{2^n}\}$. We define a corresponding operator $\operatorname{Amp} N(A)$ on $\mathcal{H}$ as the direct sum ("ampliation") of a countable number of copies of $N(A)$, with the first summand acting as a matrix with respect to the basis vectors $\{e_1, \dots, e_{2^n}\}$, the second summand acting with respect to $\{e_{2^n+1}, \dots, e_{2^{n+1}}\}$, the third summand with respect to $\{e_{2^{n+1}+1}, \dots, e_{3 \cdot 2^n}\}$, and so on. Note that

$$\| \operatorname{Amp} N(A) \| = \| N(A) \| \leq 4 \|A\|$$

(in fact it can be shown that $\|N(A)\| = 2\|A\|$), so $\operatorname{Amp} N(A)$ is bounded. Also, $(\operatorname{Amp} N(A))^2 = 0$. For each n, let $\mathcal{S}_n$ denote the set of all $\operatorname{Amp} N(A)$ such that A is a $2^{n-1} \times 2^{n-1}$ matrix. Note that $\mathcal{S}_n$ is a linear space of operators. Note also that A is a $2^{n-1} \times 2^{n-1}$ matrix, so $N(A)$ has size $2^n \times 2^n$. Therefore $\operatorname{Amp} N(N(A))$ is in $\mathcal{S}_{n+1}$ when $\operatorname{Amp} N(A)$ is in $\mathcal{S}_n$. Let $\mathcal{S} = \bigcup_{n=1}^\infty \mathcal{S}_n$.

Suppose that $n > m$. If $B \in \mathcal{S}_m$ and $C \in \mathcal{S}_n$, then BC and CB are both in $\mathcal{S}_n$, since a direct sum of 2^{n-1-m} copies of $N(B)$ is a matrix of size $2^{n-1} \times 2^{n-1}$. Hence the union, $\mathcal{S}$, is a semigroup. Moreover, every operator in $\mathcal{S}$ has the form $\operatorname{Amp}(N(A))$ for some A, and hence has square 0.

It remains to be shown that $\mathcal{S}$ is irreducible. The easiest way to do this appears to be to prove more: that $\mathcal{S}$ is dense in $B(\mathcal{H})$ in the weak operator topology. Recall that this is the topology with basis consisting of all sets of the form

$$\mathcal{U}\,(T; f_1,\ldots, f_\ell; g_1,\ldots, g_\ell; \varepsilon) = \big\{ S : \big| ((S-T)f_j, g_j)\big| < \varepsilon \big\}\ .$$

Equivalently, for a net $\{S_\alpha\}$, we define $\{S_\alpha\} \rightharpoonup T$ in the weak operator topology (the "half arrow" is standard notation for this) if $\{(S_\alpha f, g)\} \to (Tf, g)$ for all $f, g \in \mathcal{H}$. It is easily seen that $\{S_\alpha\} \rightharpoonup T$ and $\mathcal{M} \in \operatorname{Lat} S_\alpha$ for all α imply $\mathcal{M} \in \operatorname{Lat} T$ (for if $f \in \mathcal{M}$ and $g \in \mathcal{M}^\perp$, then $(S_\alpha f, g) = 0$ for all α, so $(Tf, g) = 0$). Thus any subspace invariant under a weakly dense collection of operators would be invariant under all operators, and therefore must be $\{0\}$ or $\mathcal{H}$.

To see that $\mathcal{S}$ is weakly dense, fix any weak neighborhood

$$\mathcal{U}\,(T; f_1,\ldots, f_\ell; g_1,\ldots, g_\ell; \varepsilon)$$

of any operator $T \in B(\mathcal{H})$. To get an $S \in \mathcal{S}$ in this neighborhood we will use the fact that, for any n, there is an operator $S \in \mathcal{S}$ that agrees with T on the first 2^n vectors in the given fixed basis $\{e_1, e_2, \ldots\}$ for $\mathcal{H}$. More precisely, given $\delta > 0$, if n is sufficiently large, then $\sum_{i=2^n}^{\infty} |(f_j, e_i)|^2$ and $\sum_{i=2^n}^{\infty} |(g_j, e_i)|^2$ are less than δ for all j from 1 to ℓ. Let A be the matrix with entries (Te_i, e_j) for $i, j = 1, 2, \ldots, 2^n$, and let $S = \operatorname{Amp}(N(A))$. Then $S \in \mathcal{S}_{n+1}$, and, for δ sufficiently small, S is in $\mathcal{U}(T; f_1, \ldots, f_\ell; g_1, \ldots, g_\ell; \varepsilon)$. Hence $\mathcal{S}$ is weakly dense, and, in particular, is irreducible. $\square$

There are some affirmative results for nilpotent operators.

Theorem 9.1.2. *An additive semigroup (i.e., a set closed under addition) of operators with square 0 is triangularizable.*

Proof. By the Triangularization Lemma (7.1.11), it suffices to show that every such set has a nontrivial invariant subspace. Fix any operator A in the set other than 0. Then $A^2 = 0$ implies that the closure of $A\mathcal{X}$ is not all of $\mathcal{X}$. Thus it suffices to show that $A\mathcal{X}$ is invariant under every operator in the set.

Given any B, $(A+B)^2 = 0$ and $A^2 = B^2 = 0$ imply that $BA = -AB$. Thus $B(A\mathcal{X}) \subseteq A(B\mathcal{X}) \subseteq A\mathcal{X}$. $\square$

This theorem cannot be extended to nilpotent operators of higher indices, even in finite dimensions.

Example 9.1.3. *There is a linear manifold of nilpotent 3×3 matrices that is irreducible.*

Proof. Let

$$\mathcal{L} = \left\{ \begin{pmatrix} 0 & \beta & 0 \\ \alpha & 0 & -\beta \\ 0 & \alpha & 0 \end{pmatrix} : \alpha, \beta \in \mathbb{C} \right\}.$$

It is easily verified that $\mathcal{L}$ is irreducible, and that $A^3 = 0$ for all $A \in \mathcal{L}$. $\square$

There is also a result for algebras of operators.

Theorem 9.1.4. *If $\mathcal{A}$ is a subalgebra of $B(\mathcal{X})$ and k is a natural number such that $A^k = 0$ for all $A \in \mathcal{A}$, then $\mathcal{A}$ is triangularizable.*

Proof. By the Triangularization Lemma (7.1.11), it suffices to show that $\mathcal{A}$ is reducible. If $k = 1$, there is nothing to prove, so assume that $k > 1$ and that $A^{k-1} \neq 0$ for some fixed $A \in \mathcal{A}$. For any $B \in \mathcal{A}$ and any complex number z, $(A + zB)^k = 0$. Fix B; the fact that $(A + zB)^k = 0$ for all $z \in \mathbb{C}$ implies that the coefficient of each z^m in this polynomial is 0. This is true, in particular, for the coefficient of z, so $BA^{k-1} + ABA^{k-2} + A^2BA^{k-3} + \cdots + A^{k-1}B = 0$. Hence $BA^{k-1} = -A\left(BA^{k-2} + ABA^{k-3} + \cdots + A^{k-2}B\right)$, so $B\left(A^{k-1}\mathcal{X}\right) \subseteq A\mathcal{X}$.

Since $A^k = 0$, the closure of $A\mathcal{X}$ is a proper subspace of $\mathcal{X}$, so there is a $\phi \in \mathcal{X}^*$ with $\phi \neq 0$ and $\phi(x) = 0$ for all $x \in A\mathcal{X}$. Then let y be any nonzero vector in $A^{k-1}\mathcal{X}$. Since $\mathcal{A}$ is an algebra, the closure, say $\mathcal{M}$, of $\{By : B \in \mathcal{A}\}$ is an invariant subspace of $\mathcal{A}$. Since $\mathcal{M} \subseteq A\mathcal{X}$, $\phi(v) = 0$ for all $v \in \mathcal{M}$, so $\mathcal{M} \neq \mathcal{X}$. If $\mathcal{M} = \{0\}$, then the operators in $\mathcal{A}$ all send $\{\lambda y : \lambda \in \mathbb{C}\}$ to $\{0\}$, and thus leave it invariant. In either case, then, $\mathcal{A}$ has a nontrivial invariant subspace. $\square$

Corollary 9.1.5. *A uniformly closed algebra of nilpotent operators is triangularizable.*

Proof. This will follow from the above theorem if we show that the index of nilpotence must be bounded. It is not surprising that the Baire Category Theorem does the job. To see this, let $\mathcal{A}$ denote the algebra, and, for each k, define $\mathcal{A}_k = \{A \in \mathcal{A} : A^k = 0\}$. Each $\mathcal{A}_k$ is obviously closed, and $\bigcup_{k=1}^{\infty} \mathcal{A}_k = \mathcal{A}$. Thus some $\mathcal{A}_k$ has nonempty interior; let A be in the interior of $\mathcal{A}_k$. To see that $B^k = 0$ for all $B \in \mathcal{A}$, fix any B and any bounded linear functionals ϕ on $\mathcal{B}(\mathcal{X})$, and define the polynomial p by

$$p(z) = \phi\left([A + z(B - A)]^k\right).$$

For z sufficiently small, $p(z) = 0$. Hence $p(z) \equiv 0$, and therefore $p(1) = 0$, so $\phi(B^k) = 0$. Since this is true for all linear functionals ϕ, it follows that $B^k = 0$. $\square$

If the index of nilpotence varies, however, an algebra need not be triangularizable.

Theorem 9.1.6. *There is a transitive algebra of nilpotent operators on Hilbert space.*

Proof. We construct such an algebra from the semigroups of Theorem 9.1.1. For each k, let

$$\mathcal{S}_k = \left\{ \text{Amp } N(A) : A \text{ is a } 2^{k-1} \times 2^{k-1} \text{ matrix} \right\}$$

as in the proof of Theorem 9.1.1. For each n, define $\mathcal{A}_n = \mathcal{S}_1 + \mathcal{S}_2 + \cdots + \mathcal{S}_n$ (i.e., the set of all sums $S_1 + \cdots + S_n$ with $S_j \in \mathcal{S}_j$ for all j). Then each $\mathcal{A}_n$ is a linear space, since each $\mathcal{S}_j$ is, and $\mathcal{A}_n$ is closed under multiplication, since $i < j$ implies $\mathcal{S}_i \mathcal{S}_j = \mathcal{S}_j \mathcal{S}_i = \mathcal{S}_j$. Thus each $\mathcal{A}_n$ is an algebra. Since $\mathcal{A}_{n+1} \supseteq \mathcal{A}_n$, the union $\mathcal{A} = \bigcup_{n=1}^{\infty} \mathcal{A}_n$ is an algebra. Since $\mathcal{A}$ contains $\bigcup_{n=1}^{\infty} \mathcal{S}_n$ and $\bigcup_{n=1}^{\infty} \mathcal{S}_n$ is irreducible, it follows that $\mathcal{A}$ is transitive.

It remains to be shown that every operator in $\mathcal{A}$ is nilpotent. For this, suppose $A \in \mathcal{A}_n$. Then $A = S_1 + S_2 + \cdots + S_n$ with $S_j \in \mathcal{S}_j$ for all j. Since $S_j^2 = 0$ for all j and $S_i S_j$ is in $\mathcal{S}_j$ when $i < j$, it follows that $A^2 \in \mathcal{S}_2 + \mathcal{S}_3 + \cdots + \mathcal{S}_n$. A trivial induction shows that $A^n \in \mathcal{S}_n$, so $(A^n)^2 = 0$. ☐

The following should be contrasted with Theorem 7.3.4 and with Corollary 8.4.5.

Corollary 9.1.7. *There is an algebra of nilpotent operators on Hilbert space that is transitive but has the property that every finitely generated subalgebra is triangularizable.*

Proof. We show that the irreducible algebra $\mathcal{A}$ of the previous theorem has the property that every finitely generated subalgebra is triangularizable. This is an easy consequence of the previous results, for if $\{A_1, \ldots, A_k\} \subseteq \mathcal{A}$, there is an n such that $\{A_1, \ldots, A_k\} \subseteq \mathcal{A}_n$, and $\mathcal{A}_n$ is triangularizable by Theorem 9.1.4, since $A^{2n} = 0$ for all $A \in \mathcal{A}_n$. ☐

Kolchin's Theorem (2.1.8) states that a semigroup of operators on a finite-dimensional space is triangularizable if every operator in the semigroup has spectrum $\{1\}$. This does not extend to infinite dimensions, even with the additional hypothesis that the semigroup is a group.

Corollary 9.1.8. *There is an irreducible group of operators on Hilbert space all of which have spectrum $\{1\}$.*

Proof. Let $\mathcal{A}$ be the transitive algebra of nilpotent operators constructed in Theorem 9.1.6. Define $\mathcal{S} = \{I + N : N \in \mathcal{A}\}$. To see that $\mathcal{S}$ is a group, simply note that $N^k = 0$ implies that $(I + N)^{-1} = I - N + \cdots + (-1)^{k-1}N^{k-1}$. $\qquad\qquad\qquad\qquad\qquad\qquad\qquad\qquad\qquad\qquad\square$

The irreducible group $\mathcal{S}$ of Corollary 9.1.8 is also a counterexample to several other possible conjectures: Note that $\mathcal{S}$ has permutable, submultiplicative, and sublinear spectrum. It also has the property that $AB - BA$ is nilpotent for every pair A and B in $\mathcal{S}$.

One may expect reducibility results for groups of unitary operators. For example, a unitary group with permutable spectrum is abelian, because all group commutators $ABA^{-1}B^{-1}$ have spectrum $\{1\}$ by hypothesis, so that (by the spectral theorem) $ABA^{-1}B^{-1} = I$. However, the situation is different for submultiplicative and sublinear spectra.

Example 9.1.9. *There is an irreducible group of unitary operators on which spectrum is submultiplicative and sublinear.*

Proof. Pick a bilateral orthonormal basis $\{e_n\}_{n=-\infty}^{+\infty}$ for a Hilbert space $\mathcal{H}$. Choose an aperiodic number θ of modulus one and define unitary operators U and V by

$$Ue_n = e_{n+1} \quad \text{and} \quad Ve_n = \theta^n e_n$$

for all integers n. The operator U is the bilateral shift, and V is diagonal with eigenvalues dense in $\mathbb{T} = \{z : |z| = 1\}$, so that

$$\sigma(U^m) = \sigma(V^n) = \mathbb{T}$$

for all integers m and n. Let $\mathcal{G}$ be the group generated by U and V. Since $VU = \theta UV$, we get

$$\mathcal{G} = \{\theta^r U^s V^t : r, s, t \in \mathbb{Z}\}.$$

Let $A = \theta^r U^s V^t$ be any nonscalar member of $\mathcal{G}$. Then either $s \neq 0$ or $t \neq 0$. If $s = 0$, then

$$\sigma(A) = \theta^r \sigma(V^t) = \theta^r \mathbb{T} = \mathbb{T}.$$

If $s \neq 0$, then A is similar to U^s. (It is not hard to see that every weighted shift W defined by $We_n = \lambda_n e_{n+1}$ with $|\lambda_n| = 1$ for all $n \in \mathbb{Z}$ is similar to U, so that W^s is similar to U^s.) Thus $\sigma(A) = \mathbb{T}$ in this case also.

The submultiplicativity of σ on $\mathcal{G}$ is now clear: If A and B are not both scalar, then $\sigma(A)\sigma(B) = \mathbb{T}$. Thus $\sigma(AB) \subseteq \sigma(A)\sigma(B)$.

For sublinearity of spectra on a pair $\{A, B\}$ in $\mathcal{G}$, note that it trivially holds if one is scalar. Assume that both A and B are nonscalar. It follows from $\sigma(A) = \sigma(B) = \mathbb{T}$ that, for every $\lambda \in \mathbb{C}$,

$$\sigma(A) + \lambda\sigma(B) = \{z : 1 - |\lambda| \leq |z| \leq 1 + |\lambda|\}.$$

Let $z_0 \in \sigma(A + \lambda B)$. Then

$$|z_0| \leq \|A + \lambda B\| \leq 1 + |\lambda|.$$

To show sublinearity, we must also verify that $1 - |\lambda| \leq |z_0|$. If this were not true, then the relations

$$\|A^{-1}(\lambda B - z_0 I)\| \leq \|A^{-1}\|(|\lambda| + |z_0|) = |\lambda| + |z_0| < 1$$

would imply that $I + A^{-1}(\lambda B - z_0 I)$ is invertible, and hence $A + \lambda B - z_0 I$ would be invertible, which is a contradiction.

To complete the proof we need only show that $\mathcal{G}$ is irreducible. If T is any operator commuting with every member of $\mathcal{G}$, then T is scalar, so if P is the orthogonal projection onto an invariant subspace of $\mathcal{G}$, then $AP = PA$ for every A in $\mathcal{G}$ implies $P = 0$ or $P = I$. $\square$

9.2 Commutators of Rank One

In the finite-dimensional case, Laffey's Theorem (1.3.6) states that $\{A, B\}$ is triangularizable if the rank of the commutator $AB - BA$ is 1. This does not hold in general.

Example 9.2.1. *Fix any p with $1 \leq p < \infty$. On the Banach space ℓ^p there are bounded linear operators A and B such that the rank of $AB - BA$ is 1 and the pair $\{A, B\}$ is irreducible.*

Proof. Define

$$A(x_0, x_1, \dots) = (0, x_0, x_1, \dots)$$

and

$$B(x_0, x_1, \dots) = (x_1, x_2, \dots) .$$

Then $(BA - AB)(x_0, x_1, \dots) = (x_0, 0, 0, 0, \dots)$. Thus $BA - AB$ is a rank-one projection onto the vector $(1, 0, 0, \dots)$.

It is easily seen that $\{A, B\}$ is irreducible. For if $\mathcal{M} \in \mathrm{Lat}\{A, B\}$ and $\mathcal{M} \neq \{0\}$, applying sufficiently high powers of B to a nonzero vector in $\mathcal{M}$ shows that there is an $(x_0, x_1, \dots) \in \mathcal{M}$ with $x_0 \neq 0$. Then $(AB - BA)(x_0, x_1, \dots) = (x_0, 0, 0, \dots)$ is in $\mathcal{M}$, and applying powers of A shows that, for every n, the sequence whose n-th term is x_0 and whose other terms are 0 is in $\mathcal{M}$. Thus $\mathcal{M} = \ell^p$. $\square$

Given the results of Section 9.1 above, this example is to be expected. It is more surprising that Laffey's Theorem does not even extend to pairs of operators even if they are both compact; a counterexample can be based on the following.

Lemma 9.2.2. *Let $\mathcal{H}$ be a Hilbert space and P be an orthogonal projection of rank one. Then there exist compact operators K and L such that $KL - LK = P$.*

Proof. This lemma is due to Anderson [1]; he provides a complex but explicit construction of K and L. $\qquad\square$

It follows immediately from this lemma that Laffey's Theorem does not extend to compact operators: If $\{K, L\}$ were triangularizable then $KL - LK = P$ would imply that $\sigma(P) = \{0\}$ (by the Spectral Mapping Theorem 7.2.6), and $\sigma(P) = \{0, 1\}$. A little further analysis shows that there is an irreducible pair of compact operators whose commutator has rank one.

Theorem 9.2.3. *There is an irreducible pair of compact operators on Hilbert space whose commutator is a projection of rank one.*

Proof. Let P, K, and L be as in Lemma 9.2.2. Since $\{K, L\}$ is not triangularizable, the Triangularization Lemma (7.1.11) implies that for every maximal chain $\mathcal{C}$ of common invariant subspaces of $\{K, L\}$ there is an $\mathcal{M} \in \mathcal{C}$ with the dimension of $\mathcal{M}/\mathcal{M}_-$ greater than 1. Let $K_\mathcal{M}$, $L_\mathcal{M}$, and $P_\mathcal{M}$ be the respective quotient operators. The maximality of $\mathcal{C}$ implies that $\{K_\mathcal{M}, L_\mathcal{M}\}$ is irreducible. The operator $P_\mathcal{M}$ is an orthogonal projection of rank at most 1. But $P_\mathcal{M}$ cannot be 0, for that would imply that $K_\mathcal{M}$ and $L_\mathcal{M}$ commute, which contradicts the irreducibility of $\{K_\mathcal{M}, L_\mathcal{M}\}$ (by Corollary 6.3.7). This proves the theorem. $\qquad\square$

There is an affirmative result for a large collection of compact operators, the trace-class operators on Hilbert space (which were defined in Section 6.5 and were previously used in Section 8.6). We can use the trace to produce common invariant subspaces. The fundamental theorem is the following.

Theorem 9.2.4. *If $AB - BA$ has rank 1 and AB is in the trace class, then $\{A, B\}$ is reducible.*

Proof. We use notation similar to that of Definition 7.4.1 for rank-one operators. In the case of Hilbert space, linear functionals are all represented as inner products, so a typical rank-one operator $f_0 \otimes g_0$ is defined by

$$(f_0 \otimes g_0)(f) = (f, g_0)f_0 \quad \text{for } f \in \mathcal{H} .$$

Note that the trace of $f_0 \otimes g_0$ is (f_0, g_0).

Suppose, then, that $AB - BA = f_0 \otimes g_0$. Since $BA = AB - (f_0 \otimes g_0)$, BA is also in the trace class. A fundamental property of the trace is that $\text{tr}(AB) = \text{tr}(BA)$ whenever AB is in the trace class even if neither A nor B is itself a trace-class operator. Hence $\text{tr}(f_0 \otimes g_0) = 0$. Moreover, for each pair of positive integers $\{m,n\}$, $A^n B^m = A^{n-1}(AB)B^{m-1}$ is in the trace class, as is $B^m A^n$, since the trace-class operators form an ideal. Thus $\operatorname{tr}(A^n B^m - B^m A^n) = 0$ for all m,n.

We claim that the span of

$$\{A^n B^m f_0 : m,n \ \text{natural numbers}\}$$

is a nontrivial common invariant subspace of A and B. This will be established by proving the following facts:

  (i) $A^n B^m f_0 = B^m A^n f_0$ for all m and n, and
  (ii) $(A^n B^m f_0, g_0) = 0$ for all m and n.

The first of these relations shows that the span is invariant under both A and B, while the second shows that the span is a proper subspace. We prove (i) and (ii) simultaneously, by induction on m, for all n at once. We begin with the case $m = 0$. In this case (i) is vacuous, so we need only show that $(A^n f_0, g_0) = 0$. Since $\operatorname{tr}(A^{n+1}B - BA^{n+1}) = 0$,

$$0 = \operatorname{tr}\left(A^{n+1}B - BA^{n+1}\right) = \operatorname{tr}\left(\sum_{k=0}^{n} A^{n-k}(AB - BA)A^k\right)$$

$$= \operatorname{tr}\left(\sum_{k=0}^{n} A^{n-k}(f_0 \otimes g_0)A^k\right)$$

$$= \sum_{k=0}^{n} \operatorname{tr}\left(A^{n-k}f_0 \otimes (A^*)^k g_0\right) = \sum_{k=0}^{n} (A^n f_0, g_0)$$

$$= (n+1)\,(A^n f_0, g_0)\,.$$

This proves (ii) in the case $m = 0$.

Now assume that (i) and (ii) are both established when $m < M$ and n is arbitrary. It suffices to show that they both hold for $m = M$ and n arbitrary.

We first consider (i) for $m = M$. To prove that (i) holds for all n we use induction on n. Since the case $n = 0$ is trivial, we can suppose that (i) holds for $m = M$ and $n < N$ (in addition to $m < M$ and all n) and establish the equality for $n = N$. Compute

$$B^M A^N f_0 = B(B^{M-1} A^N f_0) = B(A^N B^{M-1} f_0)$$
$$= ABA^{N-1}B^{M-1}f_0 - (AB - BA)A^{N-1}B^{M-1}f_0\,.$$

Note that

$$B^M A^N f_0 = ABA^{N-1}B^{M-1}f_0 - (f_0 \otimes g_0)A^{N-1}B^{M-1}f_0\,.$$

Now (ii) implies that

$$(f_0 \otimes g_0)A^{N-1}B^{M-1}f_0 = \left(A^{N-1}B^{M-1}f_0, g_0\right)f_0 = 0\,.$$

Thus

$$B^M A^N f_0 = ABA^{N-1}B^{M-1} f_0$$
$$= ABB^{M-1}A^{N-1}f_0 = AB^M A^{N-1}f_0$$
$$= A^N B^M f_0 \ .$$

This proves that (i) holds for $m = M$ and all n.

We now establish $\left(A^n B^M f_0, g_0\right) = 0$ for all n, using (i) for $m \leq M$ and all n. For this, compute the double sum

$$\sum_{j=0}^{M} \sum_{k=0}^{n} B^j A^{n-k}(AB - BA)A^k B^{M-j}$$

$$= \sum_{j=0}^{M} B^j \left[\sum_{k=0}^{n} A^{n-k} \left[(AB - BA)A^k \right] B^{M-j} \right]$$

$$= \sum_{j=0}^{M} B^j \left[A^{n+1}B - BA^{n+1} \right] B^{M-j}$$

$$= A^{n+1} B^{M+1} - B^{M+1} A^{n+1} \ .$$

Now, $\text{tr}\left(A^{n+1}B^{M+1} - B^{M+1}A^{n+1}\right) = 0$, so the double sum over j and k of the traces of $B^j A^{n-k}(AB - BA)A^k B^{M-j}$ is 0. For each j and k, however,

$$\text{tr}\left(B^j A^{n-k}(AB - BA)A^k B^{M-j}\right) = \text{tr}\left(B^j A^{n-k} f_0 \otimes (B^*)^{M-j}(A^*)^k g_0\right)$$

$$= \left(B^j A^{n-k} f_0, (B^*)^{M-j}(A^*)^k g_0\right)$$

$$= \left(A^k B^M A^{n-k} f_0, g_0\right) = \left(A^n B^M f_0, g_0\right) \ .$$

Since $\text{tr}\left(B^j A^{n-k}(AB - BA)A^k B^{M-j}\right)$ has the same value for all j and k, the trace of the double sum, which is 0, is also $(M+1)(n+1)\left(A^n B^M f_0, g_0\right)$. Hence $\left(A^n B^M f_0, g_0\right) = 0$, so (ii) holds by induction. $\qquad\square$

This theorem cannot easily be strengthened to conclude triangularizability without strengthening the hypothesis. Suppose, for example, that $A = A_1 \oplus A_2$ and $B = 0 \oplus B_2$ where B is a trace-class operator and the rank of $A_2 B_2 - B_2 A_2$ is 1. If there is an operator without nontrivial invariant subspaces on Hilbert space and A_1 is such an operator, then $\{A, B\}$ would be reducible, but the restriction of $\{A, B\}$ to the first summand would be irreducible. More generally, a compression of $\{A, B\}$ to some $\mathcal{M}/\mathcal{M}_-$ might be commutative even if $\text{rank}(AB - BA)$ is 1, and it is not known whether commutative families of Hilbert-space operators can be irreducible. The following result is one of several that can be formulated with hypotheses that preclude such possibilities.

Theorem 9.2.5. *If A and B are compact operators and AB is in the trace class (in particular, if A or B is a trace class operator), and if the rank of $(AB - BA)$ is 1, then $\{A, B\}$ is triangularizable.*

Proof. As indicated above, the hypothesis that the rank of $(AB - BA)$ is 1 is not necessarily inherited by quotients. However, if $\hat{A}$ and $\hat{B}$ are compressions of A and B, then the rank of $\hat{A}\hat{B} - \hat{B}\hat{A}$ is clearly at most 1; i.e., if the rank is not 1, then $\hat{A}$ and $\hat{B}$ commute. Since commutative families of compact operators are reducible (6.3.7), $\hat{A}$ and $\hat{B}$ have a nontrivial common invariant subspace if the dimension of the quotient space is greater than 1. Thus if we reformulate the hypothesis to state that the rank of $AB - BA$ is less than or equal to 1 the property is inherited by quotients and implies reducibility (by Theorem 9.2.4), so the Triangularization Lemma (7.1.11) completes the proof.   □

For the rest of this chapter we return to the general context of operators on Banach spaces. The following lemma, incorporating the essence of Laffey's Theorem (1.3.6), forms the basis for several results on triangularization of semigroups of operators whose commutators have ranks at most 1.

Lemma 9.2.6. *If A and B are operators on a Banach space and the rank of $AB - BA$ is 1, and if there is a λ such that $A - \lambda$ has nontrivial kernel and nondense range, then A and B have a common nontrivial invariant subspace.*

Proof. In the notation of Definition 7.4.1, $AB - BA = f_0 \otimes \phi_0$. If the kernel of $(A - \lambda)$ is invariant under B, we are done. If not, we show that the range of $A - \lambda$ is invariant under B, so its closure is the common invariant subspace.

Suppose that $(A - \lambda)f = 0$ but $(A - \lambda)Bf \neq 0$. Then

$$((A - \lambda)B - B(A - \lambda)) f = \phi_0(f)f_0$$

gives $(A - \lambda)Bf = \phi_0(f)f_0$. This shows that f_0 is in the range of $A - \lambda$. It follows that the range of $A - \lambda$ is invariant under B, for if $(A - \lambda)g$ is any vector in $\mathrm{ran}(A - \lambda)$, then

$$B(A - \lambda)g = (A - \lambda)Bg - \phi_0(f)f_0$$

is the sum of two vectors in $\mathrm{ran}(A - \lambda)$.   □

There are two reasons that this lemma does not immediately extend to establish triangularizability: If $\hat{A}$ and $\hat{B}$ are quotients of A and B, the rank of $\hat{A}\hat{B} - \hat{B}\hat{A}$ might be 0, and also $\hat{A} - \lambda$ may be injective and may have dense range. For certain classes of operators these problems disappear.

Definition 9.2.7. An operator A is *algebraic* if there is a polynomial p other than 0 such that $p(A) = 0$.

Every operator on a finite-dimensional space is algebraic, of course, but operators on infinite-dimensional spaces are rarely algebraic.

Theorem 9.2.8. *If A and B are algebraic operators on a Banach space, and if $\operatorname{rank}(AB - BA) \le 1$, then $\{A, B\}$ is triangularizable.*

Proof. If p is the minimal polynomial of A, let $p(z) = k \prod_{j=1}^{n}(z - \lambda_j)$. Then $\sigma(A) = \{\lambda_1, \lambda_2, \ldots, \lambda_n\}$. Each λ_j is an eigenvalue, and it is not hard to see that the closure of the range of $A - \lambda_j$ is proper. (All we really need here is the fact that this is the case for at least one λ_j; this follows immediately, for otherwise the range of $0 = \prod_{j=1}^{n}(A - \lambda_j)$ would be dense.) Thus Lemma 9.2.6 gives reducibility, and the Triangularization Lemma (7.1.11) completes the proof. $\qquad\square$

We will show that semigroups of compact or algebraic operators are triangularizable if all pairs of operators in the semigroup have commutators of rank at most 1. The following easy lemma is required.

Lemma 9.2.9. *A strong limit of a sequence of operators of rank 1 has rank at most 1.*

Proof. Suppose that $\{A_n\}$ converges strongly to A and that the rank of A_n is 1 for all n. If the rank of A is greater than 1, there are vectors f and g such that the set $\{Af, Ag\}$ is linearly independent. On the other hand, $\{A_n f, A_n g\}$ is linearly dependent for every n, so, for an infinite number of n_j, there are scalars $\{\lambda_{n_j}\}$ such that $A_{n_j} f = \lambda_{n_j} A_{n_j} g$. Then $\{A_{n_j} f\} \to Af$ and $\{A_{n_j} g\} \to Ag$ imply that $\{\lambda_{n_j}\}$ converges to some λ, so $Af = \lambda Ag$, which is a contradiction. $\qquad\square$

Theorem 9.2.10. *If S is a semigroup of compact operators on a Banach space, and if $\operatorname{rank}(AB - BA) \le 1$ for all A and B in S, then S is triangularizable.*

Proof. The hypotheses are inherited by quotients, so the Triangularization Lemma (7.1.11) implies that it suffices to show that such a semigroup is reducible. Let $\hat{S}$ be the uniform closure of $\mathbb{C}S$; then $\hat{S}$ satisfies the hypotheses of the theorem. If $\hat{S}$ were irreducible, Turovskii's Theorem (8.1.11) would imply that $\hat{S}$ contains non-quasinilpotent operators, and therefore contains nonzero operators of finite-rank, by Lemma 7.4.5.

Now let $\mathcal{F}$ denote the set of all finite-rank operators in $\hat{S}$. Since $\mathcal{F}$ is an ideal, $\mathcal{F}$ is also irreducible (Lemma 8.2.1). On the other hand, every operator in $\mathcal{F}$ is algebraic, so Theorem 9.2.8 implies that every pair of operators in $\mathcal{F}$ is triangularizable. Thus $\mathcal{F}$ is a semigroup of compact operators every pair of which is triangularizable, so $\mathcal{F}$ is triangularizable by Corollary 8.4.5. The semigroup $\mathcal{F}$ cannot be both irreducible and triangularizable, so the theorem is established.   □

We will prove the same theorem for semigroups of algebraic operators. Note that we cannot follow the outlines of the above proof, since there is a semigroup of algebraic operators that is irreducible although every pair of operators in the semigroup is triangularizable (Corollary 9.1.7). The proof in the case of algebraic operators requires the following fact, which may be of independent interest.

Theorem 9.2.11. *Suppose S is a semigroup of operators and there is a rank-one operator F such that FA is nilpotent for all $A \in S$. Then S is reducible.*

Proof. Let $F = f_0 \otimes \phi_0$. If $Af_0 = 0$ for all $A \in S$, then $\{\lambda f_0 : \lambda \in \mathbb{C}\}$ is an invariant subspace. Otherwise, we claim that the span of $\{Af_0 : A \in S\}$ is a nontrivial invariant subspace. To see this, it suffices to show that $\phi_0(Af_0) = 0$ for all $A \in S$. Note that FA nilpotent and rank $FA \leq 1$ imply that $(FA)^2 = 0$. Now, $FAf_0 = \phi(Af_0)f_0$, so

$$(FA)^2 f_0 = F\phi(Af_0)Af_0 = \phi(Af_0)\phi(Af_0)f_0.$$

Thus $(FA)^2 = 0$ implies that $\phi(Af_0) = 0$, finishing the proof.   □

Theorem 9.2.12. *A semigroup of algebraic operators on a Banach space whose commutators all have ranks at most 1 is triangularizable.*

Proof. The proof of this theorem is more involved than might be expected. We begin by recalling some basic facts about the structure of algebraic operators. If A is algebraic, there is a unique monic polynomial p of minimal degree such that $p(A) = 0$; p is the *minimal polynomial* of A, and its existence follows from the observation that the set of all polynomials q such that $q(A) = 0$ is an ideal. Consider the factorization of $p(z)$ into linear factors: $p(z) = \prod_{j=1}^{k} (z - \lambda_j)^{n_j}$ with the $\{\lambda_j\}$ distinct. Then $\sigma(A) = \{\lambda_1, \lambda_2, \ldots, \lambda_k\}$. Let $\mathcal{M}_j$ be the kernel of $(A - \lambda_j)^{n_j}$ for each j; then $\mathcal{M}_j \in \operatorname{Lat} A$. It can be shown, using an argument similar to the proof of Theorem 6.4.11 above, that the space $\mathcal{X}$ is the direct sum of the $\{\mathcal{M}_j\}$. On $\mathcal{M}_j$, let $A - \lambda_j = N_j$. Then

$$A = (\lambda_1 + N_1) \oplus (\lambda_2 + N_2) \oplus \cdots \oplus (\lambda_k + N_k).$$

We call this the *Riesz decomposition of A*. For each j, the integer n_j is the *algebraic multiplicity of λ_j*.

For each j_0, let P_{j_0} denote the projection on $\mathcal{M}_{j_0}$ along the sum of the other $\{\mathcal{M}_j\}$. Define the operators U and N by $U = \sum_{j=1}^{k} \lambda_j P_j$ and

$N = \sum_{j=1}^{k} \oplus N_j$. Then $UN = NU$, and N is nilpotent (of order the maximum of the $\{n_j\}$). This decomposition of algebraic operators is used several times in the proof, which we now begin.

By the Triangularization Lemma (7.1.11), it suffices to show that S has a nontrivial invariant subspace. We divide the proof of this into three cases, which are not disjoint but which certainly cover all possibilities:

Case (i): Every operator in S (other than 0, if $0 \in S$) is invertible and has all eigenvalues of algebraic multiplicity 1.

Case (ii): S contains an operator A with an eigenvalue λ of modulus $\rho(A)$ and algebraic multiplicity greater than 1, or S contains a nilpotent operator other than 0.

Case (iii): S contains an operator B such that all eigenvalues of B with modulus $\rho(B)$ have algebraic multiplicity 1, but B has some eigenvalues of modulus less than $\rho(B)$.

We consider the cases in order.

Proof in Case (i): If every operator in S is a scalar multiple of the identity, the result is trivial, so assume that T is in S and is not a multiple of the identity. Choose any $\lambda \in \sigma(T)$. Then T decomposes as a direct sum with respect to $\{f : Tf = \lambda f\}$ and its invariant complement. It will be convenient to represent operators with respect to this decomposition as operator matrices. (If it is preferred, the following could be stated without matrices, by writing operators S as

$$S = PSP + (1 - P)SP + PS(1 - P) + (1 - P)S(1 - P)$$

for P the projection on the eigenspace along its complement. It seems easier to understand the proof in terms of matrices.)

Thus we write $T = \begin{pmatrix} \lambda & 0 \\ 0 & X \end{pmatrix}$ where $\lambda \notin \sigma(X)$ (recall that λ has algebraic multiplicity 1). For any given $S = \begin{pmatrix} S_{11} & S_{12} \\ S_{21} & S_{22} \end{pmatrix}$ in S, the commutator is

$$ST - TS = \begin{pmatrix} 0 & S_{12}(X - \lambda) \\ (\lambda - X)S_{21} & S_{22}X - XS_{22} \end{pmatrix}.$$

This commutator has rank at most 1, and $(\lambda - X)$ is invertible, so at least one of S_{21} and S_{12} is 0. If $S_{21} = 0$ for all $S \in S$, then $\{f : Tf = \lambda f\}$ is in Lat S, and if $S_{12} = 0$ for all $S \in S$, then the complement is in Lat S. Thus it suffices to show that one of these holds.

If neither held, there would be operators A and B in S with

$$A = \begin{pmatrix} A_{11} & A_{12} \\ 0 & A_{22} \end{pmatrix} \quad \text{and} \quad B = \begin{pmatrix} B_{11} & 0 \\ B_{21} & B_{22} \end{pmatrix}$$

and $A_{12} \neq 0$, $B_{21} \neq 0$. Then $AB \in S$, and AB has the form

$$\begin{pmatrix} * & A_{12}B_{22} \\ A_{22}B_{21} & * \end{pmatrix}.$$

By the above, at least one of $A_{12}B_{22}$ and $A_{22}B_{21}$ is 0. But A_{22} and B_{22} are invertible (since A and B are, by hypothesis), so this would imply that at least one of A_{12} and B_{21} is 0, contradicting the assumption on A and B. This establishes the result in Case (i).

Proof in Case (ii): Let λ be an eigenvalue of A of modulus $\rho(A)$ that has algebraic multiplicity greater than 1. In this case, we will show that the uniform closure of $\mathbb{R}\mathcal{S}$ contains a nilpotent operator other than 0, and then that this implies reducibility. In the other subcase, if $\mathcal{S}$ contains a nontrivial nilpotent operator, the latter part of the proof applies as well.

The proof of the existence of a nilpotent operator in the closure of $\mathbb{R}\mathcal{S}$ is similar to the proof of Lemma 7.4.5. Multiply by the reciprocal of the spectral radius so that we can assume that the spectral radius of A is 1. Then collect the Riesz subspaces corresponding to eigenvalues of modulus 1, so that A can be written in the form

$$A = \begin{pmatrix} U + N & 0 \\ 0 & C \end{pmatrix},$$

where U has spectrum contained in the unit circle and is a linear combination of disjoint projections, N is a nilpotent operator commuting with U, and C has spectral radius strictly less than 1. (The summand C could be absent.) Note that $N \neq 0$ by the assumption of an eigenvalue of algebraic multiplicity greater than 1. Let k be the positive integer such that $N^k \neq 0$ and $N^{k+1} = 0$. We show that the nilpotent operator $\begin{pmatrix} N^k & 0 \\ 0 & 0 \end{pmatrix}$ is in the uniform closure of $\mathbb{R}\mathcal{S}$.

For this, first note that $UN = NU$ implies that

$$A^n = \begin{pmatrix} \sum_{j=0}^{n} \binom{n}{j} U^{n-j} N^j & 0 \\ 0 & C^n \end{pmatrix}$$

for every n. Choose a sequence $\{n_i\}$ such that $\{U^{n_i - k}\}$ converges uniformly to I. Then $\{C^{n_i}\} \to 0$, so

$$\lim_{i \to \infty} \frac{1}{\binom{n_i}{k}} A^{n_i} = \begin{pmatrix} N^k & 0 \\ 0 & 0 \end{pmatrix},$$

completing the proof that there is a nonzero nilpotent operator in the closure of $\mathbb{R}\mathcal{S}$.

Let $T = \begin{pmatrix} N^k & 0 \\ 0 & 0 \end{pmatrix}$; then $T^2 = 0$. In the other subcase, if J is in $\mathcal{S}$, $J^\ell \neq 0$, and $J^{\ell+1} = 0$, let $T = J^\ell$, so that $T^2 = 0$. We now prove that (in either case) $TST = 0$ for all $S \in \mathcal{S}$; this will then be easily seen to imply reducibility. Fix any $S \in \mathcal{S}$. By Lemma 9.2.9, $ST - TS$ has rank at most

1, so $ST - TS = f_0 \otimes \phi_0$ for some $f_0 \in \mathcal{X}$ and $\phi \in \mathcal{X}^*$. For any $f \in \mathcal{X}$, then,

$$T(ST - TS)f = \phi_0(f)Tf_0,$$

so $T^2 = 0$ yields

$$TSTf = \phi_0(f)Tf_0.$$

Similarly,

$$(ST - TS)Tf = \phi_0(Tf)f_0$$

yields $TSTf = -\phi_0(Tf)f_0$. Applying T to the last equation gives $0 = \phi_0(Tf)Tf_0$, so either $\phi_0(Tf) = 0$ or $Tf_0 = 0$. Thus $TSTf = \phi_0(f)Tf_0 = 0$ in any case.

The equation $TST = 0$ gives a nontrivial invariant subspace as follows. Fix any vector g such that $Tg \neq 0$, and let $\mathcal{M}$ be the closed linear span of

$$\{STg : S \in \mathcal{S}\} \cup \{Tg\}.$$

Then $\mathcal{M}$ is clearly in Lat $\mathcal{S}$, and is proper since it is contained in the kernel of T. This completes Case (ii).

Proof of Case (iii): In this case we show that the closure of $\mathbb{R}\mathcal{S}$ contains a nontrivial projection, and then that reducibility follows. The assumption in Case (iii) gives an operator in $\mathbb{R}\mathcal{S}$ of the form

$$B = \begin{pmatrix} V & 0 \\ 0 & D \end{pmatrix},$$

where V is a linear combination of disjoint projections, $\rho(V) = 1$, and $\rho(D) < 1$. Then suitable powers $\{V^{m_j}\}$ converge uniformly to I, so $\{B^{m_j}\}$ converges to $\begin{pmatrix} I & 0 \\ 0 & 0 \end{pmatrix}$, a nontrivial projection. Let $P = \begin{pmatrix} I & 0 \\ 0 & 0 \end{pmatrix}$. If the range of P is invariant under every operator in $\mathcal{S}$, we are done, so we can assume that there is a fixed $G \in \mathcal{S}$ with $G = \begin{pmatrix} G_{11} & G_{12} \\ G_{21} & G_{22} \end{pmatrix}$ and $G_{21} \neq 0$.

Let $F = \begin{pmatrix} 0 & 0 \\ G_{21} & 0 \end{pmatrix}$; then $F^2 = 0$. We show that the rank of F is 1, and that FX is nilpotent for all $X \in \mathcal{S}$, so that Theorem 9.2.11 implies that $\mathcal{S}$ is reducible.

To see that F has rank 1, note that Lemma 9.2.9 implies that the rank of $GP - PG$ is at most 1, and

$$GP - PG = \begin{pmatrix} G_{11} & 0 \\ G_{21} & 0 \end{pmatrix} - \begin{pmatrix} G_{11} & G_{12} \\ 0 & 0 \end{pmatrix} = \begin{pmatrix} 0 & -G_{12} \\ G_{21} & 0 \end{pmatrix}.$$

It follows that the rank of G_{21} is 1 (since we are assuming it is not 0). Thus F has rank 1.

It remains to be shown that FX is nilpotent for every $X \in \mathcal{S}$. Fix any X. Choose a sequence $\{P_n\} \subset \mathbb{R}\mathcal{S}$ that converges uniformly to P. For each n, $\{GP_n, P_nX\}$ is a triangularizable pair (by Theorem 9.2.8). Also, each commutator $(GP_n)(P_nX) - (P_nX)(GP_n)$ has rank 1, and therefore has spectrum $\{0\}$. (One way to see this is by Ringrose's Theorem (7.2.3): A triangularizing chain for $\{GP_n, P_nX\}$ also triangularizes their commutator, and each diagonal coefficient of the commutator has the form $\alpha\beta - \beta\alpha$, where α and β are complex numbers.) A rank-one quasinilpotent operator has square 0, and the uniform limit of operators whose square is 0 also has square 0, so

$$[(GP)(PX) - (PX)(GP)]^2 = 0.$$

In terms of the decomposition of the space,

$$(GP)(PX) - (PX)(GP) = \begin{pmatrix} G_{11}X_{11} - X_{11}G_{11} - X_{12}G_{21} & G_{11}X_{12} \\ G_{21}X_{11} & G_{21}X_{12} \end{pmatrix}.$$

Note that

$$(GP)(PX) = \begin{pmatrix} G_{11}X_{11} & G_{11}X_{12} \\ G_{21}X_{11} & G_{21}X_{12} \end{pmatrix},$$

and the fact that the commutator of this product with P has rank at most 1 implies that at least one of $G_{11}X_{12}$ and $G_{21}X_{11}$ is 0. Thus the matrix for $(GP)(PX) - (PX)(GP)$ is either upper or lower triangular, so the fact that its square is 0 implies, in particular, that $(G_{21}\, X_{12})^2 = 0$. But

$$FX = \begin{pmatrix} 0 & 0 \\ G_{21} & 0 \end{pmatrix} \begin{pmatrix} X_{11} & X_{12} \\ X_{21} & X_{22} \end{pmatrix} = \begin{pmatrix} 0 & 0 \\ G_{21}X_{11} & G_{21}X_{12} \end{pmatrix},$$

so $(G_{21}X_{12})^2 = 0$ implies $(FX)^4 = 0$, and Theorem 9.2.11 applies. Thus in all cases, $\mathcal{S}$ is reducible, and therefore triangularizable.   □

There is a result that applies to a pair of bounded operators that need not be algebraic or compact.

Theorem 9.2.13. *If A and B are operators on a Banach space of dimension greater than 1, and if the rank of $AB - BA$ is 1 and the ranks of $A^2B - BA^2$, $AB^2 - B^2A$, and $A^2B^2 - B^2A^2$ are each at most 1, then A and B have a nontrivial invariant subspace in common.*

Proof. By hypothesis, there is a vector $f_0 \neq 0$ and a bounded linear functional $\phi_0 \neq 0$ such that

$$(AB - BA)f = \phi_0(f)f_0 \qquad \text{for all } f.$$

We exhibit a common invariant subspace, the nature of which depends upon the relationship of ϕ_0 to A^* and B^*.

Case (i): ϕ_0 is an eigenvector for A^* and for B^*. In this case, $A^*\phi_0 = \alpha\phi_0$ implies that

$$\phi_0(Af) = (A^*\phi_0)(f) = \alpha\phi_0(f),$$

so the kernel of ϕ_0 is invariant under A. Similarly, the kernel is invariant under B.

Case (ii): ϕ_0 is not an eigenvector for either A^* or B^*. In this case we show that f_0 is a common eigenvector of A and B. For this, consider any vector g and compute

$$(A^2B - BA^2)g = A(AB - BA)g + (AB - BA)Ag$$
$$= \phi_0(g)Af_0 + \phi_0(Ag)f_0$$
$$= \phi_0(g)Af_0 + (A^*\phi_0)(g)f_0.$$

By hypothesis, $A^*\phi_0$ is not a multiple of ϕ_0, so there exist vectors h and k such that $\phi_0(h) = 0$ but $(A^*\phi_0)h \neq 0$, and $(A^*\phi_0)k = 0$ but $\phi_0(k) \neq 0$. Using the equation above yields, for any scalars α and β,

$$(A^2B - BA^2)(\alpha h + \beta k) = \alpha(A^*\phi_0)(h)f_0 + \beta\phi_0(k)Af_0.$$

The rank of $A^2B - BA^2$ is at most 1, so Af_0 is a multiple of f_0. Similarly, Bf_0 is a multiple of f_0, so the set of multiples of f_0 is a common invariant subspace in this case.

One case remains.

Case (iii): ϕ_0 is an eigenvector for exactly one of A^* and B^*. Assume that ϕ_0 is an eigenvector for B^*. Since ϕ_0 is not an eigenvector for A^*, the proof given for Case (ii) above shows that f_0 is an eigenvector for A. Let $\mathcal{M} = \bigvee_{n=0}^{\infty} \{B^n f_0\}$. Then clearly $\mathcal{M} \in \text{Lat}\, B$. We must show that $\mathcal{M}$ is nontrivial and is invariant under A.

There is a μ such that $B^*\phi_0 = \mu\phi_0$. Therefore, for any vector f,

$$(AB^2 - B^2A)f = (AB - BA)Bf + B(AB - BA)f$$
$$= \phi_0(Bf)f_0 + B\phi_0(f)f_0$$
$$= \mu\phi_0(f)f_0 + \phi_0(f)Bf_0$$
$$= \phi_0(f)(\mu + B)f_0.$$

Thus

$$(A^2B^2 - B^2A^2)f = A(AB^2 - B^2A)f + (AB^2 - BA^2)Af$$
$$= \phi_0(f)A(\mu + B)f_0 + \phi_0(Af)(\mu + B)f_0$$
$$= \phi_0(f)A(\mu + B)f_0 + (A^*\phi_0)(f)(\mu + B)f_0.$$

The rank of $A^2B^2 - B^2A^2$ is at most 1, and $A^*\phi_0$ is not a multiple of ϕ_0, so there is a scalar γ such that

$$A(\mu + B)f_0 = \gamma(\mu + B)f_0,$$

or

$$ABf_0 = \gamma(\mu + B)f_0 - \mu Af_0.$$

Since A, $Af_0 = \lambda f_0$ for some λ, we have

$$ABf_0 = \gamma(\mu + B)f_0 - \mu\lambda f_0 = \gamma Bf_0 + (\gamma - \lambda)\mu f_0.$$

But $(AB - BA)f_0 = \phi_0(f_0)f_0$, so $ABf_0 = \lambda Bf_0 + \phi_0(f_0)f_0$. Thus

$$\lambda Bf_0 + \phi_0(f_0)f_0 = \gamma Bf_0 + (\gamma - \lambda)\mu f_0 .$$

There are two possibilities: Either $\{f_0, Bf_0\}$ is one-dimensional, or $\lambda = \gamma$ and $\phi_0(f_0) = (\gamma - \lambda)\mu$. In the first situation, $\mathcal{M}$ is a one-dimensional common invariant subspace for A and B. In the other situation, we claim that $\mathcal{M}$ is contained in the kernel of ϕ_0. If $\lambda = \gamma$ and $\phi_0(f_0) = (\gamma - \lambda)\mu$, then $\phi_0(f_0) = 0$, so

$$\phi_0(B^n f_0) = ((B^*)^n \phi_0)(f_0) = \mu^n \phi_0(f_0) = 0$$

for all n, so $\mathcal{M}$ is contained in the kernel of ϕ_0, and is hence proper.

It remains to be shown that $\mathcal{M} \in \operatorname{Lat} A$ in this case. In fact, we prove that $AB^n f_0 = \lambda B^n f_0$ for all n. To see this, note first that any g in the kernel of ϕ_0 satisfies

$$(AB - BA)g = \phi_0(g)f_0 = 0,$$

so $ABg = BAg$. This holds, in particular, for $g = B^n f_0$, so $AB^{n+1}f_0 = BAB^n f_0$. Since $Af_0 = \lambda f_0$, this relation with $n = 0$ gives $ABf_0 = \lambda Bf_0$, and proceeding inductively gives $AB^n f_0 = \lambda B^n f_0$ for all n. Therefore, $\mathcal{M}$ is invariant under A.   □

Note that Example 9.2.1 shows that the hypothesis that $A^2 B^2 - B^2 A^2$ has rank at most 1 is required in the above theorem.

Corollary 9.2.14. *If the rank of $AB - BA$ is 1, and if the rank of $ST - TS$ is at most 1 for all S and T in the semigroup generated by $\{A, B\}$, then that semigroup is reducible.*

Proof. This is a special case of the theorem.   □

Note that triangularizability results can be obtained from this corollary in cases where it is known that commutative quotients are reducible. In particular, this provides an alternative proof of the corresponding theorem for compact operators (Theorem 9.2.10).

9.3 Bands

Recall that an abstract band is a semigroup whose members are all idempotent (Definition 2.3.4). We have seen that bands of compact operators (i.e., of finite-rank operators) are triangularizable. Before discussing some affirmative results in the noncompact case, we present an example showing that one cannot expect triangularizability, or even reducibility. The construction is similar to that of the irreducible semigroup of nilpotent operators (Theorem 9.1.1).

Theorem 9.3.1. *There is an irreducible semigroup of idempotent operators on a separable Hilbert space.*

Proof. For any natural number k and $k \times k$ matrices A and B, let I be the $k \times k$ identity matrix and let

$$
P_{A,B} = \begin{bmatrix} A \\ A \\ I \end{bmatrix} \begin{bmatrix} B & -B & I \end{bmatrix} = \begin{bmatrix} AB & -AB & A \\ AB & -AB & A \\ B & -B & I \end{bmatrix}.
$$

Then $P_{A,B}$ is a matrix of size $(3k) \times (3k)$; direct computation shows that $P_{A,B}^2 = P_{A,B}$. More generally, if $A, B, C,$ and D are $k \times k$ matrices, then

$$
P_{A,B} P_{C,D} = \begin{bmatrix} AB & -AB & A \\ AB & -AB & A \\ B & -B & I \end{bmatrix} \begin{bmatrix} CD & -CD & C \\ CD & -CD & C \\ D & -D & I \end{bmatrix}
$$
$$
= \begin{bmatrix} AD & -AD & A \\ AD & -AD & A \\ D & -D & I \end{bmatrix} = P_{A,D}.
$$

Now, for each pair $\{A, B\}$ the operator $T_{A,B}$ is defined on ℓ^2 as the direct sum

$$
T_0 \,\oplus\, T_1 \,\oplus\, T_2 \oplus \cdots \oplus\, T_j \,\oplus \cdots ,
$$

where $T_j = P_{A,B}$ if the ternary representation of j does not contain any 2's, and otherwise T_j is the $(3k) \times (3k)$ identity matrix. Then $T_{A,B} T_{C,D} = T_{A,D}$, by the corresponding relations on the P's, so, in particular, each $T_{A,B}$ is an idempotent.

To make the decomposition of the operators compatible, we start with matrices whose orders are powers of 3. That is, for each $n = 0, 1, 2, 3, \ldots$, let $\mathcal{S}_n$ denote the set of all $T_{A,B}$ for A and B matrices of order $(3^n) \times (3^n)$. Then each $\mathcal{S}_n$ is a semigroup of idempotents on ℓ^2, but each $\mathcal{S}_n$ is reducible. To get an irreducible semigroup, we define $\mathcal{S}$ to be the union of all the $\mathcal{S}_n$.

We first show that $\mathcal{S}$ is a semigroup. Suppose that $E \in \mathcal{S}_n$ and $F \in \mathcal{S}_m$ and that $m < n$; we show that EF and FE are both in $\mathcal{S}_n$. Now, $F = S_0 \oplus S_1 \oplus S_2 \oplus \cdots$, where each S_j is a matrix of order $3^{m+1} \times 3^{m+1}$. Let M denote the $3^n \times 3^n$ matrix $S_0 \oplus S_1 \oplus \cdots \oplus S_{3^{n-m-1}-1}$. Then F can also be written $F = C_0 \oplus C_1 \oplus C_2 \oplus \cdots$, where $C_j = M$ if j has a ternary representation with no 2's and $C_j = I$ otherwise. (Note that if j has no 2's in its ternary representation, then $3^k j$ has the same property.) Now,

$E \in \mathcal{S}_n$ implies that $E = T_{A,B}$ for some $3^n \times 3^n$ matrices A and B. Observe that

$$P_{A,B} \begin{bmatrix} M & O & O \\ O & M & O \\ O & O & I \end{bmatrix} = \begin{bmatrix} A \\ A \\ I \end{bmatrix} [BM \quad -BM \quad I] = P_{A,BM},$$

so $T_{A,B}F = T_{A,BM}$, which is in $\mathcal{S}_n$. Similarly, $FT_{A,B} = T_{MA,B}$, since

$$\begin{bmatrix} M & O & O \\ O & M & O \\ O & O & I \end{bmatrix} \begin{bmatrix} A \\ A \\ I \end{bmatrix} [B \quad -B \quad I] = \begin{bmatrix} MA \\ MA \\ I \end{bmatrix} [B \quad -B \quad I] = P_{MA,B}.$$

Thus $FT_{A,B}$ is also in $\mathcal{S}_n$. Therefore $\mathcal{S}_m \mathcal{S}_n \subseteq \mathcal{S}_n$ and $\mathcal{S}_n \mathcal{S}_m \subseteq \mathcal{S}_n$ whenever $n > m$, so $\mathcal{S}$ is a semigroup.

All that remains to be shown is that $\mathcal{S}$ is irreducible. As in the proof of Theorem 9.1.1, it seems easiest to prove the stronger assertion that $\mathcal{S}$ is weakly dense in $B(\mathcal{H})$. The proof of this is almost identical to the corresponding proof toward the end of Theorem 9.1.1. As in that proof, fix any $T \in B(\mathcal{H})$ and any weak neighborhood

$$\mathcal{U}(T; f_1, f_2, \dots, f_\ell; g_1, g_2, \dots, g_\ell; \varepsilon) = \{S : |((S - T)f_j, g_j)| < \varepsilon\}.$$

Note that

$$P_{A,I} = \begin{bmatrix} A & -A & A \\ A & -A & A \\ I & -I & I \end{bmatrix}.$$

Thus, for any n, we can find a $T_{A,I}$ in $\mathcal{S}$ such that, for $k = 1, 2, \dots, 3^n$, $T_{A,I}e_k = Te_k$, where $\{e_j\}_{j=1}^{\infty}$ is the fixed orthonormal basis of ℓ^2 with respect to which the operators $T_{A,B}$ are written. Since

$$\|P_{A,I}\|^2 \leq \left\| \begin{bmatrix} A \\ A \\ I \end{bmatrix} \right\|^2 \| [I \quad -I \quad I] \|^2 \leq (2\|A\|^2 + 1)(3),$$

and since the norm of the restriction of T to the span of $\{e_1, e_2, \dots, e_{3^n}\}$ is at most $\|T\|$, it follows that $\|T_{A,I}\|$ is at most the square root of $(2\|T\|^2 + 1)(3)$. Thus the weak density follows as in the proof of Theorem 9.1.1.   □

We need the result that a finitely generated abstract band is finite (a special case of the Green-Rees Theorem). We shall give a complete proof of this fact in Theorem 9.3.11 below. The following definitions and propositions are required for this theorem and for our operator-theoretic results.

Definition 9.3.2. Let $\mathcal{F}$ be a set. A *word in $\mathcal{F}$* is a finite expression $f_1 f_2 \dots f_n$ with $f_i \in \mathcal{F}$. The *free semigroup $\mathcal{S}$ generated by $\mathcal{F}$* is the set of all words in $\mathcal{F}$ with juxtaposition as multiplication:

$$(f_1 f_2 \dots f_n)(g_1 g_2 \dots g_n) = f_1 f_2 \dots f_n g_1 g_2 \dots g_m.$$

(Associativity is obvious.)

Proposition 9.3.3. *Let S be a free semigroup generated by a set $\mathcal{F}$. Let $\mathcal{R}$ be the collection of all relations of the form*

$$(f_1 f_2 \cdots f_n)^2 = f_1 f_2 \cdots f_n,$$

where $n \in \mathbb{N}$ and $f_i \in \mathcal{F}$. For words a and b in S, define $a\mathcal{R}b$ if a and b are the same modulo $\mathcal{R}$, that is, if a can be obtained from b by a finite number of substitutions from $\mathcal{R}$. Then

(i) *$\mathcal{R}$ is an equivalence relation, and*
(ii) *if $\{a\}$ denotes the equivalence class of a, then the operation*

$$\{a\} \cdot \{b\} = \{ab\}$$

is well-defined and makes the set of equivalence classes into a band.

Proof. That $\mathcal{R}$ is an equivalence relation is an immediate consequence of the definition. If $a\mathcal{R}a'$ and $b\mathcal{R}b'$, then $a'b'$ can be obtained from ab by the substitutions changing a to a' together with those changing b' to b. Thus $ab\mathcal{R}a'b'$, and the multiplication is well-defined. Associativity is also easy to verify. The equation $\{a\}^2 = \{a\}$ is obvious from the definition of $\mathcal{R}$, so the set of equivalence classes forms a band. □

Definition 9.3.4. The band obtained in the preceding proposition is called the *free band generated by $\mathcal{F}$.*

Definition 9.3.5. A band S is said to be *generated by a set* $\mathcal{F} \subseteq S$ if every member of S is a product of elements of $\mathcal{F}$; S is *finitely generated* if there is a finite set that generates S.

Proposition 9.3.6. *Let S be a band generated by $\mathcal{F}$. Let T be the free band (formally) generated by $\mathcal{F}$. Then there is a (semigroup) homomorphism ϕ from T onto S.*

Proof. Let $\mathcal{R}'$ be the set of all relations $f_1 f_2 \cdots f_n = g_1 g_2 \cdots g_n$ that hold in S with f_i and g_j in $\mathcal{F}$. Then $\mathcal{R}'$ includes the $\mathcal{R}$ of Proposition 9.2.3, since S is a band. Thus the identity transformation on $\mathcal{F}$ extends to a well-defined map ϕ from T onto S satisfying $\phi(ab) = \phi(a)\phi(b)$ for all a and b in T. □

Proposition 9.3.7. *Let S be an arbitrary band. For a and b in S, we define the relation $a \preceq b$ to mean $aba = a$. We also define $a \sim b$ to mean $a \preceq b$ and $b \preceq a$. Then*

(i) *$\preceq$ is a pre-order; i.e., it is reflexive and transitive,*

(ii) $\sim$ is an equivalence relation,

(iii) $a \preceq a'$ and $b \preceq b'$ implies $ab \preceq a'b'$, and

(iv) $\sim$ is a congruence: $a \sim a'$ and $b \sim b'$ imply $aa' \sim bb'$.

Proof. (i) The reflexivity is obvious. For transitivity, assume $a \preceq b \preceq c$, that is, $aba = a$ and $bcb = b$. Then $a = aba = abcba$. Hence

$$a = ab(cba) = ab(cba)^2 = a(bcb)a(cba) = aba(cba)$$
$$= a(cba) = a(cb)a,$$

so that $a \preceq cb$. Similarly,

$$a = (abc)ba = (abc)^2ba = (abc)a(bcb)a = (abc)aba$$
$$= (abc)a = a(bc)a,$$

implying $a \preceq bc$. It is clear that $cb \preceq c$. Since we have $a \preceq cb \preceq c$, we can apply the above argument to a, cb, c (instead of a, b, c) yielding $a \preceq cbc$, or $a(cbc)a = a$. Now

$$a = (acb)ca = (acb)^2ca = acb(ac)bca$$
$$= acb(ac)^2bca = (acba)c(acbca) = aca,$$

so $a \preceq c$, proving transitivity.

(ii) Since $\sim$ is symmetric by definition, it is an equivalence relation.

(iii) First note that $uv \sim vu$ for all u and v in S. We next show that if $x \preceq y$ and $x \preceq z$, then $x \preceq yz$. For $xyx = x$ and $xzx = x$ implies $x = (xzx)(xyx) = x(zxy)x$, so that

$$x \preceq zxy = (zx)(xy) \sim (xy)(zx),$$

so $x \preceq x(yz)x$, implying $x = x(xyzx)x = x(yz)x$. Finally, the assumptions $a \preceq a'$ and $b \preceq b'$ together with the obvious relations $ab \preceq a$ and $ab \preceq b$ (valid for all a and b) imply that

$$ab \preceq a' \quad \text{and} \quad ab \preceq b'.$$

Replacing x, y, and z in the argument above with ab, a', and b', respectively, we get $ab \preceq a'b'$.

(iv) This is a direct consequence of (iii). □

Definition 9.3.8. Let S be a band and let the relation $\sim$ be defined as in the preceding proposition. Each equivalence class of S relative to $\sim$ is called a *component* of S; the notation C_a is used for the component containing a. A band consisting of a single component is called a *rectangular band*. The notation $C(S)$ is used to denote the collection of all components of the band S.

Proposition 9.3.9. *Each component of a band is a band. The collection of components forms an abelian band $C(S)$ under the multiplication*

$$C_a \cdot C_b = C_{ab} \, ,$$

and the map $\phi : S \to C(S)$ defined by $\phi(a) = C_a$ is a band homomorphism. The relation $C_a \preceq C_b$ is equivalent to $a \preceq b$, and $C_a = C_b$ is equivalent to $a \sim b$.

Proof. It is easy to see that C_a is a band for each a. By Proposition 9.3.7, the multiplication $C_a \cdot C_b = C_{ab}$ is well-defined. Since this is an associative operation with $C_a^2 = C_a$ for every $a \in S$, the components form a band that is a homomorphic image of S. This band is abelian, because $ab \sim ba$ for every pair a and b in S. The remaining assertions are immediate consequences of Proposition 9.3.7. $\qquad\square$

Lemma 9.3.10. *Let S be a band generated by a finite set $\mathcal{F}$. Then the number of components of S is finite. If S is free, then each component of S is determined by a word $f_1 f_2 \cdots f_m$, where the $\{f_j\}$ are distinct members of $\mathcal{F}$.*

Proof. Proposition 9.3.6 allows us to assume, with no loss of generality, that S is free. Note that if w_1 and w_2 are words with "letters" from $\mathcal{F}$, then $w_1 \sim w_2$ if and only if w_2 is obtained from w_1 by a permutation. (Whether S is free or not, it is easily seen that permuted words are equivalent, since $C(S)$ is commutative. The converse uses the fact that S is free.) Now let C_a be a component. Then a can be taken to be a product of distinct generators, by using permutations. If $\mathcal{F}$ has n members, then there are precisely $2^n - 1$ nonempty words of this form that are mutually nonequivalent. $\qquad\square$

Theorem 9.3.11. *Every finitely generated band is finite.*

Proof. We can assume, in light of Proposition 9.3.6, that the band is free. We prove the theorem by induction. Assume that every free band with fewer than n generators is finite, and let S be a free band with n generators. By Lemma 9.3.10, it suffices to show that each component of S is finite.

Let C_a be a component. By Lemma 9.3.10 we can take $a = f_1 f_2 \cdots f_m$, where $\{f_i\}$ are distinct members of the generating set for S. Now, each word W in C_a has all the letters $\{f_1, \ldots, f_m\}$ in it. By "lengthening" W, if necessary, we can write $W = XYZ$, where each of the words X, Y, and Z has all the letters $f_1, \ldots f_m$, but

(i) the last letter of the word X appears only once in X, and
(ii) the first letter of Z appears only once in Z.

Then, by the induction hypothesis, there is only a finite number of distinct words X and Y appearing in the above representation of words in C_a. We will be done if we prove that $XYZ = XZ$ for each such representation.

Since X, Y, and Z are all in the component C_a, by construction, we have, in particular, $XZX = X$ and $Z(XY)Z = Z$. Thus
$$XYZ = (XZX)YZ = X(ZXYZ) = XZ. \qquad \square$$

Our interest in this theorem lies in the easily proven fact that a finite band of operators is triangularizable. (See Corollary 9.3.16 below.)

Proposition 9.3.12. *Every nonzero rectangular band S of operators on a Banach space is triangularizable; in fact, it has a block triangularization, at most 3×3, of the following form. The space admits a chain of invariant subspaces*
$$\{0\} = \mathcal{M}_0 \subseteq \mathcal{M}_1 \subseteq \mathcal{M}_2 \subseteq \mathcal{M}_3 = \mathcal{X}$$
for S, where the quotient bands S_i induced on $\mathcal{M}_i / \mathcal{M}_{i-1}$ satisfy
$$S_1 = \{0\}, \quad S_2 = \{I\}, \quad \text{and} \quad S_3 = \{0\},$$
and where $\mathcal{M}_1 = \cap \{\ker E : E \in S\}$ and $\mathcal{M}_2$ is the (closed linear) span of $\mathcal{M}_1$ together with the ranges of all E in S.

(Note that one or both of $\mathcal{M}_1 / \mathcal{M}_0$ and $\mathcal{M}_3 / \mathcal{M}_2$ may equal $\{0\}$; they are both trivial if $S = \{I\}$.)

Proof. Let $\mathcal{M}_1$ and $\mathcal{M}_2$ be defined as in the statement of the theorem. Clearly, $\mathcal{M}_1$ and $\mathcal{M}_2$ are invariant under S, and $S_1 = \{0\}$ and $S_3 = \{0\}$. We must show that $S_2 = \{I\}$. Let $\hat{S}$ be the quotient band on $X/\mathcal{M}_1$. If $\hat{E} \in \hat{S}$ and $\hat{E}[x + \mathcal{M}_1] = 0$ for any $[x + \mathcal{M}_1]$ in $\mathcal{X}/\mathcal{M}_1$, then $Ex \in \mathcal{M}_1$. This implies $Ex = E^2 x \in E\mathcal{M}_1 = \{0\}$. Hence if $\hat{E}[x + \mathcal{M}_1] = 0$ for all $\hat{E}$, then $Ex = 0$ for all E, yielding $x \in \mathcal{M}_1$. Thus $\hat{S}$ has trivial common kernel.

We next prove that the members of $\hat{S}$ all have the same range. Let E and F be in S and consider $\hat{E}$ and $\hat{F}$ in $\hat{S}$. To show that the range of $\hat{E}$ is invariant under $\hat{F}$, we verify that $(I - \hat{E})\hat{F}\hat{E} = 0$, which is equivalent to showing that $\hat{G}(I - \hat{E})\hat{F}\hat{E} = 0$ for all $\hat{G} \in \hat{S}$, because the common kernel of $\hat{S}$ is trivial. But
$$\hat{G}(I - \hat{E})\hat{F}\hat{E} = \hat{G}\hat{E}\hat{G}(I - \hat{E})\hat{F}\hat{E}$$
$$= \hat{G}\hat{E}\hat{G}\hat{F}\hat{E} - \hat{G}\hat{E}\hat{G}\hat{E}\hat{F}\hat{E}$$
$$= \hat{G}(\hat{E} - \hat{E}) = 0.$$

We have shown that the range of every $\hat{E}$ is invariant under every $\hat{F}$. Thus $\hat{E}\hat{F} = \hat{F}\hat{E}\hat{F} = \hat{F}$ for all $\hat{E}$ and $\hat{F}$ in $\hat{S}$. Let $\mathcal{Y}$ be the common range of the members of $\hat{S}$. Then $\hat{E}|_{\mathcal{Y}} = I$ for every $\hat{E}$, and it is easily seen that
$$\mathcal{M}_2 = \{x \in \mathcal{X} : x + \mathcal{M}_1 \in \mathcal{Y}\}.$$

Thus the quotient band $\mathcal{S}_2$ induced on $\mathcal{M}_2/\mathcal{M}_1$ is the singleton $\{I\}$.   □

By Proposition 9.3.9, the abelian band $\mathcal{C}(\mathcal{S})$ of components of a band $\mathcal{S}$ has a natural partial order inherited from the pre-order of $\mathcal{S}$. We shall show that the existence of minimal nonzero elements in this partial order implies reducibility. We start with a very simple lemma.

Lemma 9.3.13. *Let $\mathcal{S}$ be a band. For each $a \in \mathcal{S}$, the set*

$$\mathcal{J} = \{b \in \mathcal{S} : b \preceq a\}$$

is an ideal of $\mathcal{S}$.

Proof. If $s \in \mathcal{S}$, then $sa \preceq a$ and $as \preceq a$, so this follows from Proposition 9.3.7(iii).   □

Theorem 9.3.14. *If a band $\mathcal{S}$ of operators on a Banach space $\mathcal{X}$ has a minimal nonzero component C_A, then it is reducible. In fact, if $\mathcal{M}_1$ and $\mathcal{M}_2$ are the invariant subspaces given for C_A by Proposition 9.3.12, then*

(i) *every subspace $\mathcal{Y}$ of $\mathcal{X}$ satisfying $\mathcal{M}_1 \subseteq \mathcal{Y} \subseteq \mathcal{M}_2$ is invariant for $\mathcal{S}$, and*

(ii) *the quotient of every element of $\mathcal{S}$ induced on $\mathcal{M}_2/\mathcal{M}_1$ is either zero or identity.*

(The description above corresponds to a 3×3 upper-triangular block matrix, where the $(2,2)$ block is 0 or I.)

Proof. We can assume, with no loss of generality, that $0 \in \mathcal{S}$. Then $C_A \cup \{0\}$ is an ideal by Lemma 9.3.13 and the minimality of C_A. Thus the reducibility of $\mathcal{S}$ follows from Proposition 9.3.12, since a nonzero ideal of an irreducible semigroup is irreducible.

It is easily checked that if $\mathcal{J}$ is an ideal of $\mathcal{S}$, then the intersection of kernels of all E in $\mathcal{J}$ and the span of all their ranges are both invariant subspaces for $\mathcal{S}$, so that the $\mathcal{M}_1$ and $\mathcal{M}_2$ of Proposition 9.3.12, are both invariant. For $B \in \mathcal{S}$, there are two possibilities: If $AB = 0$, then the operator induced by B on $\mathcal{M}_2/\mathcal{M}_1$ is clearly zero, since $A \in C_A$ implies that the quotient induced by A is I. If AB is nonzero, then $AB \in C_A$, and thus the quotient induced by AB is I by Proposition 9.3.12.   □

In finite dimensions, the number of components of any band of operators is finite, so the following result generalizes Theorem 2.3.5. It gives a block triangularization in which the diagonal blocks of every member of $\mathcal{S}$ consist of zeros and ones.

Theorem 9.3.15. *Let S be a band of operators on a Banach space $\mathcal{X}$ and assume that its band of components is finite. Then S is triangularizable. In fact, there is a finite chain*

$$\{0\} = \mathcal{X}_0 \subseteq \mathcal{X}_1 \subseteq \cdots \subseteq \mathcal{X}_n = \mathcal{X}$$

of invariant subspaces for S such that, for each E in S, every quotient E_i of E induced on $\mathcal{X}_i / \mathcal{X}_{i-1}$ is either zero or the identity.

Proof. There is nothing to prove if $S = \{0\}$, so assume that $S \neq \{0\}$. Now, $C(S)$ contains a minimal nonzero element C_A, by finiteness. Let $\mathcal{M}_1$ and $\mathcal{M}_2$ be the invariant subspaces given by the preceding proposition. This is the start of an inductive proof on the cardinality of $C(S)$. We will be done if we can show that the induced quotient bands T_1 and T_2 on $\mathcal{M}_1 / \{0\}$ and $\mathcal{X} / \mathcal{M}_2$, respectively, have fewer components than $C(S)$.

If $\hat{E}$ and $\hat{F}$ are in T_1 (with E and F in S), then $\hat{E}\hat{F}\hat{E} \neq \hat{E}$ certainly implies $EFE \neq E$. Thus distinct components of T_1 come from distinct components of S itself, and the cardinality of $C(T_1)$ is at most that of $C(S)$. On the other hand, since the operator $\hat{A}$ induced by A on T_1 is zero, no nonzero component of T_1 is induced by the component C_A. This shows that $C(T_1)$ has fewer elements than $C(S)$. The proof for T_2 is similar. (When we run out of nonzero components, all the remaining induced bands are zero.) $\qquad\square$

Corollary 9.3.16. *Every finitely generated band of operators on a Banach space is triangularizable and has a block triangularization as described in Theorem 9.3.15.*

Proof. Such a band is finite by Theorem 9.3.11 and thus has only a finite number of components. $\qquad\square$

Corollary 9.3.17. *Let S be an arbitrary band of operators on a Banach space $\mathcal{X}$. Then*

(i) *the algebra $\mathcal{A}(S)$ generated by S consists of algebraic operators,*
(ii) *$AB - BA$ is nilpotent for every A and B in $\mathcal{A}(S)$, and*
(iii) *spectrum is submultiplicative and permutable on $\mathcal{A}(S)$.*

Proof. (i) Since each A in $\mathcal{A}(S)$ is a linear combination of a finite number $E_1, \ldots, E_m$ of members of S, we can consider A as an element in $\mathcal{A}(S_0)$, where S_0 is the subband generated by the $\{E_i\}$. But S_0 is finite by Theorem 9.3.11, and thus has a finite chain of invariant subspaces $\mathcal{X}_i$ as described in Theorem 9.3.15. It follows that the corresponding quotients A_i of A induced on the spaces $\mathcal{X}_i/\mathcal{X}_{i-1}$ are all scalars. If these scalars are α_i, then

$$(A - \alpha_i)(A - \alpha_2) \cdots (A - \alpha_n) = 0,$$

so A is algebraic.

(ii) If A is a linear combination of $\{E_1, \ldots, E_m\}$ and B is another operator in $\mathcal{A}(\mathcal{S})$, then B is a linear combination of some $\{F_1, \ldots, F_k\}$ in $\mathcal{S}$, so that A and B both belong to $\mathcal{A}(\mathcal{S}_1)$, where $\mathcal{S}_1$ is the finite subband generated by $\{E_i\} \cup \{F_j\}$. Using Theorem 9.3.15 again, we see that A and B have a simultaneous finite chain of invariant subspaces with scalar quotients. Thus $AB - BA$ has zero quotients and is therefore nilpotent.

(iii) It is clear from the preceding paragraph that $\sigma(AB) \subseteq \sigma(A)\sigma(B)$ for A and B in $\mathcal{A}(\mathcal{S})$. The proof of permutability of σ is similar: Any triple $\{A, B, C\}$ in $\mathcal{A}(\mathcal{S})$ belongs to $\mathcal{A}(\mathcal{S}_2)$ for some finite subband $\mathcal{S}_2$.   □

The corollary above, together with Theorem 9.3.1, shows the existence of an irreducible band $\mathcal{S}$ such that σ is permutable and submultiplicative on $\mathcal{A}(\mathcal{S})$ and the norm closure of $\mathcal{A}(\mathcal{S})$ contains no nonzero compact operators.

Corollary 9.3.18. *Let $\mathcal{S}$ be a band of operators on a Banach space and let $\mathcal{A}(\mathcal{S})$ be the algebra generated by $\mathcal{S}$. If the norm closure of $\mathcal{A}(\mathcal{S})$ contains a nonzero compact operator, then $\mathcal{S}$ is reducible.*

Proof. Let $\mathcal{J}$ be the ideal of compact operators in the norm closure of $\mathcal{A}(\mathcal{S})$. Since spectrum is continuous at any compact operator (Theorem 7.2.13), it follows from the preceding corollary that σ is submultiplicative on $\mathcal{J}$. Since $\mathcal{J}$ is reducible by Theorem 8.3.1, so is $\mathcal{A}(\mathcal{S})$ by Lemma 8.2.1.   □

An easy consequence of this corollary is the following, which generalizes the fact that a band of finite-rank operators is triangularizable (Theorem 8.3.9).

Corollary 9.3.19. *Let $\mathcal{S}$ be a band of operators on a Banach space such that for each $E \in \mathcal{S}$ either E or $I - E$ is of finite rank. Then $\mathcal{S}$ is triangularizable.*

Proof. Since the property assumed for $\mathcal{S}$ is clearly inherited by quotients, we need only verify the reducibility; the proof is then completed by the Triangularization Lemma (7.1.11). If we assume that I belongs to $\mathcal{S}$, which we can do with no loss of generality, then reducibility follows from the preceding corollary.   □

The following easy result is somewhat surprising.

Theorem 9.3.20. *Let S be any band of operators on a Banach space $\mathcal{X}$. Then*

(i) $(EF - FE)^3 = 0$ *for all pairs E and F in S, and*
(ii) *if $(EF - FE)^2 = 0$ for all pairs in S, then S is triangularizable.*

Proof. (i) Actually, more is true for an arbitrary band: For all pairs E and F in S,

$$E(EF - FE)^2 = E(EF + FE - EFE - FEF)$$
$$= EF + EFE - EFE - EF = 0,$$

and similarly, $(EF - FE)^2 E = 0$.

(ii) Now assume $(EF - FE)^2 = 0$ for all pairs. Since this property is inherited by quotients of S, the Triangularization Lemma (7.1.11) implies that we need only show the reducibility of S. We adjoin 0 and I to S, with no loss of generality. Assume that S has a member E with $E \neq 0$, $E \neq I$. By Lemma 9.3.13, the set $\mathcal{J} = \{F \in \mathcal{J} : F \preceq E\}$ is an ideal. Since $E \neq 0$, this ideal is nonzero. We show that it is reducible.

Since $(FE - EF)^2 = 0$, we get

$$(I - E)(FE - EF)^2(I - E)$$
$$= (I - E)(FE + EF - EFE - FEF)(I - E)$$
$$= (I - E)(FE - FEF)(I - E) + (I - E)(EF - EFE)(I - E)$$
$$= (I - E)FEF(I - E) = 0.$$

But $FEF = F$, so that $(I - E)F(I - E) = 0$ for all F in S. Since $E \neq I$, we can use Lemma 8.2.10 with $P = I - E \neq 0$ to conclude that $\mathcal{J}$ has a nontrivial invariant subspace. The reducibility of S now follows by Lemma 8.2.1.   □

9.4 Nonnegative Operators

In this section we merely touch upon decomposability problems for semigroups of nonnegative bounded operators. The setting is still that of Section 8.7: The underlying space will be $\mathcal{X} = \mathcal{L}^p(X, \mu)$, and all definitions and terminology are as in that section. Although there are indecomposable semigroups of nonnegative quasinilpotent operators—even singly generated ones (see the notes and remarks)–certain affirmative results hold. The presence of some compact operators is, of course, expected to yield standard invariant subspaces, as in Theorem 8.7.7, but there are also different hypotheses (e.g., on the measure space X) that imply decomposability and sometimes complete decomposability.

The following simple lemma is quite useful.

Lemma 9.4.1. *If $0 \le A \le B$, then $\|A\| \le \|B\|$ and $\rho(A) \le \rho(B)$.*

Proof. Since $A^n \le B^n$ for all natural n, the spectral radius formula (6.1.10) will give the result if we verify that $0 \le A \le B$ implies $\|A\| \le \|B\|$. Since $Af \le Bf$ for all positive f, and since $0 \le f \le g$ implies $\|f\| \le \|g\|$, it suffices to show that the norm of every nonnegative operator A is achieved on nonnegative vectors. But

$$\|A\| = \sup\{\|Af\| : f \text{ a simple function}, \|f\| = 1\},$$

and, since every simple function f can be written as $\Sigma \lambda_i \chi_i$, with the χ_i characteristic functions of mutually disjoint Borel sets, we get

$$|Af| = |A\Sigma\lambda_i\chi_i| = |\Sigma\lambda_i A\chi_i|$$
$$\le \Sigma|\lambda_i|A\chi_i = A\Sigma|\lambda_i|\chi_i = A|f|.$$

Hence

$$\|Af\| = \|(|Af|)\| \le \|(A|f|)\|,$$

which shows, since the simple function $|f|$ also has norm one, that $\sup\|Af\|$ can be taken over nonnegative simple functions to yield $\|A\|$. $\square$

Recall (Definition 8.4.9) that T is called a strong quasiaffinity if $\overline{T\mathcal{M}} = \mathcal{M}$ for every invariant subspace $\mathcal{M}$ of T. If $\mathcal{M}$ and $\mathcal{N}$ are invariant for such a T, with $\mathcal{M} \subseteq \mathcal{N}$, then the operator induced by T on $\mathcal{N}/\mathcal{M}$ is a strong quasiaffinity.

Theorem 9.4.2. *Let $\mathcal{S}$ be a semigroup of nonnegative quasinilpotent operators. If a member A of $\mathcal{S}$ satisfies $0 \le K \le A$ for a nonzero compact K, then $\mathcal{S}$ is decomposable. If K is a strong quasiaffinity, then $\mathcal{S}$ is completely decomposable.*

Proof. Let

$$\hat{\mathcal{S}} = \{T : 0 \le T \le S \quad \text{for some} \quad S \text{ in } \mathcal{S}\}.$$

Each member of $\hat{\mathcal{S}}$ is quasinilpotent by Lemma 9.4.1. Also, $\hat{\mathcal{S}}$ is a semigroup, for if T_1 and T_2 are in $\hat{\mathcal{S}}$, then $T_1 \le S_1$ and $T_2 \le S_2$ with S_1 and S_2 in $\mathcal{S}$, so that $0 \le T_1T_2 \le S_1S_2$, and $T_1T_2 \in \hat{\mathcal{S}}$. Since this enlarged semigroup contains K, it is decomposable by Corollary 8.7.10.

Now assume that K is a strong quasiaffinity. If $\mathcal{C}$ is a maximal chain of standard invariant subspaces for $\hat{\mathcal{S}}$, and if $\mathcal{M} \in \mathcal{C}$ with $\mathcal{M}_- \ne \mathcal{M}$, then the induced operator $K_\mathcal{M}$ on $\mathcal{M}/\mathcal{M}_-$ is a strong quasiaffinity, so that the induced semigroup $\mathcal{S}_\mathcal{M}$ is decomposable by the preceding paragraph. (Note that $K \le S$ for $S \in \mathcal{S}$ implies $K_\mathcal{M} \le S_\mathcal{M}$ for the corresponding induced operators.) Thus $\mathcal{M}/\mathcal{M}_-$ has dimension one, and the proof is complete by Lemma 8.7.8. $\square$

Conditions on the underlying space $\mathcal{L}^p(X, \mu)$ itself sometimes imply decomposability for an arbitrary semigroup of quasinilpotent operators. Recall that a subset Y of X is called an atom (relative to the measure μ) if $\mu(Y) \neq 0$ and every subset Z of Y has measure $\mu(Y)$ or zero. By the Lindelöf property of X and regularity of μ, both assumed at the outset, an atom Y contains a minimal closed atom. Also, if Y is an atom, we can replace it with one point $y \in Y$ (so that X can be reduced to $(X \backslash Y) \cup \{y\}$).

Note that if Y is an atom, then the standard subspace $\mathcal{L}^p(Y, \mu)$ is one-dimensional. If (X, μ) is atomic, (i.e., X is the union of atoms, necessarily countable in number by the σ-finiteness assumption), then $\mathcal{L}^p(Y, \mu)$ is a weighted l^p space.

Theorem 9.4.3. *Let S be any semigroup of nonnegative quasinilpotent operators on $\mathcal{X} = \mathcal{L}^p(X, \mu)$. If (X, μ) has an atom, then S is decomposable.*

Proof. Let Y be an atom. Let E be the rank-one idempotent with range $\mathcal{L}^p(Y)$ and kernel $\mathcal{L}^p(X \backslash Y)$. Note that both E and $I - E$ are nonnegative operators on $\mathcal{X}$. Next observe that $ESE \leq S$ for every S in $\mathcal{S}$, because

$$S - ESE = (I - E)SE + ES(I - E) + (I - E)S(I - E) \geq 0.$$

Thus it follows from Lemma 9.4.1 that ESE is quasinilpotent. Since ESE is a multiple of the rank-one idempotent E, we deduce that it is zero. Hence $ESE = \{0\}$, and the decomposability of S follows from Lemma 8.7.6 (iii). ☐

Corollary 9.4.4. *If (X, μ) is atomic, then every semigroup of nonnegative quasinilpotent operators on $\mathcal{L}^p(X, \mu)$ is completely decomposable.*

Proof. Let $\mathcal{C}$ be a maximal chain of standard invariant subspaces for $\mathcal{S}$. If $\mathcal{M} \in \mathcal{C}$ and $\mathcal{M}_- \neq \mathcal{M}$, we must show that $\mathcal{M}/\mathcal{M}_-$ is one-dimensional. Let $\mathcal{M} = \mathcal{L}^p(Y, \mu)$ and $\mathcal{M}_- = \mathcal{L}^p(Z, \mu)$, so that the induced semigroup $\mathcal{S}_\mathcal{M}$ can be identified with a semigroup of nonnegative operators on $\mathcal{L}^p(Y \backslash Z)$. Since $Y \backslash Z$ is atomic, we will be done by Theorem 9.4.3 and Lemma 8.7.8 if we prove that $\mathcal{S}_\mathcal{M}$ consists of quasinilpotent operators. To this end, let E be the nonnegative idempotent with range $\mathcal{L}^p(Y \backslash Z)$ and kernel $\mathcal{L}^p((X \backslash Y) \cup Z)$. Then E and $I - E$ are both nonnegative, and an argument similar to the one used in the proof of the preceding theorem shows that $ESE \leq S$ for every S in $\mathcal{S}$, and thus, since S is quasinilpotent, ESE is quasinilpotent by Lemma 9.4.1. ☐

Corollary 9.4.5. *Let $\mathcal{X} = l^p$ and denote its standard basis by $\mathcal{B}$. Let $\mathcal{S}$ be a semigroup of quasinilpotent operators, with nonnegative matrices relative to $\mathcal{B}$. Then there is a total ordering of the basis such that if α, β in $\mathcal{B}$ with $\alpha < \beta$, then the (α, β) entry of the matrix of S is zero for every S in $\mathcal{S}$.*

Proof. This is a direct consequence of the preceding corollary.   □

It is interesting that in the atomic case we do not need quasinilpotence for this discrete triangularization, so long as all diagonal entries are zero.

Theorem 9.4.6. *Let S be any semigroup of operators on l^p whose matrices are nonnegative. If the diagonal of every S in S consists of zeros, then the standard basis B of l^p has a total ordering such that the (α, β) entry of every S in S is zero whenever $\alpha < \beta$.*

Proof. As in the proof of Corollary 9.4.4, we need only show decomposability. (Then the result can be applied to induced semigroups on a maximal chain of standard invariant subspaces, and Lemma 8.7.8 completes the proof.) But the decomposability of S follows from Lemma 8.7.6 (iii).   □

Let A be the backwards shift on l^p: $Ae_1 = 0$ and $Ae_n = e_{n-1}$ for $n \geq 2$, where $\{e_n\}$ is the standard basis. Let S be the semigroup generated by A. Then S satisfies the hypothesis of the preceding theorem and yet contains no quasinilpotent elements.

The next two results concern semigroups of integral operators. Recall that an operator on $\mathcal{L}^p(X, \mu)$ is a *nonnegative integral operator* if there exists a nonnegative-valued, measurable function $\mathcal{K}_T$ on $(X \times X, \mu \times \mu)$, called a *kernel function* for T, such that, for all $f \in \mathcal{L}^p(X, \mu)$,

$$(Tf)(t) = \int \mathcal{K}_T(t, s) f(s) d\mu(s).$$

(The assumption that for almost all t the right-hand side is defined is part of the definition.) It is not hard to verify that if S and T are nonnegative integral operators, then $S \leq T$ if and only if $\mathcal{K}_S \leq \mathcal{K}_T$ almost everywhere.

We need a lemma.

Lemma 9.4.7. *Let X_0 be a Borel subset of X with $0 < \mu(X_0) < \infty$, and let F be a bounded, nonnegative measurable function on $X_0 \times X_0$. If T is the integral operator with kernel function F, then T is compact in $\mathcal{L}^p(\mathcal{X}_0, \mu)$ if $1 < p < \infty$, and T^2 is compact on $\mathcal{L}^1(\mathcal{X}_0, \mu)$.*

Proof. For $p > 1$, this is a special case of the theorem of Hille-Tamarkin [1] (which applies to a class of kernels that properly includes the bounded ones); see also Zaanen [2, p. 320] or Jörgens [1, Theorem 11.6]. In our special case, it is not hard to show that T is a uniform limit of integral operators whose kernels are linear combinations of characteristic functions of measurable rectangles, and the latter integral operators have finite rank.

In the case $p = 1$, T need not be compact, as shown by an example due to von Neumann (see Hille-Tamarkin [1] or Zaanen [2, p. 322]). It has been shown by Zaanen [2, pp. 323-326] that T^2 is compact (under a more general hypothesis than the boundedness of F). □

For singly generated S , the next theorem is a specialization to $\mathcal{L}^p$ of the Ando-Krieger Theorem on Banach lattices; see Section 9.5.

Theorem 9.4.8. *Let S be a semigroup of nonnegative, quasinilpotent operators on $L^p(X, \mu)$ with $p > 1$. If S contains a nonzero integral operator, then S is decomposable.*

Proof. Let A be a nonzero integral operator in S with kernel function $\mathcal{K}_A$. By the regularity and σ-finiteness assumptions on μ, there exists a Borel subset X_0 of X with $0 < \mu(X_0) < \infty$ such that the restriction F_0 of $\mathcal{K}_A$ to $X_0 \times X_0$ is bounded and is not almost everywhere zero. By Lemma 9.4.7, there is a nonzero, compact operator T_0 on $L^p(X_0)$ with kernel function F_0. Extend T_0 to an operator T on $L^p(X_0)$ by defining $Tf = 0$ for all $f \in L^p(X \setminus X_0)$. Also, extend F_0 to a function F on $X \times X$ by defining $F(t, s) = 0$ for all (t, s) off $X_0 \times X_0$. Then clearly, T is a nonnegative operator with kernel function F, and $0 \leq F \leq \mathcal{K}_A$ implies $T \leq A$. Thus the proof is completed by Theorem 9.4.2. □

Theorem 9.4.9. *A semigroup of nonnegative, quasinilpotent integral operators on $\mathcal{L}^p(X, \mu)$ is completely decomposable.*

Proof. We first prove decomposability. In view of Theorem 9.4.8, we need only consider the case $p = 1$. It is not hard to verify that if A is any integral operator with $A^2 \neq 0$ and with nonnegative kernel function $\mathcal{K}_A$, then we can choose a restriction F_0 as in the proof of Theorem 9.4.8 such that the corresponding operator T_0 also satisfies $T_0^2 \neq 0$. (Of course T_0 may not be compact, but T_0^2 is, by Lemma 9.4.7.) Thus the decomposability of S follows if it contains a member A with $A^2 \neq 0$. We now assume that every member of S has square zero.

Pick a nonzero member T of S. If $TST = \{0\}$, we are done by Lemma 8.7.6 (ii). Hence assume that $TBT \neq 0$ for some $B \in S$. Now consider the integral operator $A = TB + T$. Since

$$A^2 = (TB + T)^2 = (TB)^2 + T^2 + T^2B + TBT = 0 + TBT \neq 0,$$

we can find an integral operator T_0 as in the preceding paragraph such that $T_0 \leq A$ and T_0^2 is compact and nonzero. Since $T_0^2 \leq TBT \in S$, the proof of decomposability is completed by Theorem 9.4.2

If $\mathcal{C}$ is a maximal chain of standard invariant subspaces for S, and $\mathcal{M} \in \mathcal{C}$, then $\mathcal{M} = \mathcal{L}^p(X_1)$ and $\mathcal{M}_- = \mathcal{L}^p(X_2)$ with $X_2 \subseteq X_1$. If $\mathcal{M} \neq \mathcal{M}_-$, we must show that $\mathcal{M}/\mathcal{M}_-$ is one-dimensional. For $S \in \mathcal{S}$ with kernel function $\mathcal{K}_S$, the induced operator $S_\mathcal{M}$ on $\mathcal{M}/\mathcal{M}_-$ is naturally identifiable with an integral operator; the restriction of $\mathcal{K}_s$ to $(X_1 \setminus X_2) \times (X_1 \setminus X_2)$ is then a kernel function for $S_\mathcal{M}$. The semigroup $\{S_\mathcal{M} : S \in \mathcal{S}\}$ is decomposable by what we have already proved. Thus $\mathcal{M}/\mathcal{M}_-$ is one-dimensional, and the proof is completed by Lemma 8.7.8. $\qquad\qquad\square$

Our next results concern bands of nonnegative operators. The first lemma below extends Lemma 8.7.12 (which was stated for a singleton band of finite rank).

Lemma 9.4.10. *Let $\mathcal{S}$ be a rectangular band of nonnegative operators on $\mathcal{X} = \mathcal{L}^p(X, \mu)$.*

(i) *Assume that $Sh = 0$ with $0 \leq h \in \mathcal{X}$ implies $h = 0$ and that $\phi|_{S\mathcal{X}} = 0$ with $0 \leq \phi \in \mathcal{X}^*$ implies $\phi = 0$. If for some Borel set $Y \subseteq X$, $\mathcal{L}^p(Y)$ is invariant under $\mathcal{S}$, then so is $\mathcal{L}^p(X \setminus Y)$.*

(ii) *In general, $\mathcal{X}$ has a decomposition $\mathcal{L}^p(Y_1) \oplus \mathcal{L}^p(Y_2) \oplus \mathcal{L}^p(Y_3)$ relative to which the matrix of each member of $\mathcal{S}$ is of the form*

$$\begin{pmatrix} 0 & A_S F_S & A_S F_S B_S \\ 0 & F_S & F_S B_S \\ 0 & 0 & 0 \end{pmatrix},$$

where $A_S \geq 0$, $B_S \geq 0$, and $\mathcal{S}_0 = \{F_S : S \in \mathcal{S}\}$ is a rectangular band satisfying (i) above.

Proof. (i) Relative to the direct sum decomposition $\mathcal{L}^p(Y) \oplus \mathcal{L}^p(X \setminus Y)$, a typical member of $\mathcal{S}$ has a matrix

$$\begin{pmatrix} E_1 & A \\ 0 & E_2 \end{pmatrix};$$

we must show that $A = 0$. For arbitrary F and G in $\mathcal{S}$,

$$FEG = \begin{pmatrix} F_1 & B \\ 0 & F_2 \end{pmatrix} \begin{pmatrix} E_1 & A \\ 0 & E_2 \end{pmatrix} \begin{pmatrix} G_1 & C \\ 0 & G_2 \end{pmatrix}$$

$$= \begin{pmatrix} F_1 E_1 G_1 & F_1 E_1 C + F_1 A G_2 + B E_2 G_2 \\ 0 & F_2 E_2 G_2 \end{pmatrix}.$$

Setting $D = F_1 E_1 C + F_1 A G_2 + B E_2 G_2$ and using the fact that $(FEG)^2 = FEG$, we get $F_1 E_1 G_1 D + D F_2 E_2 G_2 = D$, so

$$F_1 E_1 G_1 (F_1 E_1 G_1 D + D F_2 E_2 G_2) F_2 E_2 G_2 = F_1 E_1 G_1 (D) F_2 E_2 G_2,$$

which yields $F_1 E_1 G_1 (D) F_2 E_2 G_2 = 0$. Noting that $0 \leq F_1 A G_2 \leq D$, we deduce that

$$F_1 E_1 G_1 (F_1 A G_2) F_2 E_2 G_2 = (F_1 E_1 G_1 F_1) A (G_2 F_2 E_2 G_2) = 0,$$

or, by rectangularity, that $F_1 A G_2 = 0$ for all F and G in $\mathcal{S}$. Now, as in the proof of Lemma 8.7.12, the hypothesis implies first that $A G_2 = 0$ for all $G \in \mathcal{S}$, and then $A = 0$.

(ii) The argument is very similar to that of the proof of (ii) in Lemma 8.7.12: Let $\mathcal{L}^p(Y_1)$ be the maximal standard subspace annihilated by all the members of $\mathcal{S}$ and let $\mathcal{L}^p(Y_1 \cup Y_2)$ be the minimal standard subspace containing $\mathcal{L}^p(Y_1)$ and the ranges of all S in $\mathcal{S}$. We then get the 3×3 block matrix for each S claimed above, and the set $\mathcal{S}_0$ is certainly a band of nonnegative operators on $\mathcal{L}^p(Y_2)$. It is also easily seen to be rectangular. Now, if $h \in \mathcal{L}^p(Y_2)$ with $\mathcal{S}_0 h = \{0\}$, then $F_S h = 0$ implies $A_S F_S h = 0$ and thus $Sh = 0$ for all $S \in \mathcal{S}$, yielding $h = 0$ by the maximality of $\mathcal{L}^p(Y_1)$. The fact that $\phi|_{\mathcal{S}_0 \mathcal{L}^p(Y_2)} = 0$ with $0 \le \phi \in \mathcal{L}^p(Y_2)^*$ implies that $\phi = 0$ can be proven in a similar fashion. $\qquad\square$

The above lemma does not imply decomposability in the case (i). For example, a rank-one idempotent $f \otimes \phi$ with f and ϕ strictly positive and $\phi(f) = 1$ is indecomposable (as we have already seen).

Observe that if $\mathcal{S}$ is a band, and if $A, B \in \mathcal{S}$ with $A \preceq B$ in the notation of Section 9.3, then rank $A \le$ rank B, where infinite values for the rank are permitted. (This is an easy consequence of rank $A =$ rank $(ABA) \le$ rank B.) Thus each component of $\mathcal{S}$ has constant (finite or infinite) rank.

Definition 9.4.11. The *rank* of a rectangular band $\mathcal{S}$ is the rank of any member of $\mathcal{S}$.

Theorem 9.4.12. *Let $\mathcal{S}$ be a band of nonnegative operators on $\mathcal{L}^p(X, \mu)$. If $\mathcal{S}$ has a minimal nonzero component of rank at least 2, then $\mathcal{S}$ is decomposable.*

Proof. By hypothesis, there is a nonzero $E \in \mathcal{S}$ such that for every nonzero S in $\mathcal{S}$ we have $ESE = E$. Consider the singleton (and thus rectangular) band $\{E\}$ and apply Lemma 9.4.10 to it to get $\mathcal{L}^p(X) = \mathcal{L}^p(Y_1) \oplus \mathcal{L}^p(Y_2) \oplus \mathcal{L}^p(Y_3)$ and

$$E = \begin{pmatrix} 0 & AE_0 & AE_0 B \\ 0 & E_0 & E_0 B \\ 0 & 0 & 0 \end{pmatrix},$$

where E_0 is an idempotent on $\mathcal{L}^p(Y_2)$. Observe that rank $E_0 =$ rank E. Now, the proof of the fact that E_0 is decomposable is precisely that given in Lemma 8.7.12 (which did not use the finiteness of r). Thus, by Lemma 9.4.10, $\mathcal{L}^p(Y_2)$ has complementary subspaces $\mathcal{L}^p(Z_1)$ and $\mathcal{L}^p(Z_2)$, both invariant under E_0, such that

$$E_0|_{\mathcal{L}^p(Z_1)} \ne 0 \quad \text{and} \quad E_0|_{\mathcal{L}^p(Z_2)} \ne 0.$$

With respect to the decomposition $\mathcal{L}^p(X) = \mathcal{L}^p(Y_1 \cup Z_1) \oplus \mathcal{L}^p(Z_2 \cup Y_3)$, the matrix of E is of the form

$$\begin{pmatrix} E_1 & C \\ 0 & E_2 \end{pmatrix},$$

where E_1 and E_2 are both nonzero.

We now use the relation $ESE = E$ for all nonzero S in $\mathcal{S}$. Writing the matrix of $S \neq 0$ in the new decomposition of E, we have

$$\begin{pmatrix} E_1 & C \\ 0 & E_2 \end{pmatrix} \begin{pmatrix} S_{11} & S_{12} \\ S_{21} & S_{22} \end{pmatrix} \begin{pmatrix} E_1 & C \\ 0 & E_2 \end{pmatrix} = \begin{pmatrix} E_1 & C \\ 0 & E_2 \end{pmatrix},$$

which implies $E_2 S_{21} E_1 = 0$. This equation holds for $S = 0$ also. Thus for all S in $\mathcal{S}$,

$$\begin{pmatrix} 0 & 0 \\ 0 & E_2 \end{pmatrix} \begin{pmatrix} S_{11} & S_{12} \\ S_{21} & S_{22} \end{pmatrix} \begin{pmatrix} E_1 & 0 \\ 0 & 0 \end{pmatrix} = 0,$$

yielding the decomposability of $\mathcal{S}$ by Lemma 8.7.6 (ii). $\qquad\square$

The following corollary includes the case of bands on finite-dimensional spaces.

Corollary 9.4.13. *Let $\mathcal{S}$ be a band of nonnegative operators on $\mathcal{X} = \mathcal{L}^p(X, \mu)$. Either of the following conditions implies decomposability of $\mathcal{S}$:*

(i) *The number of components in $\mathcal{S}$ is finite, and each component has rank at least 2.*

(ii) *The minimal rank in $\mathcal{S}$ is finite and greater than one.*

Proof. (i) Since there are minimal components by hypothesis, the assertion follows directly from Theorem 9.4.12.

(ii) Pick a member E of minimal rank r and consider the component $\mathcal{C}_E$ of $\mathcal{S}$. Note that $\mathcal{C}_E$ is minimal, because if $F \in \mathcal{S}$ and $EF \neq 0$, then EFE has the same range and same kernel as E, and thus $EFE = E$. Conclusion (ii) thus also follows from Theorem 9.4.12. $\qquad\square$

The following result gives the structure of a rectangular band. It is particularly interesting in the "nondegenerate" case, which yields a "direct integral" of rank-one idempotents for each member of the band.

Theorem 9.4.14. *Let $\mathcal{S}$ be a rectangular band of nonnegative operators on $\mathcal{X} = \mathcal{L}^p(X, \mu)$ and let rank $\mathcal{S} = r \leq \infty$. Then $\mathcal{S}$ has a chain*

$$\{0\} = \mathcal{M}_0 \subset \mathcal{M}_1 \subset \cdots \subset \mathcal{M}_r = \mathcal{X}$$

of distinct, standard invariant subspaces. Furthermore, the following hold:

(i) *If $\mathcal{S}h = 0$ with $0 \leq h \in \mathcal{X}$ implies $h = 0$ and also $\phi|_{\mathcal{S}\mathcal{X}} = 0$ with $0 \leq \phi \in \mathcal{X}^*$ implies $\phi = 0$, then there is an s with $0 \leq s \leq r$ and standard subspaces*

$$Y, \mathcal{Z}_1, \mathcal{Z}_2, \ldots, \mathcal{Z}_s,$$

all invariant under S, whose direct sum is $\mathcal{X}$, such that $S|_\mathcal{Y}$ is completely decomposable with a continuous chain and $S|_{\mathcal{Z}_i}$ is an indecomposable, rectangular band of rank one for each i.

(ii) With the hypothesis of the preceding paragraph, if r is finite, then $s = r$ and $\mathcal{Y} = \{0\}$.

Proof. By the second part of Lemma 9.4.10, it suffices to prove (i) and (ii) only.

Suppose the hypothesis of (i) holds. We first treat the case $r = 1$. We must show that S is indecomposable (so that $s = 1$ and $\mathcal{Y} = \{0\}$). Assuming that a nontrivial $\mathcal{L}^p(Z)$ is invariant under S, then so is $\mathcal{L}^p(X \setminus Z)$ (by Lemma 9.4.10). By hypothesis, S has members E and F with matrices

$$E = \begin{pmatrix} E_1 & 0 \\ 0 & E_2 \end{pmatrix} \quad \text{and} \quad F = \begin{pmatrix} F_1 & 0 \\ 0 & F_2 \end{pmatrix},$$

where $E_1 \neq 0$ and $F_2 \neq 0$. Since $EFE = E$ implies $E_1 F_1 E_1 = E_1$, we conclude that $F_1 \neq 0$. Then rank $F =$ rank F_1+ rank $F_2 \geq 2$, which is a contradiction. This disposes of the case $r = 1$.

Now assume $r \geq 2$. Let Ω be a maximal collection of mutually disjoint nontrivial Borel subsets Z of X such that $S|_{\mathcal{L}^p(Z)}$ is indecomposable (and therefore of rank one by Theorem 9.4.12). Then Ω is countable and may, of course, be empty. Let $\Omega = \{Z_i\}$ and $\mathcal{Z}_i = \mathcal{L}^p(Z_i)$ for each i. If $Y = X \setminus \cup Z_i$, then the standard subspace $\mathcal{Y} = \mathcal{L}^p(Y)$ is invariant, by Lemma 9.4.10, and $S|_\mathcal{Y}$ is a rectangular band satisfying the hypothesis in (i).

Next let $\mathcal{C}$ be a maximal chain of standard invariant subspaces for $S|_\mathcal{Y}$. If $\mathcal{M} \in \mathcal{C}$, then $\mathcal{M} = \mathcal{L}^p(X_1)$ and $\mathcal{M}_- = \mathcal{L}^p(X_2)$ with $X_2 \subseteq X_1$. We must show that $\mathcal{M} = \mathcal{M}_-$ (or, equivalently, $\mu(X_1 \setminus X_2) = 0$). Suppose otherwise. Applying Lemma 9.4.10 twice, we see that $\mathcal{L}^p(X_2)$, $\mathcal{L}^p(X_1 \setminus X_2)$ and $\mathcal{L}^p(Y \setminus X_2)$ are all invariant under $S|_\mathcal{Y}$ and, relative to the corresponding decomposition of $\mathcal{L}^p(Y)$, every $S \in S|_\mathcal{Y}$ has a matrix

$$S = \begin{pmatrix} S_1 & 0 & 0 \\ 0 & S_2 & 0 \\ 0 & 0 & S_3 \end{pmatrix},$$

where S_2 is naturally identified with the induced operator $S_\mathcal{M}$ on $\mathcal{M}/\mathcal{M}_-$. If $X_1 \setminus X_2$ has positive measure, then the set $\{S_2 : S \in S\}$ is an indecomposable, rectangular band by construction. The existence of nontrivial $\mathcal{L}^p(X_1 \setminus X_2)$ thus contradicts the maximality of Ω. Hence $\mathcal{M} = \mathcal{M}_-$ for every $\mathcal{M}$ in $\mathcal{C}$, and $\mathcal{C}$ is a continuous chain.

To prove (ii), observe that if $\mathcal{Y} \neq \{0\}$, then the existence of the continuous chain $\mathcal{C}$ on it implies that it is infinite-dimensional; hence, we can pick distinct members $\mathcal{M}_i$ of $\mathcal{C}$ with

$$\mathcal{M}_0 \subset \mathcal{M}_1 \subset \mathcal{M}_2 \subset \cdots \subset \mathcal{M}_t$$

and $t > s$. Let $\mathcal{M}_i = \mathcal{L}^p(Y_i)$ for each i and note that each $\mathcal{L}^p(Y_i \setminus Y_{i-1})$ is invariant under $\mathcal{S}$, so each S in $\mathcal{S}$ restricts to a nonzero idempotent on $\mathcal{L}^p(Y_i \setminus Y_{i-1})$ for each i. This contradicts the fact that S has rank r. Thus $\mathcal{Y} = \{0\}$. The claim $r = s$ now follows easily. $\square$

The following example shows that the hypotheses of Theorem 9.4.12 and Corollary 9.4.13 are essential, even in the case of abelian bands on atomic spaces.

Example 9.4.15. *There exists an abelian, indecomposable band of non-negative operators with no minimal component other than zero.*

Proof. Let $\mathcal{X}$ be the Hilbert space l^2, and let $\{e_i\}$ be the standard basis for $\mathcal{X}$. We present our operators by describing their matrices in this basis. Fix the 2×2 idempotent matrix $P = \begin{pmatrix} 1/2 & 1/2 \\ 1/2 & 1/2 \end{pmatrix}$ and denote by $P^{(n)}$ the n-fold tensor product $P \otimes P \otimes \cdots \otimes P$. For each n, let S_n be the ampliation of $P^{(n)}$; i.e.,

$$S_n = P^{(n)} \oplus P^{(n)} \oplus \cdots .$$

Then $\mathcal{S} = \{S_1, S_2, \dots\}$ is an abelian band. To see this, note that S_n is the ampliation of the $2^n \times 2^n$ matrix whose elements are all $1/2^n$; thus $m \leq n$ implies $S_m S_n = S_n S_m = S_n$. It follows that $S_n S_m S_n = S_n$ if and only if $m \leq n$. Hence each component of $\mathcal{S}$ is a singleton $\mathcal{C}_n = \{S_n\}$, with $\mathcal{C}_n \leq \mathcal{C}_m$ if and only if $m \leq n$.

Every S_n clearly has infinite rank. The indecomposability of $\mathcal{S}$ follows from the fact that $\mathcal{S}$ contains, for arbitrary n, operators with $n \times n$ upper left sections all of whose entries are positive. $\square$

9.5 Notes and Remarks

The transitive algebra problem was first raised in the early 1960s by R.V. Kadison. Although there are some affirmative results under various special hypotheses (see Radjavi-Rosenthal [2, Chap. 8]), the general problem remains elusive. With respect to invariant subspaces of individual operators, there are several existence theorems in addition to those related to Lomonosov's. Scott Brown [1], answering a long-standing question raised by Paul Halmos, proved that subnormal operators (operators that are restrictions of normal operators to invariant subspaces) have nontrivial invariant subspaces. This powerful theorem has been substantially generalized to apply to operators on Hilbert space of norm 1 whose spectra include the unit circle (Brown-Chevreau-Pearcy [1], [2]). There are also results for some operators that are perturbations of certain normal operators; see Radjavi-Rosenthal [2, Chap. 6].

The construction given in Theorems 9.1.1 and 9.1.6 are from Hadwin-Nordgren-Radjabalipour-Radjavi-Rosenthal [1]. These constructions are slightly generalized in Hadwin-Nordgren-Radjabalipour-Radjavi-Rosenthal [2], which also contains Theorems 9.1.2 and 9.1.4 and Corollary 9.1.5. (The proof that the index of nilpotence is bounded is from Grabiner [1].) Grabiner [1] also shows that a weakly closed algebra of operators that contains a maximal abelian self-adjoint algebra is triangularizable if it is commutative modulo its radical. Corollary 9.1.8 is in Cigler-Drnovšek-(Kokol-Bukovsek)-Laffey-Omladič-Radjavi-Rosenthal [1], which also contains an affirmative result for semigroups of operators with spectra $\{1\}$ such that the collection of translates by $-I$ forms a set of nilpotent operators of bounded index. Example 9.1.9 is from Radjavi [6]. It is shown in Radjavi-Rosenthal-Shulman [1] that a strongly compact group of operators on a Banach space is abelian if $AB - BA$ is quasinilpotent for all A and B in the group.

Lemma 9.2.2 is due to Anderson [1]; it solved several problems on commutators that had been open for some time. The application of Anderson's lemma to Theorem 9.2.3 is from Laurie-Nordgren-Radjavi-Rosenthal [1], in which Theorems 9.2.4 and 9.2.5 are also established. Lemma 9.2.6 is from Choi-Laurie-Radjavi [1]. Theorems 9.2.12 and 9.2.13 were proven by a group of people at a workshop in Bled in May 1996, and appear in Cigler-Drnovšek-(Kokol-Bukovsek)-Laffey-Omladič-Radjavi-Rosenthal [1].

The counterexample given in Theorem 9.3.1 is due to Drnovšek [2]; it settles a question that was raised several years earlier, after affirmative results in the case of finite-rank operators had been obtained. The results leading to Theorem 9.3.11 are essentially in several texts on abstract semigroups, including Petrich [1]; the arrangement here was made in collaboration with L. Livshits, G. MacDonald, and B. Mathes. Theorem 9.3.11 is a special case of results of Green-Rees [1], which studies the following more general problem. Let r be a fixed positive integer and let $\mathcal{S}$ be a semigroup such that $a^r = a$ for every a in $\mathcal{S}$. If $\mathcal{S}$ is finitely generated, must it necessarily be finite? It is proved in Green-Rees [1] that this problem is equivalent to the corresponding Burnside problem: Is a finitely generated group of exponent $r - 1$ necessarily finite? The latter question has been answered negatively in general, but affirmatively for exponents up to 4. It seems to be unknown for $r = 5$. For a good historical account, see Lam [2]. A nice exposition of the Burnside and "restricted Burnside" problems is in Zelmanov [1]. In the case $r = 2$ (that is, the band case), Green-Rees [1] gives a recursive formula for the number of elements in a band generated by n members.

Theorem 9.3.12 is an adaptation to the Banach space setting of a result of Livshits-MacDonald-Mathes-Radjavi [1], as are Theorems 9.3.14, 9.3.15, 9.3.16, and 9.3.17. Corollary 9.3.18 is from Livshits-MacDonald-Mathes-Radjavi [2], which also contains Corollary 9.3.19 and Theorem 9.3.20. There are other results in the last two papers. The second paper, for example, also treats the case of bands on a vector space (without topology). It turns out that, in this case, every band is triangularizable; i.e., irreducible representations of abstract bands on vector spaces are trivial. A more recent paper, Drnovšek-Livshits-MacDonald-Mathes-Radjavi-Šemrl [1], proves the rather surprising fact that for each $\varepsilon > 0$ there exists an irreducible band on Hilbert space that is uniformly bounded by $1 + \varepsilon$. (Recall that a band of uniform norm 1 on Hilbert space is necessarily abelian and thus highly reducible.) In the Banach-space case the bound can even be 1.

The singly generated case of Theorem 9.4.2 is a special case of results of Abramovich-Aliprantis-Burkinshaw [2]; the general case is part of Drnovšek [3]. Corollaries 9.4.4 and 9.4.5 are essentially contained in Choi-Nordgren-Radjavi-Rosenthal-Zhong [1]. These results and Theorem 9.4.6 have been generalized in Drnovšek [5]. Theorems 9.4.8 and 9.4.9 are true in a more general setting: Drnovšek [4] shows that Turovskii's Theorem, together with the Abramovich-Aliprantis-Burkinshaw results, implies that a semigroup of nonnegative abstract kernel operators on any Dedekind-complete Banach lattice is decomposable. The well-known Ando-Krieger Theorem (Schaefer [1, p. 335]) is the singly generated version of this assertion. A short proof of the Ando-Krieger Theorem is given in Grobler [2].

Not every quasinilpotent, nonnegative operator on $\mathcal{L}^p(X, \mu)$ is decomposable; there is a counterexample in Schaefer [2]. Interesting examples are given in Zhong [2] of semigroups of nilpotent, nonnegative operators on $\mathcal{L}^2[0, 1]$ that are not only indecomposable, but actually irreducible (see also Theorem 9.1.1). Other connections with reducibility are explored in Zhong [1]; it is proved, for example, that a semigroup of nonnegative compact integral operators on $\mathcal{L}^2(X, \mu)$ on which ρ is submultiplicative is reducible. Jahandideh ([1], [2]) includes several extensions of the Abramovich-Aliprantis-Burkinshaw results in the setting of Banach lattices; a sample result in a special case is that a quasinilpotent, nonnegative operator on $C_0(X)$, where X is a locally compact Hausdorff space, is completely decomposable.

There are some (necessarily weak) extensions of the Perron-Frobenius Theorem to the noncompact case. A theorem of Krein (see Krein-Rutman [1]) states that the adjoint of a positive operator on $C(X)$, with X compact Hausdorff, has a positive eigenvector corresponding to a nonnegative eigenvalue. A result of Krein-Rutman [1] asserts that for any positive operator T on any Banach lattice, $\rho(T) \in \sigma(T)$. A nice exposition of these ideas is in Grobler [1]. A thorough discussion of peripheral spectra can be found in Schaefer [1, pp.322–333].

The structure of rectangular bands of nonnegative operators (Lemma 9.4.10) is essentially contained in Marwaha ([1], [2]), as are most of the other results of this section on bands; the proofs, originally given for $\mathcal{L}^2$, have been adapted here to the $\mathcal{L}^p$ case.

References

Y.A. Abramovich, C.D. Aliprantis, and O. Burkinshaw.

1. *Invariant subspace theorems for positive operators*, J. Funct. Anal. **124** (1994),95-111.
2. *The invariant subspace problem: some recent advances*, Rend. Inst. Mat. Univ. Triste **29** (1998),3-79.

C.D. Aliprantis and O. Burkinshaw.

1. *Positive Operators*, Academic Press, Orlando, 1985.

G.D. Allen, D.R. Larson, J.D. Ward and G. Woodward

1. *Similarity of nests in L_1*, J. Funct. Anal. **92** (1990), 49-76.

N.T. Andersen.

1. *Compact perturbations of reflexive algebras*, J. Funct. Anal. **38** (1980), 366-400.

J. Anderson.

1. *Commutators of compact operators*, J. Reine Angew. Math. **291** (1977), 128-132.

T. Ando

1. *Positive operators in semi–ordered linear spaces*, J. Fac. Sci. Hokkaido Univ. Ser. I **13** (1957), 214-228.

N. Aronszajn and K.T. Smith.

1. *Invariant subspaces of completely continuous operators*, Ann. of Math. **60** (1954), 345-350.

W.B. Arveson.

1. *A density theorem for operator algebras*, Duke Math J. **34** (1967), 635-648.
2. *Analyticity in operator algebras*, Amer. J. Math. **89** (1967), 578-642.
3. *Operator algebras and invariant subspaces*, Ann. of Math. (2) **100** (1974), 433-532.
4. *Ten lectures on operator algebras*, CBMS Regional Conference Series in Mathematics, 55, American Mathematical Society, Providence, 1984.
5. *Notes on extensions of C*-algebras*, Duke Math. J. **44** (1977), 329-355.

H. Auerbach

1. *Sur les groupes bornés de substitutions linéaires*, C. R. Acad. Sci. Paris **195** (1932), 1367-1369.

B. Aupetit.

1. *Caractérisation spectrale des algèbres de Banach commutative*, Pacific J. Math. **63** (1976), 23-35.
2. *Propriétés spectrales des algèbres de Banach*, Lecture Notes in Math. 735, Springer, Berlin, 1979.
3. *A Primer on Spectral Theory*, Universitext, Springer-Verlag, New York, 1991.

E.A. Azoff.

1. *On finite rank operators and preannihilators*, Mem. Amer. Math. Soc. **64** (1986).

S. Banach.

1. *Théorie des Opérations Linéaire* , Monografje Matematyczne, Warsaw, 1932.

R.B. Bapat and T.E.S. Raghavan.

1. *Nonnegative Matrices and Applications*, Encyclopedia of Mathematics and its Applications 64, Cambridge Univ. Press, Cambridge, 1997.

G.P. Barker, L.Q. Eifler, and T.P. Kezlan.

1. *A non-commutative spectral theorem*, Linear Algebra and Appl. **20** (1978), 95-100.

B.A. Barnes.

1. *Density theorems for algebras of operators and annihilator Banach algebra*, Mich. Math. J. **19** (1972), 149-155.

B.A. Barnes and A. Katavolos.

1. *Properties of quasinilpotents in some operator algebras*, Proc. Roy. Irish Acad. Sect. A **93** (1993) 155-170.
2. *Corrigendum to: "Properties of quasinilpotents in some operator algebras,"* Proc. Roy. Irish Acad. Sect. A **95** (1995), 249-250.

H. Bart and R.A. Zuidwijk.

1. *Simultaneous reduction to triangular forms after extension with zeroes*, Linear Algebra Appl. **281** (1998), 105-135.

W. Barth and G. Elencwajg.

1. *Concernant la cohomologie des fibres algebriques stables sur $P_n(C)$*, Variétés analytiques compactes (Colloq., Nice, 1977), Lecture Notes in Math., 683, Springer, Berlin, 1978, 1-24.

R. Bellman.

1. *Introduction to Matrix Analysis*, McGraw–Hill, New York, 1970.

H. Bercovici.

1. *Notes on invariant subspaces*, Bull. Amer. Math. Soc. **23** (1990), 1-33.

M.A. Berger and Y. Wang.

1. *Bounded semigroups of matrices*, Linear Algebra Appl. **166** (1992), 21-27.

A. Berman and R.J. Plemmons.

1. *Nonnegative Matrices in Mathematical Sciences*, Academic Press, New York, 1979.

A.R. Bernstein and A. Robinson.

1. *Solution of an invariant subspace problem of K. T. Smith and P. R. Halmos*, Pacific J. Math. **16** (1966), 421-431.

R. Bhatia.

1. *Matrix Analysis*, Springer-Verlag, New York, 1997.

B. Bollobás.

1. *Linear Analysis. An Introductory course*, Cambridge Univ. Press, Cambridge, 1990.

F.F. Bonsall.

1. *Operators that act compactly on an algebra of operators*, Bull. London Math. Soc. **1** (1969), 163-170.

N. Bourbaki.

1. *Lie Groups and Lie Algebras. Chapters 1-3*, Elements of Mathematics, Springer-Verlag, Berlin-Heidelberg-New York, 1989.

M. Boyle and D. Handelman.

1. *The spectra of nonnegative matrices via symbolic dynamics*, Ann. of Math. (2) **133** (1991), 249-316.

M.S. Brodskiĭ.

1. *Triangular and Jordan Representations of Linear Operator*, Translations of Mathematical Monographs, Vol. 32, American Mathematical Society, Providence, 1971.

L.G. Brown, R.G. Douglas, and P.A. Fillmore.

1. *Unitary equivalence modulo the compact operators and extensions of C*-algebras*, Proceedings of a Conference on Operator Theory (Dalhousie Univ., Halifax, N.S., 1973), Lecture Notes in Math., Vol. 345, Springer, Berlin, 1973, pp. 58-128.
2. *Extensions of C*-algebras and K-homology*, Ann. of Math. (2) **105** (1977), 265-324.

S.W. Brown.

1. *Some invariant subspaces for subnormal operators*, Integral Equations Operator Theory **1** (1978), 310-333.
2. *Lomonosov's theorem and essentially normal operators*, New Zealand J. Math. **23** (1994), 11-18.

S.W. Brown, B. Chevreau, and C. Pearcy.

1. *Contractions with rich spectrum have invariant subspaces*, J. Operator Theory **1** (1979), 123-136.
2. *On the structure of contraction operators. II*, J. Funct. Anal. **7** (1988), 30-55.

J.J. Buoni.

1. *A trace inequality for functions of triangular Hilbert–Schmidt operators*, Czechoslovak Math. J. **25 (100)** (1975), 4-7.

W. Burnside.

1. *On the condition of reducibility of any group of linear substitutions*, Proc. London Math. Soc. **3** (1905), 430-434.
2. *On criteria for the finiteness of the order of a group of linear substitutions*, Proc. London Math. Soc. **3** (1905), 435-440.
3 *Theory of Groups of Finite Order*, Second Edition, Dover, New York, 1955.

D. Carlson and S. Pierce.

1. *Common eigenvectors and quasicommutativity of sets of simultaneously triangularizable matrices*, Linear Algebra Appl. **71** (1985), 49-55.

288   References

V. Caselles.

1. *An extension of Ando-Krieger's Theorem to ordered Banach spaces*, Proc. Amer. Math. Soc. **103** (1988), 1070-1072.

B. Chevreau, W.S. Li and C. Pearcy.

1. *A new Lomonosov lemma*, J. Operator Theory **40** (1998), 409-417.

B. Chevreau and C. Pearcy.

1. *On the structure of contraction operators. I*, J. Funct. Anal. **76** (1988), 1-29.

M.D. Choi, C. Laurie and H. Radjavi.

1. *On commutators and invariant subspaces*, Linear and Multilinear Algebra, **9** (1980/81), 329-340.

M.D. Choi, E.A. Nordgren, H. Radjavi, P. Rosenthal and Y. Zhong.

1. *Triangularizing semigroups of quasinilpotent operators with nonnegative entries*, Indiana Univ. Math. J. **42** (1993), 15-25.

E. Christensen.

1. *On invertibility preserving linear mappings, simultaneous triangularization and property L*, to appear.

G. Cigler, R. Drnovšek, D. Kokol-Bukovšek, T. J. Laffey, M. Omladič, H. Radjavi, and P. Rosenthal.

1. *Invariant subspaces for semigroups of algebraic operators*, J. Funct. Anal. **160** (1998), 452-465.

J. M. Clauss.

1. *Elementary chains of invariant subspaces of a Banach space*, Canad. J. Math. **47** (1995), 290-301.

J. B. Conway.

1. *A Course in Functional Analysis*, Second Edition, Springer-Verlag, New York, 1990.

J. Daughtry.

1. *An invariant subspace theorem*, Proc. Amer. Math. Soc. **49** (1975), 267-268.

K. Davidson.

1. *Nest algebras.*, Pitman, New York, 1988.
2. *Similarity and compact perturbations of nest algebras*, J. Reine. Angew. Math. **348** (1984), 286–294.

A.M. Davie.

1. *The approximation problem for Banach spaces*, Bull. London Math. Soc. **5** (1973), 261-266.

C. Davis.

1. *Generators of the ring of bounded operators*, Proc. Amer. Math. Soc. **6** (1955), 970-972.

B. de Pagter.

1. *Irreducible compact operators*, Math. Z. **192** (1986), 149-153.

M.A. Dehghan and M. Radjabalipour.

1. *Matrix algebras and Radjavi's trace condition*, Linear Algebra Appl. **148** (1991), 19-25.

E. Deutsch and H. Schneider.

1. *Bounded groups and norm Hermitian matrices*, Linear Algebra Appl. **9** (1974), 9-27.

J. Dieudonné.

1. *Foundations of Modern Analysis*, Academic Press, New York, 1969.

W.F. Donoghue.

1. *The lattice of invariant subspaces of a completely continuous quasi-nilpotent transformation*, Pacific J. Math. **7** (1957), 1031-1035.

M.P. Drazin.

1. *Some generalizations of matrix commutativity*, Proc. London. Math. Soc. (3) **1** (1951), 222-231.

M.P. Drazin, J. W. Dungey, and K. W. Gruenberg.

1. *Some theorems on commutative matrices*, J. London Math. Soc. **26** (1951), 221-228.

R. Drnovšek.

1. *On reducibility of semigroups of compact quasinilpotent operators*, Proc. Amer. Math. Soc. **125** (1997), 2391-2394.
2. *An irreducible semigroup of idempotents*, Studia Math. **125** (1997), 97-99.
3. *Common invariant subspaces for collections of operators*, preprint.
4. *A generalization of the Ando-Krieger theorem*, preprint.

R. Drnovšek, L. Livshits, G. MacDonald, B. Mathes, H. Radjavi, and P. Šemrl.

1. *On operator bands*, preprint.

N. Dunford and J.T. Schwartz.

1. *Linear Operators. Part I: General Theory*, Interscience, New York, 1957.
2. *Linear Operators. Part II: Spectral Theory*, Interscience, New York, 1963.
3. *Linear Operators. Part III: Spectral Operators*, Interscience, New York, 1971.

P. Enflo.

1. *A counterexample to the approximation problem in Banach spaces*, Acta Math. **130** (1973), 309-317.
2. *On the invariant subspace problem in Banach spaces*, Acta Math. **158** (1987), 213-313.

J.A. Erdos.

1. *Operators of finite rank in nest algebras*, J. London Math. Soc. **43** (1968), 391-397.
2. *On the trace of a trace class operator*, Bull. London Math. Soc. **6** (1974), 47-50.

P. Fillmore, G. MacDonald, M. Radjabalipour, and H. Radjavi.

1. *Towards a classification of maximal unicellular bands*, Semigroup Forum **49** (1994), 195-215.
2. *On principal-ideal bands*, Semigroup Forum, to appear.

I. Fredholm.

1. *Sur une classe d'équations fonctionelles*, Acta Math. **27** (1903), 365-390.

G. Frobenius.

1. *Über vertauschbare Matrizen*, Sitzungsberichte der Preussischen Akademie der Wissenschaften zu Berlin (1896), 601-614.
2. *Über Matrizen aus nicht negativen Elementen*, Sitzungsberichte der Preussischen Akademie der Wissenschaften zu Berlin (1912), 456-477.

G. Frobenius and I. Schur.

1. *Über die Äquivalenz der Gruppen linearer Substitutionen*, Sitzungsber. Preuss. Akad. Wiss. (1906), 209-217.

L.S. Goddard and H. Schneider.

1. *Pairs of matrices with a non-zero commutator*, Proc. Cambridge Philos. Soc. **51** (1955), 551-553.

I.C. Gohberg and M.G. Kreĭn.

1. *Introduction to the Theory of Linear Non-selfadjoint Operators*, Translations of Mathematical Monographs, Vol. 18, American Mathematical Society, Providence, 1969.
2. *Theory and Applications of Volterra Operators in Hilbert Space*, Translations of Mathematical Monographs, Vol. 24, American Mathematical Society, Providence, 1970.

J.K. Goldhaber and G. Whaples.

1. *On some matrix theorems of Frobenius and McCoy*, Canadian J. Math. **5** (1953), 332-335.

S. Grabiner.

1. *The nilpotency of Banach nil algebras*, Proc. Amer. Math. Soc. **21** (1969), 510.

J.A. Green and D. Rees.

1. *On semigroups on which $x^r = x$*, Proc. Cambridge Philos. Soc. **48** (1952), 35-40.

J.J. Grobler.

1. *Spectral theory in Banach lattices*, Operator theory in function spaces and Banach lattices, Birkhäuser, Basel, 1995, pp. 133-172.
2. *A short proof of the Ando-Krieger theorem*, Math. Z. **174** (1980), 61-62.

L. Grunenfelder, M. Omladič, and H. Radjavi.

1. *Jordan analogs of the Burnside and Jacobson density theorems*, Pacific J. Math. **161** (1993), 335-346.

L. Grunenfelder, R. Guralnick, T. Košir, and H. Radjavi.

1. *Permutability of characters on algebras*, Pacific J. Math. **178** (1997), 63-70.

P.S. Guinand.

1. *On quasinilpotent semigroups of operators*, Proc. Amer. Math. Soc. **86** (1982), 485-486.

R.M. Guralnick.

1. *Triangularization of sets of matrices*, Linear and Multilinear Algebra **9** (1980), 133-140.

A. Haar.

1. *Der Massbegriff in der Theorie der kontinuierlichen Gruppen*, Annals of Math. **34** (1933), 147-169.

D. Hadwin.

1. *Radjavi's trace condition for triangularizability*, J. Algebra **109** (1987), 184-192.
2. *An operator still not satisfying Lomonosov's hypothesis*, Proc. Amer. Math. Soc. **123** (1995), 3039-3041.

D. Hadwin, E. Nordgren, M. Radjabalipour, H. Radjavi, and P. Rosenthal.

1. *A nil algebra of bounded operators on Hilbert space with semisimple norm closure*, Integral Equations Operator Theory **9** (1986), 739-743.
2. *On simultaneous triangularization of collections of operators*, Houston J. Math. **17** (1991), 581-602.

D.W. Hadwin, E.A. Nordgren, H. Radjavi, and P. Rosenthal.

1. *An operator not satisfying Lomonosov's hypothesis*, J. Funct. Anal. **38** (1980), 410-415.

P.R. Halmos.

1. *Measure Theory*, D. Van Nostrand Company, New York, 1950.
2. *Introduction to Hilbert Space and the Theory of Spectral Multiplicity*, Second Edition, Chelsea Publishing Company, New York, 1957.
3. *A Hilbert Space Problem Book*, Second Edition, Springer–Verlag, New York, 1982.

P.R. Halmos and V.S. Sunder.

1. *Bounded Integral Operators on L^2 Spaces*, Springer-Verlag, New York, 1978.

I. Halperin and P. Rosenthal.

1. *Burnside's theorem on algebras of matrices*, Amer. Math. Monthly **87** (1980), 810.

D.A. Herrero.

1. *Triangular operators*, Bull. London Math. Soc. **23** (1991), 513-554.

D.A. Herrero, D.R. Larson, and W.R. Wogen.

1. *Semitriangular operators*, Houston J. Math. **17** (1991), 477-499.

D. Hilbert.

1. *Grunzüge einer allgemeinen Theorie der linearen Integralgleichungen IV*, Nachr. Akad. Wiss. Göttingen. Math.–Phys. Kl. (1906), 157-227.

R.D. Hill.

1. *Inertia theory for simultaneously triangulable complex matrices*, Linear Algebra and Appl. **2** (1969), 131-142.

E. Hille and J.D. Tamarkin.

1. *On the theory of linear integral equations II*, Annals of Math. **35** (1934), 445-455.

J. Holbrook, E. Nordgren, H. Radjavi and P. Rosenthal

1. *On the operator equation $AX = XAX$*, Linear Algebra Appl. **295** (1999), 113-116.

Y.P. Hong and R.A. Horn.

1. *On simultaneous reduction of families of matrices to triangular or diagonal form by unitary congruences*, Linear and Multilinear Algebra **17** (1985), 271-288.

R.B. Honor.

1. *Density and transitivity results on l^{∞} and l^{1}*, J. London Math. Soc. (2) **32** (1985), 521-527.

N. Jacobson.

1. *Lectures in Abstract Algebra II: Linear Algebra*, Van Nostrand, Princeton, 1953.
2. *Lie Algebras*, Interscience, New York, 1962.

A.A. Jafarian, H. Radjavi, P. Rosenthal, and A.R. Sourour.

1. *Simultaneous triangularizability, near commutativity and Rota's theorem*, Trans. Amer. Math. Soc. **347** (1995), 2191-2199.

M.T. Jahandideh.

1. *On the ideal-triangularizability of positive operators on Banach lattices*, Proc. Amer. Math. Soc. **125** (1997), 2661-2670.
2. *On the ideal-triangularizability of semigroups of quasinilpotent positive operators on $C(\mathcal{K})$* Canad. Math. Bull. **41** (1998), 298-305.

K. Jörgens.

1. *Linear Integral Operators*, Pitman, Boston, 1982.

R.V. Kadison and I.M. Singer.

1. *Triangular operator algebras. Fundamentals and hyperreducible theory*, Amer. J. Math. **82** (1960), 227-259.

I. Kaplansky.

1. *Completely continuous normal operators with property L*, Pacific J. Math. **3** (1953), 721-724.
2. *The Engel–Kolchin theorem revisited*, Contributions to Algebra, H. Bass, P.J. Cassidy and J. Kovaciks (eds.), Academic Press, New York, 1977, pp. 233-237.
3. *Fields and Rings*, Second Edition, University of Chicago Press, Chicago, 1972.
4. *Commutative Rings*, Allyn & Bacon, Boston, 1970.

A. Katavolos and H. Radjavi.

1. *Simultaneous triangularization of operators on a Banach space*, J. London Math. Soc. (2) **41** (1990), 547-554.

A. Katavolos and C. Stamatopoulos.

1. *Commutators of quasinilpotent operators and invariant subspaces*, Studia Math. **128** (1998), 159-169.

T. Kato.

1. *Perturbation Theory for Linear Operators*, Second Edition, Springer-Verlag, Berlin-New York, 1976.

H.W. Kim, C. Pearcy, and A.L. Shields.

1. *Sufficient conditions for rank-one commmutators and hyperinvariant subspaces*, Mich. Math. J. **23** (1976), 235-243,

E. Kolchin

1. *On certain concepts in the theory of algebraic matric groups*, Ann. of Math. (2) **49** (1948), 774-789.

T. Košir, M. Omladič, and H. Radjavi.

1. *Maximal semigroups dominated by 0–1 matrices*, Semigroup Forum **54** (1997), 175-189.

V.P. Kostov.

1. *A generalization of the Burnside theorem and of Schur's lemma for reducible representantions*, J. Dynam. Control Systems **1** (1995), 551-580.

M.G. Kreĭn and M.A. Rutman.

1. *Linear operators leaving invariant a cone in a Banach space*, Uspehi Matem. Nauk (N.S.) **3** (1948), 3-95 (Russian); *Amer. Math. Soc. Translation*, no. 26, 1950 (English).

A.G. Kurosh.

1. *The Theory of Groups*, Vol. I, Second Edition, Chelsea Publishing Company, New York, 1960.
2. *The Theory of Groups*, Vol. II, Second Edition, Chelsea Publishing Company, New York, 1960.

T.J. Laffey.

1. *Simultaneous triangularization of a pair of matrices*, J. Algebra **44** (1977), 550-557.
2. *Simultaneous quasidiagonalization of a pair of 3×3 complex matrices*, Rev. Roumaine Math. Pures Appl. **23** (1978), 1047-1052.
3. *Simultaneous triangularization of matrices-low rank cases and the nonderogatory case*, Linear and Multilinear Algebra **6** (1978/79), 269-305.
4. *Simultaneous triangularization of a pair of matrices whose commutator has rank two*, Linear Algebra Appl. **29** (1980), 195-203.

T.Y. Lam.

1. *A theorem of Burnside on matrix rings*, Amer. Math. Monthly **105** (1998), 651-653.
2. *Representations of finite groups: a hundred years, part II*, Notices Amer. Math. Soc. **45** (1998), 465-474.

M. Lambrou, W.E. Longstaff, and H. Radjavi.

1. *Spectral conditions and reducibility of operator semigroups*, Indiana Univ. Math. J. **41** (1992), 449-464.

S. Lang.

1. *Algebra*, Second Edition, Addison-Wesley, Reading, Mass., 1984.

D.R. Larson.

1. *A solution to a problem of J. R. Ringrose*, Bull. Amer. Math. Soc. (N.S.) **7** (1982), 243-246.
2. *Nest algebras and similarity transformations*, Ann. of Math (2) **121** (1985), 409-427.
3. *Triangularity in operator algebras*, Surveys of some recent results in operator theory, J.B. Conway and B.B. Morrel (eds.), Pitman, New York, 1988.

4. *Reflexivity, algebraic reflexivity and linear interpolation*, Amer. J. Math. **110** (1988), 283-299.
5. *On similarity of nests in Hilbert space and in Banach spaces*, Functional analysis (Austin, TX, 1986-86), Springer, Berlin-New York, 1988, pp. 179-194.
6. *Some recent progress in nest algebras*, Operator theory: operator algebras and applications, Part 1 (Durham, NH, 1988), American Mathematical Society, Providence, RI, 1990, pp. 333-346.

D.R. Larson and W.R. Wogen.

1. *Some problems on triangular and semi-triangular operators*, Selfadjoint and nonselfadjoint operator algebras and operator theory (Fort Worth, TX, 1990), American Mathematical Society, Providence, RI, 1991, pp, 97-100.

C. Laurie, E. Nordgren, H. Radjavi, and P. Rosenthal.

1. *On triangularization of algebras of operators*, J. Reine Angew. Math. **327** (1981), 143-155.

W.C. Lee and F. Ma.

1. *Simultaneous triangularization of the coefficients of linear systems*, Trans. ASME J. Appl. Mech. **64** (1997), 430-432.

J. Levitzki.

1. *Über nilpotente Unterrringe*, Math. Ann. **105** (1931), 620-627.

V.B. Lidskiĭ.

1. *Non-selfadjoint operators with a trace*, Dokl. Akad. Nauk SSSR **125** (1959), 485-487 (Russian); Transl. Amer. Math. Soc. (2) **47** (1965), 43-46, (English).

S. Lie and F. Engel.

1. *Theorie der Transformationsgruppen*, 3 vol., Leipzig (Teubner), 1888-1893.

R.S. Lipyanskiĭ.

1. *Triangulability of bounded algebras and Lie algebras*, Algebra and discrete mathematics, Latv. Gos. Univ., Riga, 1984, pp. 65-80 (Russian).

L. Livshits, G. MacDonald, B. Mathes, and H. Radjavi.

1. *Reducible semigroups of idempotent operators*, J. Operator Theory **40** (1998), 35-69.
2. *On band algebras*, preprint.

V. Lomonosov.

1. *Invariant subspaces for the family of operators commuting with compact operators*, Funkcional. Anal. i Priložen. **7** (1973), 55-56 (Russian); Funct. Anal. and Appl. **7** (1973), 213-214 (English).
2. *An extension of Burnside's theorem to infinite-dimensional spaces*, Israel J. Math. **75** (1991), 329-339.

R.R. London and H.P. Rogosinski.

1. *Decomposition theory in the teaching of elementary linear algebra*, Amer. Math. Monthly **97** (1990), 478-485.

W.E. Longstaff and H. Radjavi.

1. *On the equation $AB = BX$ in collections of matrices*, Linear and Multilinear Algebra **25** (1989), 173-184.
2. *On permutability and submultiplicativity of spectral radius*, Canad. J. Math. **47** (1995), 1007-1022.

T.W. Ma.

1. *On rank one commutators and triangular representations*, Canad. Math. Bull. **29** (1986), 268-273.

A. Marwaha.

1. *Decomposability and structure of nonnegative bands in $M_n(\mathbb{R})$*, Linear Algebra Appl. **291** (1999), 63-82.
2. *Decomposability and structure of nonnegative bands in infinite dimensions*, preprint.

B. Mathes, M. Omladič, and H. Radjavi.

1. *Linear spaces of nilpotent matrices*, Linear Algebra Appl. **149** (1991), 215-225.

N.H. McCoy.

1. *On quasi-commutative matrices*, Trans. Amer. Math. Soc. **36** (1934), 327-340.
2. *On the characteristic roots of matric polynomials*, Bull. Amer. Math. Soc. **42** (1936), 592-600.

H. Minc.

1. *Nonnegative matrices*, John Wiley & Sons, New York, (1988).

T.S. Motzkin and O. Taussky.

1. *Pairs of matrices with property L*, Trans. Amer. Math. Soc. **73** (1952), 106-114.

  2. *On representations of finite groups*, Nederl. Akad. Wetensch. Proc. Ser. A 55 = Indagationes Math. **14** (1952), 511-512.
  3. *Pairs of matrices with property L.* II, Trans. Amer. Math. Soc **80** (1955), 387-401.

G.J. Murphy.

  1. *Triangularizable algebras of compact operators*, Proc. Amer. Math. Soc. **84** (1982), 354-356.

M.A. Naĭmark

  1. *Normed Rings*, P. Noordhoff N.V., Groningen, Netherlands, 1959.

J.D. Newburgh.

  1. *The variation of spectra*, Duke Math. J. **18** (1951), 165-176.

E.A. Nordgren, M. Radjabalipour, H. Radjavi, and P. Rosenthal.

  1. *Quadratic operators and invariant subspaces*, Studia Math **88** (1988), 263-268.

E. Nordgren, H. Radjavi, and P. Rosenthal.

  1. *A geometric equivalent of the invariant subspace problem*, Proc. Amer. Math. Soc. **61** (1976), 66-69.
  2. *Triangularizing semigroups of compact operators*, Indiana Univ. Math. J. **33** (1984), 271-275.
  3. *On density of transitive algebras*, Acta Sci. Math. (Szeged) **30** (1969), 175-179.

J. Okniński.

  1. *Triangularizable semigroups of matrices*, Linear Algebra Appl. **262** (1997), 111-118.
  2. *Semigroups of matrices*, World Scientific, Singapore, 1998.

M. Omladič, M. Radjabalipour, and H. Radjavi.

  1. *On semigroups of matrices with traces in a subfield*, Linear Algebra Appl. **208/209** (1994), 419-424.

M. Omladič and H. Radjavi.

  1. *Irreducible semigroups with multiplicative spectral radius*, Linear Algebra Appl. **251** (1997), 59-72.

M. Petrich.

  1. *Lectures in Semigroups*, Academic-Verlag, John Wiley & Sons Ltd., Berlin, 1977.

O. Perron.

1. *Zur Theorie der Matrizen*, Math. Ann. **64** (1907), 248-263.
2. *Über Stabilität und asymptotisches Verhalten der Integrale von Differentialgleichungssystemen*, Math. Z. **29** (1929), 129-160.

J. S. Ponizovskii

1. *On matrix semigroups over a field* $\mathbb{K}$ *conjugate to matrix semigroups over a proper subfield of* $\mathbb{K}$, Semigroups with Applications (Oberwolfach, 1991), 1-5, World Sci. Publishing, River Edge, N. J., 1992

S.C. Power.

1. *Another proof of Lidskiĭ's theorem on the trace*, Bull. London Math. Soc. **15** (1983), 146-148.

C. Procesi.

1. *Rings with Polynomial Identities*, Marcel Dekker, New York, 1973.

A.J. Pryde.

1. *Inequalities for the joint spectrum of simultaneously triangularizable matrices*, Miniconference on probability and analysis (Sydney, 1991), Proc. Centre Math. Appl. Austral. Nat. Univ., 29, Austral. Nat. Univ., Canberra, 1992, 196-207.

M. Radjabalipour.

1. *Simultaneous triangularization of algebras of polynomially compact operators*, Canad. Math. Bull. **34** (1991), 260-264.

M. Radjabalipour and H. Radjavi.

1. *A finiteness lemma, Brauer's Theorem and other irreducibility results*, Comm. Algebra **27 (1)** (1999), 301-319.

H. Radjavi.

1. *On the reduction and triangularization of semigroups of operators*, J. Operator Theory **13** (1985), 63-71.
2. *A trace condition equivalent to simultaneous triangularizability*, Canad. J. Math. **38** (1986), 376-386.
3. *The Engel-Jacobson theorem revisited*, J. Algebra **111** (1987), 427-430.
4. *On reducibility of semigroups of compact operators*, Indiana Univ. Math. J. **39** (1990), 499-515.
5. *Invariant subspaces and spectral conditions on operator semigroups*, Linear Operators (Warsaw, 1994), Banach Center Publ., 38, Polish Acad. Sci., Warsaw, 1997, pp. 287-296.

6. *Sublinearity and other spectral conditions on a semigroup*, Canad. J. Math., to appear.
7. *The Perron-Frobenius Theorem revisited*, J. Positivity, to appear.
8. *Isomorphisms of transitive operator algebras*, Duke Math. J. **41** (1974), 555-564.

H. Radjavi and P. Rosenthal.

1. *On invariant subspaces and reflexive algebras*, Amer. J. Math. **91** (1969), 683-692.
2. *Invariant Subspaces*, Springer-Verlag, Berlin-Heidelberg-New York, 1973.
3. *On transitive and reductive operator algebras*, Math. Ann. **209** (1974), 43-56.
4. *The invariant subspace problem*, Math. Intelligencer **4** (1982), 33-37.
5. *From local to global triangularization*, J. Funct. Anal. **147** (1997), 443-456.
6. *A sufficient condition that an operator algebra be self-adjoint*, Canad. J. Math. **23** (1971), 588-597.

H. Radjavi, P. Rosenthal, and V.S. Shulman.

1. *Operator semigroups with quasinilpotent commutators*, Proc. Amer. Math. Soc., to appear.

C.J. Read.

1. *A solution to the invariant subspace problem on the space l_1*, Bull. London Math. Soc. **17** (1985), 305-317.
2. *A short proof concerning the invariant subspace problem*, J. London Math. Soc. (2) **34** (1986), 335-348.
3. *Quasinilpotent operators and the invariant subspace problem*, J. London Math. Soc. (2) **56** (1997), 595-606.

F. Riesz.

1. *Über lineare Funktionalgleichungen*, Acta Math. **41** (1917), 71-98.
2. *Les systèmes d'équations linéaires à une infinité d'inconnues*, (Paris, 1913).

J.R. Ringrose.

1. *Algebraic isomorphisms between ordered bases*, Amer. J. Math. **83** (1961), 463-478.
2. *Super-diagonal forms for compact linear operators*, Proc. London Math. Soc. (3) **12** (1962), 367-384).
3. *On some algebras of operators*, Proc. London Math. Soc. (3) **15** (1965), 61-83.

4. *On some algebras of operators II*, Proc. London Math. Soc. (3) **16** (1966), 385-402.

5. *Compact Non-self-adjoint Operators*, Van Nostrand Reinhold Co., New York, 1971.

E. Rosenthal.

1. *A remark on Burnside's theorem on matrix algebras*, Linear Algebra Appl. **63** (1984), 175-177.

P. Rosenthal.

1. *Applications of Lomonosov's lemma to non-selfadjoint operator algebras*, Proc. Royal Irish Acad. **74** (1974), 271-281.

2. *Equivalents of the invariant subspace problem*, Paul Halmos. Celebrating 50 Years of Mathematics, J. H. Ewing and F. W. Gehring (eds.), Springer-Verlag, New York, (1991), 179-188.

P. Rosenthal and A. Soltysiak.

1. *Formulas for the joint spectral radius of non-commuting Banach algebra elements*, Proc. Amer. Math. Soc. **12** (1995), 2705-2708.

G.-C. Rota and W.G. Strang.

1. *A note on the joint spectral radius*, Indag. Math. **22** (1960), 379-381.

W.E. Roth.

1. *On the characteristic values of the matrix $f(A, B)$*, Trans. Amer. Math. Soc. **39** (1936), 234-243.

2. *On k-commutative matrices*, Trans. Amer. Math. Soc. **39** (1936), 483-495.

W. Rudin.

1. *Real and Complex Analysis*, Third Edition. McGraw-Hill, New York, 1987.

2. *Functional Analysis*, Second Edition. McGraw-Hill, New York, 1991.

A.A. Sagle and R.E. Walde.

1. *Introduction to Lie Groups and Lie Algebras*, Academic Press, New York, 1973.

D. Sarason.

1. *Invariant subspaces and unstarred operator algebras*, Pacific J. Math. **17** (1966), 511-517.

H.H. Schaefer.

1. *Banach Lattices and Positive Operators*, Springer-Verlag, New York-Heidelberg, 1974.
2. *Topologische Nilpotentz irreduzibler Operatoren*, Math. Z. **117** (1970), 135-140.

O.J. Schmidt.

1. *Über Gruppen, deren sämtliche Teiler spezielle Gruppen sind*, Math. Sbornik **31** (1924), 366-372.

H. Schneider.

1. *An inequality for latent roots applied to determinants with dominant principal diagonal*, J. London Math. Soc. **28** (1953), 8-20.
2. *A pair of matrices with property P*, Amer. Math. Monthly **62** (1955), 247-249.

H. Schneider and M.H. Schneider.

1. *Max–balancing weighted directed graphs and matrix scaling*, Math. Oper. Res. **16** (1991), 208-222.

I. Schur.

1. *Über die characteristischen Wurzeln einer linearen Substitution mit einer Anwendung auf die Theorie der Integral Gleichungen*, Math. Ann. **66** (1909), 488-510.

P. Šemrl.

1. *Isomorphisms of standard operator algebras*, Proc. Amer. Math. Soc. **123** (1995), 1851-1855.

J.P. Serre.

1. *Linear Representations of Finite Groups*, Springer-Verlag, New York, 1977.

H. Shapiro.

1. *Simultaneous block triangularization and block diagonalization of sets of matrices*, Linear Algebra Appl. **25** (1979), 129-137.

V.S. Shulman.

1. *On invariant subspaces of Volterra operators*, Funktsional. Anal. i Prilozhen **18** (1984), 85-86 (Russian).

V.S. Shulman and Y.V. Turovskii.

1. *Joint spectral radius, operator semigroups and a problem of W. Wojtyński*, preprint.

2. *Invariant subspaces and spectral mapping theorems*, Functional Analysis and Operator Theory (Warsaw, 1992), Banach Center Publ., 30, Polish Acad. Sci., Warsaw, 1994, pp. 313-325.

B. Simon.

1. *Trace Ideals and their Applications*, London Mathematical Society Lecture Note Series, 35, Cambridge University Press, Cambridge, 1979.

A. Simonič.

1. *Matrix groups with positive spectra*, Linear Algebra Appl. **173** (1992), 57-76.
2. *A construction of Lomonosov functions and applications to the invariant subspace problem*, Pacific J. Math. **175** (1996), 257-270.
3. *An extension of Lomonosov's techniques to non-compact operators*, Trans. Amer. Math. Soc. **348** (1996), 975-995.

W.S. Sizer.

1. *Triangularizing semigroups of matrices over a skew field*, Linear Algebra Appl. **16** (1977), 177-187.

D.A. Sprunenko.

1. *Matrix Groups*, Translations of Mathematical Monographs, Vol. 45, American Mathematical Society, Provindence, 1976.

G. Szép

1. *Simultaneous triangularization of projector matrices*, Acta Math. Hungar. **48** (1986), 285-288.

O. Taussky.

1. *Commutativity of finite matrices*, Amer. Math. Monthly **64** (1957), 229-235.
2. *Some results concerning the transition from the L- to the P-property for pairs of finite matrices (2)*, Linear and Mult. Alg. **2** (1974), 195-202.

A. Thue.

1. *Ueber unendliche Zeichenreih*, Selected Mathematical Papers of Axel Thue, Universitetsforlaget, Oslo-Bergen-Tromsø, 1977, pp. 139-158.

Y. Tong.

1. *Quasinilpotent integral operators*, Acta Math. Sinica **32** (1989), 727-735 (Chinese).

304   References

V.G. Troitsky.

1. *Lomonosov's theorem cannot be extended to chains of four operators*,
    Proc. Amer. Math. Soc., to appear.

Y.V. Turovskiĭ.

1. *Spectral properties of certain Lie subalgebras and the spectral radius
    of subsets of a Banach algebra*, Spectral theory of operators and its
    applications, No. 6 "Élm", Baku, 1985, 144-181. (Russian)
2. *Volterra semigroups have invariant subspaces*, J. Funct. Anal. **162**
    (1999), 313-322.

K. Vala.

1. *On compact sets of compact operators*, Ann. Acad. Sci. Fenn. Ser.
    A I **351** (1964), 1-8.

V.S. Varadarajan.

1. *Lie Groups, Lie Algebras and their Representations*, Prentice–Hall
    Inc., Englewood Cliffs, New Jersey, 1974.

M. Vaughan-Lee.

1. *On Zelmanov's solution of the restricted Burnside problem*, J. Group
    Theory **1** (1998), 65-94.

E.B. Vinberg.

1. *Linear Representations of Groups*, Birkhäuser, Basel-Boston-Berlin,
    1989.

D. Voiculescu.

1. *A non-commutative Weyl-von Neumann theorem*, Rev. Roumaine
    Math. Pures Appl. **21** (1976), 97-113.

D.B. Wales and H.J. Zassenhaus

1. *On L-groups* Math. Ann. **198** (1972), 1-12.

J.F. Watters.

1. *Block triangularization of algebras of matrices*, Linear Algebra Appl.
    32 (1980), 3-7.

B.A.F. Wehrfritz.

1. *On the Lie-Kolchin-Mal'cev theorem*, J. Austral. Math. Soc. Ser.
    A **26** (1978), 270-276.

H. Wielandt.

1. *Lineare Scharen von Matrizen mit reellen Eigenwerte*, Math. Z. **53** (1950), 219-225.

J. Williamson

1. *The simultaneous reduction of two matrices to triangle form*, Amer. J. Math. **57** (1935), 281-293.

W. Wojtyński.

1. *Engel's theorem for nilpotent Lie algebras of Hilbert-Schmidt operators*, Bull. Acad. Polon. Sci. Sér. Sci. Math. **24** (1976), 797-801.
2. *Associative algebras generated by Lie algebras of compact operators*, Bull. Acad. Polon. Sci. Sér. Sci. Math. **28** (1980), 237-241.

M.R. Yahaghi.

1. *Recent developments in reducibility of operator semigroups*, M.Sc. Thesis, Dalhousie University, 1998.

A.C. Zaanen

1. *Riesz Spaces II*, North-Holland, Amsterdam, 1983.
2. *Linear Analysis*, Third Edition, North-Holland, Amsterdam, 1964.

H.J. Zassenhaus.

1. *On L-semi-groups*, Math. Ann. **198** (1972) 13-22.

E.I. Zelmanov

1. *On the restricted Burnside problem*, Fields Medalists' Lectures, World Scientific, River Edge, NJ, (1997), 623-632.

J. Zemànek.

1. *Properties of the spectral radius in Banach algebras*, Spectral Theory (Warsaw, 1977), Banach Center Publ., 8, PWN, Warsaw, 1982, pp. 579-595.

Y. Zhong

1. *Functional positivity and invariant subspaces of semigroups of operators*, Houston J. Math. **19** (1993), 239-262.
2. *Irreducible semigroups of functionally positive nilpotent operators*, Trans. Amer. Math. Soc. **347** (1995), 3093-3100.

Notation Index

Author Index

Subject Index

Universitext   *(continued)*

Meyer: Essential Mathematics for Applied Fields
Mines/Richman/Ruitenburg: A Course in Constructive Algebra
Moise: Introductory Problems Course in Analysis and Topology
Morris: Introduction to Game Theory
Polster: A Geometrical Picture Book
Porter/Woods: Extensions and Absolutes of Hausdorff Spaces
Radjavi/Rosenthal: Simultaneous Triangularization
Ramsay/Richtmyer: Introduction to Hyperbolic Geometry
Reisel: Elementary Theory of Metric Spaces
Rickart: Natural Function Algebras
Rotman: Galois Theory
Rubel/Colliander: Entire and Meromorphic Functions
Sagan: Space-Filling Curves
Samelson: Notes on Lie Algebras
Schiff: Normal Families
Shapiro: Composition Operators and Classical Function Theory
Simonnet: Measures and Probability
Smith: Power Series From a Computational Point of View
Smoryski: Self-Reference and Modal Logic
Stillwell: Geometry of Surfaces
Stroock: An Introduction to the Theory of Large Deviations
Sunder: An Invitation to von Neumann Algebras
Tondeur: Foliations on Riemannian Manifolds
Wong: Weyl Transforms
Zhang: Matrix Theory: Basic Results and Techniques
Zong: Sphere Packings
Zong: Strange Phenomena in Convex and Discrete Geometry